T0261472

WEYERHAEUSER ENVIRONMENTAL BOOKS

William Cronon, Editor

Weyerhaeuser Environmental Books explore
human relationships with natural environments
in all their variety and complexity. They seek to
cast new light on the ways that natural systems
affect human communities, the ways that people
affect the environments of which they are a part,
and the ways that different cultural conceptions
of nature profoundly shape our sense
of the world around us.

THE LOST WOLVES OF JAPAN

BRETT L. WALKER

Foreword by William Cronon

UNIVERSITY OF WASHINGTON PRESS

Seattle and London

The Lost Wolves of Japan has been published with
the assistance of a grant from the Weyerhaeuser
Environmental Books Endowment, established
by the Weyerhaeuser Company Foundation,
members of the Weyerhaeuser family,
and Janet and Jack Creighton.

© 2005 by the University of Washington Press
Printed in the United States of America
Designed by Pamela Canell
Maps by Dale Martin
12 11 10 09 08 07 06 05   5 4 3 2 1

ISBN 13: 978-0-295-98814-6

University of Washington Press
P.O. Box 50096, Seattle, WA 98145
www.washington.edu/uwpress

Library of Congress Cataloging-in-Publication Data
can be found at the back of this book.

The paper used in this publication is acid-free and 90 percent
recycled from at least 50 percent post-consumer waste. It meets
the minimum requirements of American National Standard
for Information Sciences—Permanence of Paper for Printed
Library Materials, ANSI Z39.48-1984. ♾ ♼

FOR MY PARENTS,
LINDA HARBERS
AND NELSON WALKER

CONTENTS

FOREWORD

A Strange Violent Intimacy

William Cronon

Wolves and other big predators have long loomed large in the history, literature, and folklore of human interactions with the natural world. Although there is wide variation from culture to culture in the ways these creatures are depicted, they rarely surface in our collective memory without standing in one way or another for the fierce autonomy of nature in resisting the visions of order that people in the modern era like to imagine they can impose on the world around them. Wolves in particular occupy a peculiarly fraught place in our cultural conceptions of nature, since their intelligence, their communication skills, their sociability, their cooperative hunts, and, not least, their close kinship to the domesticated dogs that are arguably our closest animal companions make them seem uniquely familiar to us. These and other qualities enable us to gaze on these creatures and see in them powerful reflections of some of our own most cherished human qualities.

And yet wolves also contest our dominion of the landscapes from which we wrest our livelihoods, threatening our comfortable conviction that we are safely in control of nature. They kill our animals and, under rare circumstances, can even feast on human flesh to prove that we too can become prey to chains of predation that lead not just to our own stomachs, but to the stomachs of other animals as well. No doubt because of this strange violent intimacy between ourselves and wolves, the history of our relationship is bathed in blood, so much so that nation after nation over the past two hundred years has eventually succeeded in driving wolves to the brink of local extinction and beyond. It is no accident that they have come to symbolize the wildest places on earth, for until very recently we have systematically extirpated them from almost everywhere else.

Much has been written about the environmental history of wolves in the United States, in part because of the striking reversal in American attitudes toward predators during the middle decades of the twentieth century. Aldo Leopold's classic essay "Thinking Like a Mountain" usually serves as the symbolic benchmark for this change, describing as it does a landscape that underwent massive environmental degradation as deer herds exploded with the removal of large predators like wolves. The essay was a famous mea culpa for Leopold himself, since he had devoted considerable energy early in his career to promoting predator control in the American Southwest, realizing too late that his own efforts were helping destroy the very wilderness he hoped to preserve. Although Leopold's conclusions about the role of wolves in controlling deer populations have been challenged in subsequent years—he likely exaggerated the scale of the famous Kaibab deer disaster of the 1920s, and probably overemphasized the role of wolves in controlling deer herds while underemphasizing the importance of Native American hunters in performing that same function—the effects of his arguments on modern environmentalist attitudes toward predators have nonetheless been profound. Wolves, cougars, and grizzlies are now typically regarded as among the most important keystone species in North American ecosystems, so that their presence or absence has become one of the most widely embraced biological measures of wilderness in the United States and Canada.

But wolves have played critical roles in the human histories of many other parts of the world as well, which is why we can be grateful to Brett L. Walker for producing this highly original and thought-provoking study, *The Lost Wolves of Japan.* Walker is a scholar who specializes in Japanese environmental history at Montana State University. Living in Bozeman enabled him to observe at first hand the reintroduction of wolves into the Greater Yellowstone Ecosystem in the 1990s. So moved was Walker by his encounters with Yellowstone wolves in the wild—experiences that he narrates as evocative autobiographical sketches in this book—that he embarked on an exploration of wolf biology, folklore, and history in the Japanese archipelago that finally yielded this ambitious and boldly speculative book. What he found will be of great interest to anyone who cares about human relationships with wolves—and the rest of wild nature besides—anywhere on earth.

The question at the center of Walker's book is deceptively simple: why did Japanese wolves become extinct? To answer it, he begins by sketching the origin of wolves in Siberia a million or more years ago, and their eventual migrations into Japan, where they enter history as two distinct subspecies traditionally labeled *yamainu* (mountain dog) and *ōkami* (wolf). He introduces us to medieval and modern debates over the taxonomic status of these separately named creatures, depicting their eventual merger in the writings of modern naturalists as the single canid "Japanese wolf" as evidence of the ways in which natural variation and complexity become homogenized when viewed through the lens of modernity. Walker sees this impulse toward homogenization and control as a defining characteristic of the modern world that would eventually spell doom for the Japanese wolf.

Before mustering evidence on behalf of this argument, however, Walker first introduces us to a complex array of folk traditions and spiritual beliefs that governed Japanese interactions with wolves for centuries before the dawn of the modern era. He describes roadside shrines for the worship of wolves in rituals seeking protection of crops from grain-eating animals like deer and wild boars. He tells of talismans with depictions of wolves as magical charms effective in warding off all manner of calamities. Although wolves were hunted, they were also treated with immense respect, so that wolves and people coexisted for many centuries in a magically animate world that the two shared equally with one another.

Things began to change in the eighteenth century, in the wake of an astonishingly virulent rabies epidemic that suddenly converted these formerly reclusive animals into crazed man-killers whose random violence was terrifying to behold. The Japanese response was a series of ceremonial hunts designed to attack a creature that had suddenly become demonic, beginning the long-term destruction of wolf populations throughout the islands. By the nineteenth century, wolves were being redefined simply as noxious pests worthy of destruction because they threatened the domesticated grazing animals of a modernizing Japanese agriculture. For such destruction to be possible, a prior universe in which humans saw themselves sharing their islands with countless animal spirits—many of them very powerful and even dominant over humans—had to give way to a far more mechanistic and alienated universe in which human control of the natural world came to seem not only feasible but desirable. Expert wolf hunters were welcomed to Japan from North America, bringing with them the modern tools of guns and bounty laws and strychnine to slaughter the last wolves of Japan. By 1905, that effort had achieved its goal.

But Walker's story does not end with the final death of the last Japanese wolf. Instead, he traces the efforts by Japanese ecologists to understand the meaning of this event both for Japanese ecosystems and for the Japanese nation. Theories of wolf extinction become an unexpectedly revealing window into the founding of ecological science in Japan in the early twentieth century, enabling Walker to return to one of the most important themes of his book. Although one might assume that material competition between humans and wolves for control of the same habitat—both desiring to consume the same large herbivores, for instance—made their collision inevitable, Walker wishes us to consider two other possibilities that are subtler and potentially more profound in their implications.

First, he wants us to remember that premodern societies were not nearly so instrumentalist in their attitudes toward the fellow beings with whom humans share this planet. For wolves to become the objects of a sustained campaign directed at their total destruction, they had to be stripped of the animist magic and empathetic identifications that had rendered them spiritual companions for earlier Japanese communities—dangerous and frightening companions, to be sure, but also sentient beings possessed of a spiritual autonomy that commanded respect as well as fear. In the West, we sometimes attribute to the Cartesian logic of the Enlightenment what the environmental historian Carolyn Merchant has called "the death of nature," and one of Walker's core arguments is that this same logic reproduced itself with the spread of modernity to other parts of the world as well.

But Walker adds to this claim about modernity another more speculative insight that is even closer to his own heart. It came to him from his experiences in Yellowstone as he felt the cold chill along his spine listening to the howl of wolves, and watching them tear into the flesh of the bloody carcass they were consuming. Although it would be difficult ever to prove this claim with finality, Walker believes that the violence we have directed against these creatures reflects both our recognition and our denial of the close kinship we feel for them. We see in them sentient beings whose emotions are closer to our own than we dare acknowledge. If this is so, then the animist traditions of premodern Japan express truths about wolves and people that we moderns have willfully chosen to forget. And in that forgetting, we have put at risk not only the wolves, but the world we share with them—and a vital part of ourselves as well.

PREFACE

When it comes to our relationship with wolves, we have both come a long way and gone nowhere at all. While I was revising the manuscript of this book, a disturbing incident occurred near Bozeman that reminded me of this fact. In March 2004, wolves in the Madison Valley attacked several head of livestock. In one frightening episode, a rancher watched helplessly as wolves mauled his dog to death; his wife noticed that wolves had urinated on a snowbank near where their children play. Within a week of the wolf attacks, the headline of the *Bozeman Daily Chronicle* read, "Killer Wolf Packs Marked for Death," after Montana politicians began clamoring for the wolves to be destroyed. Governor Judy Martz was among those politicians who made hay from wolf hides. "Families are afraid to allow their children to play outdoors," she claimed in a letter to Ed Bangs, wolf recovery leader for the U.S. Fish and Wildlife Service.

Bangs swiftly ordered the eradication of all wolves in the Madison Valley. He also noted to reporters, however, that the trouble had started when poachers had shot three adult members of the Sentinel pack. Without these adults, he explained, "you basically have a bunch of teenagers walking through cattle." Within days, federal trappers had shot these teenage wolves from helicopters while some of them stood snared in steel traps at baiting stations in the Madison Range; some ranchers were also given permission to shoot wolves on sight. Federal trappers even kept alive one badly injured collared wolf, apparently shot illegally, in the hopes that it would somehow limp along, despite its loss of blood, and lead them to any remaining wolves so that trappers could shoot them as well. Because the Sentinel and Ennis Lake wolves had killed two steers, one heifer, and a stock dog, they were eradicated.

As I watched this sad event unfold I reflected on the book I had just written, and I realized that the Madison Valley incident was replete with the same intense emotions that always seem to surface when people deal with, write about, or interact with wolves. I have tried to capture some of those emotions in these pages. When it comes to wolves, rarely is there ever a right or wrong path to follow or a good or bad decision to make, despite the fact that so many people think otherwise; science, the reality of wolf ecology, and the economy of ranching always surrender to anger, anxiety, fear, and passion. I've never met a person who, when asked for his or her opinions about wolf recovery, responded with indifference. Even among supposedly wilderness-loving Montanans, people just lose their heads when it comes to wolves (see the *Bozeman Daily Chronicle* online edition).

I come from a Montana family and so I sympathize with these ranchers at many levels. Mainly, I know that the odds are already stacked against farming families and so I do not blame them for being angry at one more obstacle being placed in front of them. But I also sympathize with wolves. The brutality of their hunting and killing that so offends the sensibilities of some today will never even approach the cruelty with which we have tortured their kind throughout the ages. History tells us that ours is the disturbing species, not theirs: had humans, in some twist of evolutionary fate, not inherited the earth, it would surely belong to wolves, and they probably would have proved better stewards.

In retrospect, what brought me to admire wolves, to cherish them as a precious part of our national heritage, to value them as an integral part of our wildlands, and to sympathize with their sad plight was researching the history of Japan's extinct wolves, which is the subject of this book. But Japanese ranchers and upland hunters are also an important part of this story, and for them as well I have gained a new level of respect. Wolves truly are, as some Native Americans believe, teachers of the highest order; they can tell us something about the health of our planet and of ourselves. But I believe they can teach us something about our history as well. So to the Sentinel and Ennis Lake wolves and to their brethren throughout the ages and around the world I can only apologize for our arrogance and cruelty and for our failure to learn our lessons.

When writing this book, I have tried to heed the lessons taught by Japan's wolves and the hunters and ranchers who interacted with them. I also relied heavily on the knowledge of friends and colleagues, as well as on the kindness of the staffs of libraries and archives and of shrines and temples in the United States and Japan. Some colleagues even suffered through painfully early versions of this manuscript, and for that I am forever grateful. In particular, Bill Cronon utterly changed the nature of this project during a single conversation in the Denver airport. After that conversation, we had several lengthy discussions about writing history that opened new doors for me intellectually. Conrad Totman read the entire manuscript at least twice and added countless insights I otherwise would have missed. Douglas Smith kindly allowed me to participate as an observer in the Yellowstone Wolf Project Winter Study. The Yellowstone wolves simply could not have a better spokesperson. After studying the Druids with the Wolf Project, I learned that when I want perspective on life I travel to the Lamar Valley in the dead of winter. Julidta Tarver, editor at the University of Washington Press, had the faith to gracefully shepherd this project along, despite its many strange incarnations.

I also received advice in various forms and forums from Akizuki Toshiyuki, Amano Tetsuya, Sam Anderson, Gordon Brittan, Pamela Bruton, Rob Campbell, Diane Cattrell, Jennifer Chrisman, John DeBoer, Tak Fujitani, Jim Halfpenny, Rosanne Halouf, Jeff Hanes, Valerie Hansen, Hara Michiko (for the generous hospitality in Japan), Peter Harbers, David Howell, Inoue Katsuo, Ishihara Makoto, Susan Jones, Stephen Kohl, Tim LeCain, Lawrence Marceau, Dale Martin (who drew the marvelous maps), Michelle Maskiell, Rick McIntyre, L. David Mech, Raymond Mentzer, Ian Miller, Mary Murphy, Ronald Nowak, Greg Pflugfelder, Michael Reidy, Luke Roberts, Adam Rome (and anonymous readers for *Environmental History*), Robert Rydell, Lynda Sexson, Aaron Skabelund, Billy Smith, Tanabe Yasuichi (who selflessly shared years' worth of his own archival research), Julia Adeney Thomas, Ron Toby, Tozawa Shirō, Tsuchiya Tatsuhide, Anne Walthall, Kären Wigen, Donald Worster, Marcia Yonemoto, and anonymous readers for the University of Washington Press. Parts of chapter 4 first appeared in the journal *Environmental History,* and I appreciate their permission to reproduce them. Finally, Yuka Hara, my wife and partner, read through the manuscript, remained supportive throughout the research and writing process, elegantly put up with my constant absentmindedness, and helped in countless other ways, as did my inspirations Shō and Taiyō. I want to sincerely thank all of these souls because writing history can be such a collaborative effort; but I also want to acknowledge that the mistakes in the text are mine to keep.

Writing this book would not have been possible without the extremely generous economic support of a Vice President for Scholarship, Creativity, and Technology Transfer Research Grant, Montana State University; a National Science Foundation EPSCoR Research Grant, Montana State University; a Japan Foundation Research Grant, The Japan Foundation, Tokyo, Japan; a College of Letters and Science Research and Creativity Grant, Montana State University; and constant support from the Department of History and Philosophy at Montana State University. The talented staffs at the following centers made researching this project relatively painless: the Resource Collection for Northern Studies (Hoppō Shiryō Shitsu), Hokkaido University Library, Sapporo; the Hokkaido Prefectural Archives (Hokkai Dōritsu Monjokan), Sapporo; the Montana State University Library and Special Collections, Bozeman, Montana. And, finally, cheers to the interlibrary loan staff at Montana State University.

A NOTE TO THE READER

As is customary with scholarly writing on Japan, Japanese names are given in the Japanese order, with the surnames preceding the given names— that is, unless written otherwise, as in the case of an author of an English-language scientific publication, for example. I have supplied the appropriate diacritical marks for Japanese words and names, except for common words or place-names such as Hokkaido and Tokyo. I have provided translations of all Japanese titles cited in this book. Whenever an English title was appended to a Japanese title in the original publication, however, I used that, sometimes with slight modifications. Place-names, mountain ranges, shrine locations, and other spatial information appear in the maps so masterfully drawn by Dale Martin. All Chinese is written with the pinyin system, and the diacritical marks have been excluded. There are only a few Ainu names, but they are written as they appear in the sources. Except for such names, I have standardized the Ainu language according to Kayano Shigeru's *Ainugo jiten* (A dictionary of the Ainu language) (Tokyo: Sanseidō, 1996).

Months written as, for example, "the third month" of a year refer to the lunar calendar that was employed in Japan prior to "the twelfth month" of 1872, when Japan adopted the Gregorian calendar, but all years have been estimated according to the modern Western calendar. In the original documents, Japanese dating used imperial reign names and the Chinese calendar. The equivalents for weights and measures have been translated in the text.

THE LOST WOLVES OF JAPAN

If we carefully observe the countless varieties of birds and beasts, even tiny insects, we shall discover that they love their children, long to be near their parents, that husband and wife remain together, that they are jealous, angry, greedy, self-seeking, and fearful for their own lives to an even worse degree than men because they lack intelligence. How can we not feel pity when pain is inflicted on them or people take their lives?

A man who can look on sentient creatures without feeling compassion is no human being.

KENKŌ, *Tsurezuregusa* (Essays in idleness; c. 1330)

Forty years ago, there were very few Japanese in Hokkaido, which, under the uninviting name of Yezo, was considered as an island cold in the extreme and full of danger from wild beasts. If you should go to Hokkaido now—which you can easily in two days from Tokyo—it would be hard to see an Ainu hut, and so far as a bear or a wolf is concerned, I am afraid you would never make his acquaintance in his home. The old haunts of these animals are now turned into plowed fields; and where they once roamed in unmolested freedom, you find in their stead children playing; where two decades ago you heard the hungry howl of wolves and the angry growl of bears, you hear the sweet notes of school songs.

NITOBE INAZŌ, May 1906

The wolves' hunting noises were always far off, back north in the river bottoms. In the eerie clarity of the white nights they seemed to cry from inexpressible distances, faint and musical and clear, and he might have been tempted to think of them as something not earthly at all, as creatures immune to cold and hunger and pain, hunting only for the wolfish joy of running and perhaps not even visible to human eyes, if he had not one afternoon ridden through a coulee where they had bloodied half an acre with a calf.

WALLACE STEGNER, *Wolf Willow*

INTRODUCTION

In central Japan in 1905, near Washikaguchi, a rural timber community nestled along the banks of the Takami River, a lone American traveler checked into the Hōgetsurō Inn. Malcolm Anderson had come to Japan under the auspices of an English zoologist, the Duke of Bedford, to collect exotic animal specimens from East Asia for the London Zoological Society and the British Museum of Natural History. One suspects that only rarely did Westerners such as Anderson venture to this remote location in Nara Prefecture, and even less frequently did they venture there to gather what local Japanese surely saw as the filthy carcasses of dead animals. Not surprisingly, Anderson intrigued the villagers, and some even brought specimens to the inn.

On January 23, 1905, while Anderson reportedly labored to preserve a dead mouse, three men hauled in the stiff carcass of a wolf. After bickering over the price with the help of a translator, who later wrote about the incident, Anderson eventually purchased the wolf. At least one account explains that, two days earlier, the hapless wolf had strayed near a log pile while chasing deer, where hunters had promptly killed it. Initially, they had thrown the carcass away; but after hearing that Anderson had come to the village to buy dead animals, they decided to bring it to the Hōgetsurō Inn to see if it might fetch a price. After buying it, Anderson shipped the pelt from this wolf, along with a ghoulish menagerie of other specimens (including the Eurasian wild boar, Japanese deer, and even the rare Japanese serow), to London, where curators eventually placed them in the British Museum of Natural History. The Japanese wolf proved to be the last of its kind.[1]

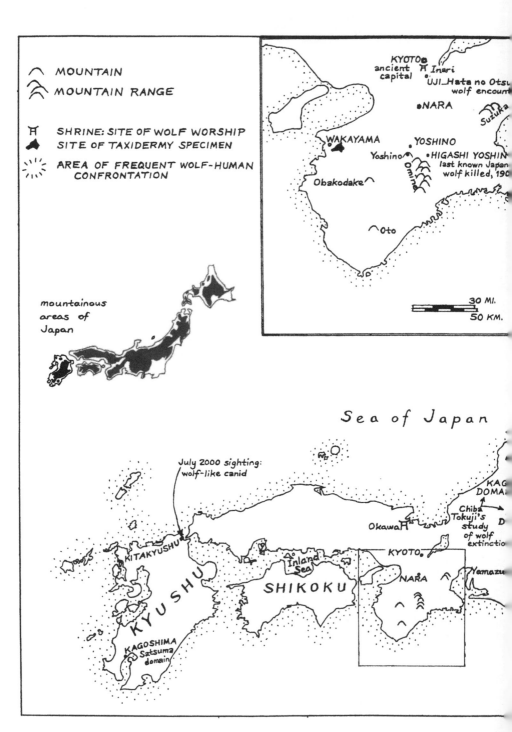

MAP 1. A wolf map of Japan.

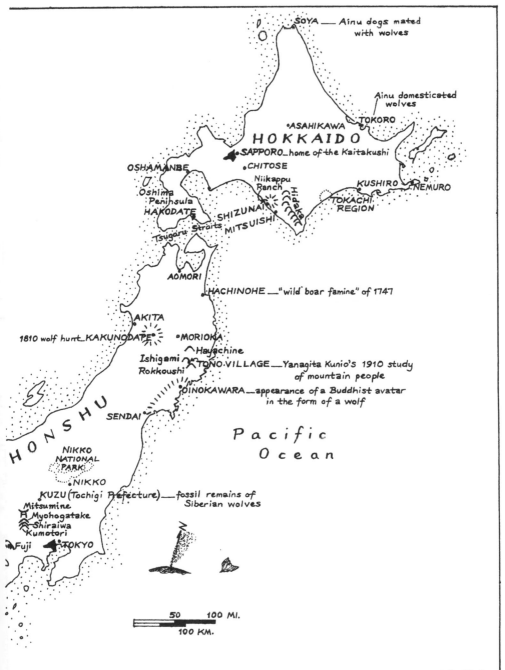

SOYA — Ainu dogs mated with wolves

Ainu domesticated wolves

TOKORO

ASAHIKAWA

HOKKAIDO

SAPPORO — home of the Kaitakushi

CHITOSE

Niikappu Ranch

KUSHIRO

NEMURO

OSHAMANBE

Oshima Peninsula

HAKODATE

SHIZUNAI

MITSUISHI

Hidaka

TOKACHI REGION

Tsugaru Straits

AOMORI

HACHINOHE — "wild boar famine" of 1747

AKITA

1810 wolf hunt — KAKUNODATE

MORIOKA

Hayachine

Ishigami Rokkoushi

TONO VILLAGE — Yanagita Kunio's 1910 study of mountain people

OINOKAWARA — appearance of a Buddhist avatar in the form of a wolf

SENDAI

HONSHU

NIKKO NATIONAL PARK

NIKKO

KUZU (Tochigi Prefecture) — fossil remains of Siberian wolves

Mitsumine

Myohogatake

Shiraiwa

Kumotori

Fuji

TOKYO

Pacific Ocean

N

50   100 MI.

100 KM.

D·MARTIN

Almost a century later, in March 2002, while researching and traveling in Japan, I visited the site where hunters had killed this last Japanese wolf. Outside my inn and down a narrow path, a clear stream rushed through deposits of large boulders; deep eddies, which spiraled away like small galaxies from the swift current, proved bountiful fishing holes for locals with their simple poles. Thinking back, the delightful aroma of what upland Japanese call "mountain cuisine" still makes my mouth water, particularly when I recall the salted trout and wild boar stew that became my nightly fare while I stayed at this and other locations. Today, at Washikaguchi (now called Higashi Yoshino), only a simple memorial stands where the last Japanese wolf once stood, a memorial hardly alluring enough to attract motorists and logging trucks as they speed by on the narrow roads that wind precariously around Yoshino's partially denuded mountainsides (see figs. 1 and 2). I remember thinking when at the memorial that the low grumbling sound of heavy machinery in the adjacent lumberyard, when combined with the rhythmic metallic chopping of a helicopter lifting cedar logs off a nearby steep mountainside, served as more appropriate statements than the bland one on the memorial about the cruel fate of the Japanese wolf and the landscape where it once lived.[2] As I poked around this sleepy village, my imagination got the best of me, and I began to wonder what it must have been like to be the last Japanese wolf—the last of one's kind.

The staff at the Higashi Yoshino administrative office eagerly copied for me information related to the memorial and to the last Japanese wolf; but they appeared more interested in the fact that I spoke Japanese and that I had come all the way from Montana to visit the wolf memorial. "Few Japanese people ever stop to see the memorial," explained one office worker. Of the many wolf sites I visited during my trip to Japan that winter, the stay at Higashi Yoshino to see the wolf memorial left me feeling the most perplexed. The visit exposed a troubling historical mystery that gripped my imagination and came to determine the direction and focus of my research on Japan's wolves.

I teach Japanese history, and in my heart, I have long believed that most preindustrial Japanese, because of their experiences with Shinto and Buddhist theologies—and because of the basic commonsense fact that most folks just lived closer to the land back then—empathized more with wild animals, even if these same Japanese did routinely hunt, kill, and eat them. But if this age-old stereotype about a Japanese "oneness" with nature was even remotely true, then what had happened to Japan's wolves? I mean, think of it this way: it is well known that most cattle ranchers in the American

FIG 1. Erected in March 1987, the Japanese Wolf Memorial stands near the banks of the Takami River in Higashi Yoshino, Nara Prefecture. The statue is a replica of the last wolf (a male), which hunters killed nearby in 1905.

West hated wolves and killed them throughout the nineteenth and twentieth centuries, even pushing them, at one point, to the brink of extinction with their poisons, traps, and guns. Yet miraculously, North American wolves survived the slaughter and once more inhabit—with the protection of the same U.S. government that once persecuted them, no less—my home state of Montana.[3] Japanese farmers, by contrast, once worshiped wolves as sacred animals, and upland villagers, like those from Higashi Yoshino, even went so far as to leave ceremonial dishes of red beans and rice next to wolf dens, humble offerings to these neighborhood "large-mouthed" gods (as Japanese sometimes called them) and their pups. Nonetheless, it was the once-worshiped wolves of Japan, not the relentlessly persecuted wolves of North America, that had become extinct by the early twentieth century.

FIG 2. The Japanese Wolf Memorial in Higashi Yoshino.

Later, during the train ride to Kyoto from the memorial, my mind could not help but wander back to the mountains of Higashi Yoshino. I thought, if history serves to commemorate past events and experiences, then perhaps a book such as this one, on Japan's extinct wolves, is a better way than the wolf memorial to remember these wolves and the hardy mountain people who interacted with them and to celebrate the life and death of a wild animal and vanishing culture that have been basically erased from the Japanese Archipelago.[4] Weeks later, however, once back in Bozeman, I read through my travel notes once more and reflected further on these initial reactions to the wolf memorial. I came to realize that a book such as this one serves as much more than a simple remembrance to wolves, because wolves, as living and breathing creatures who experienced from the inside Japan's eighteenth- and nineteenth-century transformation into an industrialized nation, represent a powerful subject with which to explore the many lessons, both cultural and ecological, hidden within this particular moment of Japan's natural and human history. Simply, over the centuries, Japanese culture has changed dramatically and so too have Japanese attitudes toward wolves and other animals. In part, these changes led to wolf extinction, a fact that exposes the ways in which culture shapes the relationship people have with the natural world, a relationship that should constrain human behavior, but when it does not, the implications for other creatures and the landscape can be devastating. Indeed, with Anderson's acquisition of the dead wolf in Higashi Yoshino, the last wolf in Japan—a member of one of Japan's two wolf subspecies: the Japanese wolf (*Canis lupus hodophilax* Temminck, 1839) or the Hokkaido wolf (*Canis lupus hattai* Kishida, 1931)—disappeared from Japan, and the world, forever.

What makes this story so compelling is that many Japanese once revered the wolf as Ōguchi no Magami, or Large-Mouthed Pure God. Grain farmers worshiped wolves at shrines, beseeching this elusive canine to protect their crops from the sharp hooves and voracious appetites of wild boars and deer. At Shinto shrines such as the one at Mitsumine in Saitama Prefecture, talismans adorned with images of wolves guarded worshipers from fire, disease, and other calamities; charms with images of wolves with their pups supposedly brought fertility to surrounding agrarian communities and to couples hoping to have children. Unlike biblical interpreters in the West who cast the wolf as symbolic of wilderness and hence as the devilish inhabitant of a "place without God," in the highly syncretic world of Japanese religion wolves emerged as animals who protected grain farmers from various hardships. Northward, on the island of Hokkaido, the Ainu, at least before being stripped of their culture after 1800, believed that their people were born from the union of a wolflike canine and a goddess. Much like some Native American cultures, Ainu folklore and epic poetry celebrated the wolf, and some Ainu communities even sacrificed wolves, along with bears and owls, in elaborate ceremonies called *iomante*, the "sending away." They believed wolves to be gods and their ancestors.

Two of the three major ethnic groups on the Japanese Archipelago (the third being the Okinawan people, who lived where wolves were not found) integrated wolves into their cultural landscapes and views of the natural world, and so, on a cultural level, an exploration of the extinction of wolves in Japan promises to expose how the process of nation building and industrialization in Japan's eighteenth and nineteenth centuries reconfigured relationships with the natural world in ways that led to the disappearance of these canines. This book focuses on the cultural shifts that accompanied Japan's modern transformation, an extended historical moment when animals, and even the mountains and forests they inhabited, lost their otherworldly status in the Japanese imagination and became "noxious animals" who needed to be killed and primitive landscapes that needed to be developed, in the context of Japan's burgeoning industrial order.

On an ecological level, by contrast, this book attempts to transcend the confining national boundaries of "Japanese" history and seeks to explain why one species—our species, *Homo sapiens*—has worked so tirelessly to destroy another. For this reason, this book is highly comparative, both historically and methodologically, and attempts to incorporate select theories regarding human and nonhuman ecologies into a broader analysis of interspecific conflict and animal extinction. In doing so, my point is in fact a simple one: if we speak of animals as exclusively creatures of instinct, or of humans as exclusively creatures of culture, we risk doing neither species justice and belying the complexity of human interaction with other animals and vice versa. To write on human interaction with wolves and not speak of biology and ecology is akin to writing about driving and not mentioning automobiles and roads.

THE EMOTIONAL LIVES OF ANIMALS

During the 1990s, canine specialist and historian Hiraiwa Yonekichi (1898–1986) published several important animal histories and ecologies. Many of the topics covered by Hiraiwa stemmed naturally from his intense passion for raising dogs, wolves, and other animals, and when such passions intersected with his desire to understand their ecology and history, it exposed the importance of personal experience, scientific inquiry, and historical analysis in formulating creative historical ideas regarding the place of animals in the past. Of Hiraiwa's many books, *Ōkami: Sono seitai to rekishi* (The wolf: Its ecology and history; 1992) is the most intriguing.[5] For me, Hiraiwa's research into Japan's wolf history provided a textual road map of sorts through the centuries of extremely complex historical documentation, folklore, biology, and even canine pathology.

From Hiraiwa we learn how, for example, rabies contributed to wolf attacks in the eighteenth century and how such epizootic concerns contributed to wolf extinction. For Hiraiwa—as for myself—Japan's wolves were more than relics of Japan's past and present cultural cartographies; they were living and breathing creatures with agency, experiences, and stories of their own. Hiraiwa tried to focus on what animals, particularly advanced social carnivores like wolves, might have to say, at least insofar as we can discern, about our writing of their histories.[6] Hiraiwa's passion for observing, researching, and interacting with wolves imbues his scholarship with an authority that transcends most other Japanese animal histories and cultural studies.[7] Simply put, when it came to wolves, he just knew what he was talking about because he took the time to get to know these animals as best he could. Other Japanese scholars often handle textual evidence with more theoretical savvy; Hiraiwa, however, possessed the scientific knowledge and personal experiences required to present a more balanced history that, at a basic level, took into account the behavioral or ecological agency of wolves.

This book, too, attempts to retrieve some statement from Japan's wolves, an extinct creature who harbors a yet unheard perspective on that country's past and who needs to be seen as another casualty of Japan's transformation into an industrialized nation. In the chapters to come, where I treat wolves as complex beings with emotional lives and historical agency, I was inspired by an 1872 work by Charles Darwin, *The Expression of the Emotions in Man and Animals*. At this point, having introduced the topic of animal cognition and emotion—the potential for emotional lives and historical agency among wolves—let us wade together into the dangerous waters of writings related to the emotional lives of animals, waters in which, as we all know, scholarly and scientific sharks have been swimming for well over a century. I promise that it is worth taking the risk, however, because understanding the emotional lives of animals materially and, perhaps more important philosophically, made writing this book possible.

In *Expression of the Emotions*, Darwin argued that evolution (by this time the hallmark of his thought) shaped the behavior, not just the physiological characteristics, of a given species, and so behavioral consistencies just as reliably identify a given species as do physiological ones. Behavior, that is, unites a species, such as wolves or humans, as part of recognizable taxonomic categories. But it was Darwin's blurring of the line between the emotional lives of humans and animals that drew me toward his research on emotional expression. Darwin traced the origin of some human expressions to earlier stages in our evolution, pushing us closer, philosophically speaking, to our animal cousins. "With mankind some expressions," wrote Darwin, "such as the bristling of the hair under the influence of extreme terror, or the uncovering of the teeth under that of furious rage, can hardly be understood, except on the belief that man once existed in a much lower and animal-like condition."[8] If emotions such as love and hate or courage and fear—the basic building blocks for the culture we celebrate as a species in our histories and hagiographies—are related to emotions shared by other advanced animals, then clearly animals need to be treated by historians less as cultural relics or instinctual automatons and more as historical beings with stories of their own.

Darwin's study of expression linked humans to a nonhuman past; but it also integrated many animals into a more human-like emotional present. To begin with, Darwin argued that physical expression serves as one means of communication among advanced social animals, including humans and wolves. "With social animals," Darwin explained, "the power of intercommunication between the members of the same community,—and with other species, between the opposite sexes, and as well as between the young and the old,—is of the highest importance to them." To these ends, both humans and nonhumans employ "inarticulate cries, gestures, and expressions" for intercommunication; what sets humans apart is their use of complex languages. So, when archival documents discussed in chapter 4 tell of wolves howling near settlements on Hokkaido, such "inarticulate cries," cries no doubt quite articulate to wolf communities, need to be seen as voices that expressed some kind of lupine emotion, even though loathed, feared, and misinterpreted by humans. Much like Japanese settlers, perhaps these wolves were afraid too, and they said as much; or perhaps they vocalized notions of territorial competition. Human beings cry out when calling for aid, when in intense pain, and when afraid: so too do animals such as wolves. Darwin wrote: "I have often recognized, from a distance on the Pampas, the agonized death-bellow of the cattle, when caught by the lasso and hamstrung. It is said that horses, when attacked by wolves, utter loud and peculiar screams of distress." Indeed, cries of terror appeared "almost universal" to Darwin. In the expression of emotions, in other words, humans and wolves share similar means of articulation, creating a common emotional space for human and nonhuman animals. Darwin explained: "He who will look at a dog preparing to attack another dog or a man, and at the same animal when caressing his master, or will watch the countenance of a monkey when insulted, and when fondled by his keeper, will be forced to admit that the movements of their features and their gestures are almost as expressive as those of man."[9]

I have heard wolves howl on many occasions, usually when the sun begins to set and the temperature begins to drop in the Lamar Valley in Yellowstone National Park. Darwin would have been delighted to learn that it indeed raised every hair on the back of my neck. On at least one occasion in chapter 4, I have tried to tap into my own natural impulses (the same ones that made my hair stand on end) to inform explanations of the human hatred of wolves; but rather than try to narrate the emotional responses of wolves to such hatred, I instead draw on the work of animal scientists most sensitive to the notion that advanced animals do have complex mental lives. Still, even though I refrain from too much speculation on the emotions and motivations of Japanese wolves, the sheer probability of such emotional lives made writing this book philosophically possible. As a historian trying to piece together the whole picture, these traces of wolf voices, so lifelessly recorded in archival documents, still haunt me. I can never feel that I have done justice to this topic without having tried to comprehend what those voices said. To me, it would be akin to writing Japanese history and not reading the Japanese language.

Darwin believed that evolution, and a similar facial physiology, helped explain why humans and nonhumans share so many emotional expressions. He went into excruciating detail to explain the physiology of a child weeping, including photographs of his own child doing so. He then went on to describe the physiological similarities in the weeping of elephants and humans, such as the contracting of the orbicular muscles, implying that most animals, including humans, possess the same muscular hardware for the expression of emotions. Hence, if nonhuman animals share with humans the same hardware for expressing emotions, then they must share, at least in part, the potential for similar emotional lives. Darwin wrote that human laughter resembled similar expressions in advanced primates. When laughing, the "lower jaw often quivers up and down, as is likewise the case with some species of baboons, when they are much pleased." Even the expression of love is not isolated to the human species. Darwin submitted that just as the expression of love in human cultures around the world includes a "strong desire to touch the beloved person," with animals "we see the same principle of pleasure derived from contact in association with love."[10] Simply stated, if our emotional lives are an important part of our past experiences and histories—that is, if our emotional lives are part of who we are as individuals and as members of nations and broader civilizations—then the implications of Darwin's study can be quite profound for those writing animal histories.

As I see them, the implications of Darwin's work are that advanced social animals such as wolves also experience historical lives, and so, as feeling beings with limited culture and vast experiences of their own, they too deserve actual histories, because they represent part of the greater cultural and ecological milieu that defines not just human histories but our shared natural histories. Moreover, the expression of animal emotions, expression that might be recorded by the discerning eye of the naturalist, can be read by the historian as a kind of "text" with which to give animals greater agency in historical narrative (just as historians might use anthropology to retrieve the lost voices of some nonliterate human cultures) and to help historians better understand why interspecific conflicts take the form they do.

Conversely, if humans share with animals the complex web of biology and culture that defines the basic nature of our being and behavior, then historians might read their own instincts and impulses as another kind of "text" that, if such emotional responses as the bristling of hair are "universal" to our species, links us to both our human and our nonhuman subjects in a profoundly organic way. In this sense, chapter 4 will take the topic of subjectivity in history to an entirely new level: perhaps we can, at certain moments, tap into our "animal" past to transcend culturally generated behavior and better understand our "human" present and vice versa.

Like Darwin, Konrad Lorenz also wrote about the emotional lives of animals. In *King Solomon's Ring*, Lorenz recalled, with both charm and scientific authority, his personal experiences with the myriad animals that he and his wife kept at their home, ranging from greylag geese and the larvae of the water beetle *Dytiscus* to ravens and wolflike dogs. A pioneer in the study of animal behavior, Lorenz wrote about the emotional lives and personal histories of animals; but a strong desire to convey to readers what he saw as the "infinite beauty of our fellow creatures and their life" also motivated Lorenz's life work. Lorenz wrote, "I take very seriously the task of awakening, in as many people as possible, a deeper understanding of the awe-inspiring wonder of Nature and I am fanatically eager to gain proselytes." Lorenz, again like Darwin, drew comparisons between the behavior of humans and nonhumans, such as when he likened the "war-dance" of fighting fish to the ceremonial dances of Javanese and other Indonesian peoples. More to the point, however, Lorenz wrote about the expression of emotions in animals and how it resembled, but also differed from, the expression of emotions in humans.[11]

"In the higher vertebrates," explained Lorenz, "as also in insects, particularly in the socially living species of both great groups, every individual has a certain number of innate movements and sounds for expressing feelings." Lorenz believed that "animal utterances" and "human languages" share only superficial similarities but that both humans and nonhumans do share the "mimetic signs" that communicate their moods. Following the development of language, Lorenz speculated, the need for highly specialized "mood-convection" diminished in humans, but animals continued to require such "minute intention-displaying movements." Likewise, Lorenz believed that animals remained better equipped to receive mood-convection signals. In a discussion of his relationship with the several dogs he owned, Lorenz wrote of the ability of these animals to read, with uncanny accuracy, innate signals, ones articulated through expression well before they ever took the form of spoken commands. Lorenz believed he was able to communicate with many of his animals, especially with his dogs and with his raven (ravens and crows are highly intelligent creatures and are discussed in chapter 5 of this book).[12]

If, as Darwin and Lorenz argued, nonhuman animals have complex mental lives and hence create cultures and experiences—if, in their own ways, they can love and hate, experience joy and suffering; and if, when horses cry out or elephants shed tears, they are articulating such emotional experiences—historians need to revise the ways they write about animals. Interestingly, power engenders a peculiar kind of ignorance: dominant humans almost never take the time to really get to know the peoples, plants, and animals they subordinate. One might speculate that historians ignore animals at the peril of ignoring part of the wholeness of our history and our selves, however. It is precisely this point that the controversial scholar Jeffrey Moussaieff Masson and his colleague Susan McCarthy addressed in *When Elephants Weep* (1995). Masson and McCarthy argued that animals do experience complex emotional lives, but that, because of the specter of anthropomorphism, scientists have avoided exploring this emotional world in the context of ethology and other studies of animal behavior. In effect, scientists have ignored the best available road map for exploring the emotional lives of animals: their own emotional lives.

Masson and McCarthy submitted that science has become content to simply describe what animals do and to rehash the old Neo-Darwinian line about how such actions fit into theories of "ultimate causation" or how different species obtain reproductive success. Emotions, or what Masson and McCarthy described as those subjective experiences that define our being, should also be taken into account when people interact with, or study and write about, the lives of animals. "Biographies without grief, sadness, and nostalgia would appear unreal" if written about people, wrote Masson and McCarthy. They continue:

> An ordinary person's life in which no one loves, is loved, or wants to be loved; in which no one fears anything; in which no one becomes angry or makes anyone else angry; in which the depths of despair remain unfathomed; in which no one feels pride in anything they do; in which no one is ever ashamed to do anything or feels guilty if they do—this would be an unnatural, unrealistic, paltry description. It would be neither believable nor accurate. It would be called inhuman. To describe the lives of animals without including their

emotions may be just as inaccurate, just as superficial and distorted, and may strip them of their wholeness just as profoundly.[13]

Masson and McCarthy identified the social and biological constructions of power as explanations for why humans treat other animals with such indifference: "Dominant human groups have long defined themselves as superior by distinguishing themselves from groups they are subordinating," and so, depriving animals of emotional lives, much as white slave-owners once deprived African Americans of rational lives, serves as a tool with which to subordinate them. Linking animal subordination to identity formation, Masson and McCarthy explained, "People define themselves as distinct from animals, or similar when convenient or entertaining, in order to keep themselves dominant over them." As a consequence, humans can, with little guilt, benefit from the only slightly bridled exploitation of other creatures: "Human beings presumably benefit from treating animals the way they do—hurting them, jailing them, exploiting their labor, eating their bodies, gaping at them, and even owning them as signs of social status."[14] To remedy this situation, philosophers and legal scholars such as Peter Singer, Gordon Brittan, Steven Wise, and others have made cases for the existence of mental lives among nonhuman animals and therefore better ethical treatment and legal rights for them; but historians, though in the vanguard of retrieving the voices of women, ethnic minorities, and other underrepresented groups around the globe, have largely ignored the voices of nonhuman animals when narrating the past.[15] Yet, all the while, wolves still howled and people heard and interpreted their calls.

Despite the diverse methodologies employed in these pages, and as tempting as it might have been, I have not tried to narrate the emotional lives of wolves when telling their histories. When speculating on the roots of certain instances of conflict between humans and wolves, I rely on wolf science, ethology, animal pathology, and the basic commonsense assumption that human and nonhuman behavior is the result of ecology and culture. I wanted to give wolves agency and voices—credible agency and voices. Stanley Young and others have attempted to narrate the emotional lives of some rather famous wolves. In my opinion, however, the results were disastrous. In particular, Young, a division chief in the U.S. Bureau of Biological Survey who killed wolves for a living, celebrated the killing of some of the most famous "outlaw" wolves in the American West by narrating their experiences and emotions in the *The Last of the Loners* (1970). He tried to reproduce what it felt like to be hunted relentlessly by humans.

According to Young, when these wolves killed cattle or evaded the poisoned carcasses left by federal wolf hunters (called "wolfers"), these wolves—such as "Old Three Toes" and "Rags, the Digger"—were not simply feeding themselves in a changing environment or fleeing sites of known or perceived danger. Rather, "Old Three Toes" and her mate, a collie named Shep, experienced "the spirits of the wild, they played in open sunny places, and then on dark nights trotted through the dusk, slaughtering and gorging in a bacchanalian orgy of blood feast." We learn that "Rags, the Digger," relished "[k]illing, slaughter, blood, warm and steamy with the victim burbling out its dying breath as he ripped at the steaming carcass." For Young, such wolves came to resemble canine partisans defiantly fighting the most noble kind of battle—a losing battle—over the fate of their species: "They were great leaders, superb outlaws, these last loners. They deserved and received the profound respect of those men who finally conquered them. Defiant, they were striking back at man, playing a grim game but never acknowledging defeat until fate had called the last play." Of course, the only respect Young and his wolfers ever afforded these wolves was to kill them, and so his sentimental drivel should be ignored; but the notion of a war between humans and wolves is not as ridiculous as it might sound, as we shall see in the pages ahead.[16]

Besides Stanley Young's books on wolves, several excellent wolf histories have been written in the United States (though only a precious few have demonstrated an interest in exploring the forces that motivate wolf behavior). However, because I refer to this literature so frequently in the narrative, discussing it here risks later redundancies.

SELF, SUBJECT, AND SUBJECTIVITY IN HISTORY

In *Desert Solitaire*, Edward Abbey wrote about the harsh but beautiful landscape of Arches National Park near Moab, Utah. Abbey never intended his writings to be confused with a tour guide, and so he cautioned his readers against trying to reproduce his experiences with this unforgiving landscape by just driving around Moab. "Do not jump into your automobile next June and rush out to the canyon country hoping to see some of that which I have attempted to evoke in these pages," Abbey wrote from a bar in Hoboken. "In the first place you can't see *anything* from a car; you've got to get out of the goddamned contraption and walk, better yet crawl, on hands and knees, over the sandstone and through the thornbush and cactus. When traces of blood begin to mark your trail you'll see something, maybe."[17]

I admit that the hidden voices of the extinct wolves of Japan insisted much the same thing of me: "Get out of the goddamned archives and go see Japan's few remaining wild places and Montana's wolves for yourself," they said. And so, on arriving at Montana State University in Bozeman, I began to research and write under the assumption that, when dealing with animals and the natural world in historical narrative, some experiences simply cannot be discovered in the stained folders of an archive. Though unruly to be sure, Abbey's preaching applied to me because, considering the speed with which I narrate swaths of historical time and geographic space in this book, in a metaphoric sense at times I too needed to get out of my automobile-like historical mind-set and just walk, or sometimes even crawl, through some of the places where wolves once lived, or do live, and where Japanese once worshiped and killed them. That is, I came to equate traditional historical methodologies to a kind of automobile, an impersonal metal shell racing down the highways of the human past at mind-boggling speeds.

Usually, historians see only things strategically placed at the side of the road, things like state- and industry-sponsored advertising billboards in the form of "official" written documents, and even these pass by with blurring speed. With this book, I wanted to do more than simply look out the windows of the impersonal metal shell that history can be. I wanted to get out of that automobile, even if only in a limited fashion, and crawl through some of Japan's most obscure historical documents, rugged forests, and sacred shrines, not to mention some of Montana's wildest places where wolves live. Most of the shrines discussed in chapter 2, for example, particularly Mitsumine Shrine, once represented sacred sites for practitioners of Shugendō, or "mountain asceticism," and so I ventured into these same forests and mountains, touching and smelling the same waterfalls, viewing the same rugged peaks, all along trying to conjure a past human vision that saw this as a landscape alive with wolves and other otherworldly deities.

In contrast, I did not have to conjure a landscape alive with wolves when I did research with the Yellowstone Wolf Project Winter Study, on the other side of the planet from the extinct wolves of Japan. Subjecting myself to a landscape alive with wolves was an attempt to probe the human expression of emotions as they relate to wolf conflicts, an attempt to trigger something inside me that might better help me explain the creation of a culture of wolf hatred. One night in March 2004, I even guarded calving pastures with a couple who are friends and ranchers in the Tom Miner Basin near Yellowstone, a majestic location notable for its jigsaw-puzzle-like jurisdictions of private, national-forest, and national-park lands; but superimposed over all of these human forms of landownership is also a wolf one. The Tom Miner Basin belongs to the Chief Joseph wolves as well. When we shined our flashlights into the aspen groves surrounding the pasture, several pairs of yellow and red eyes stared back, and I realized, for the first time, how difficult it must have been for Japanese ranchers to defend their livestock from hungry wolves.

I was born in Montana and have spent a great deal of time in rural and northernmost Japan, and so this book naturally took on a more personal tone. The process of writing this book resembled at times the experiences narrated by Wallace Stegner in *Wolf Willow*, although nothing in these pages is as fictitious or as gracefully written. Stegner wrote about his native ground, on the Canada-Montana border, from both a personal and a historical vantage point; but he also offered important insights into the role of personal narrative in historical writings, particularly ones that strike close to home. "Easing watchfully back into the past" is how Stegner described his journey home and his research into his home's rich past, and so I too decided to ease into a different kind of past, or at times an ecological present, one alive with howling wolves, eviscerated elk, Japanese mountain communities, and shrines devoted to wolf worship. Stegner came to believe that an "ancient, unbearable recognition" of a place can be realized through scents and shapes, a recognition more authoritative than the dreamlike world of memory and history, and so in this spirit I decided to take in deeply the odor of life and death when, with Yellowstone researchers, I went to the sites of wolf kills, ripped the jaws from devoured elk to measure teeth, and stained my hands with elk blood and wolf scat. Like the plains embraced by Stegner, the "colors and shapes that evoke my deepest pleasure" are in Montana and the hinterlands of Japan, and so these places, quite naturally, became the subject of this book.[18]

A BLIP IN NATURAL TIME

It should be obvious by now that this book draws on a multifaceted array of sources and intellectual traditions to solve the historical mystery of why Japan's once-worshiped wolves became extinct. For this reason, I have organized the chapters according to topical and interpretive themes rather than string them together as a chronological narrative.

Chapter 1 is a history of Japanese taxonomic traditions, as well as a natural history of Japan's two subspecies of wolf. It covers the emergence of the first wolves of Siberia about a million years ago through the evolution of Japan's subspecies. This chapter also probes the ways in which Japanese naturalists have, from ancient times to the present, interpreted wolf behavior and taxonomically categorized them; I pay special attention to the nineteenth century and the adoption of the Linnaean system. Forcing wolves into Linnaean categories was complicated by the fact that prior to the nineteenth century Japanese commonly used two names to describe wolflike canines: *yamainu* (mountain dog) and *ōkami* (wolf). On the one hand, medieval and early modern Japanese naturalists debated whether these two canines were distinct animals, and they mustered fascinating physiological and behavioral evidence to make their respective cases. Modern naturalists, on the other, ignored these earlier debates and argued that these two names referred to one canine—the Japanese wolf—exposing how the Linnaean system homogenized the complexities of Japan's ever evolving wildlife. This discussion reveals how taxonomic debates mirrored the birth of the modern Japanese nation in the late nineteenth century; many scientists glossed debates regarding "mountain dogs" and "wolves" in order to create the singular *Nihon ōkami*—the "Japanese wolf." This is admittedly a somewhat difficult and detailed chapter to start off with, one that analyzes topics ranging from contemporary craniometries to Neo-Confucian encyclopedias, but my hope is that it will elucidate how Japanese naturalists have invented and imagined wolves throughout the ages. To be frank, just using the word "wolf" in this book is tricky, and I wanted to make this fact as translucent as possible.

Chapter 2 departs the worlds of East Asian and Linnaean taxonomies to explore the place of wolves in the mountain traditions of both the ethnic Japanese and the Ainu of northern Japan. Wolf worship occurred at the site of ancient imperial authority and within the context of traditions of nature worship throughout Japan, from the early Shinto and Buddhist theologies of the Japanese to the animistic order of the Ainu. And so this chapter promises to take us on a journey to some of Japan's most ancient shrines, sacred sites, and wild places to trace how wolves became deities in the Japanese and Ainu imagination. Wolves served as divine messengers at Shinto shrines in some Japanese communities and protected the crops of peasants from wild boars and deer, while the Ainu viewed wolves as the high-ranking deity Horkew Kamuy, and they sacrificed wolves in elaborate ceremonies called the "sending away." Not only do these two distinct traditions of wolf worship resemble each other (thus bridging the vast cultural divide separating these two peoples), but both traditions attempted, in their own ways, to create natural and supernatural realities to understand and control the decidedly complex and uncontrollable habits of such intelligent carnivores as wolves (a theme I will return to in the Epilogue). Ultimately, creating and killing wolves represent two approaches to controlling them.

Even though Japanese worshiped wolves, they also killed them, particularly after the spread of rabies in the eighteenth century. In a bizarre ecological episode unique to global wolf history, in the eighteenth century wolves became rabid man-killers in many parts of Japan, and wolf hunts, designed to cleanse the landscape of what many Japanese saw as demons, often looked more like ceremonies than hunts. The point I want to emphasize in chapter 3 is that although Japanese killed wolves in the medieval and early modern periods, these canines still remained, with some exceptions, part of preindustrial natural and supernatural realities. Even as wolves killed and ate Japanese travelers in the eighteenth century, most Japanese still viewed wolves as deities who lived in the otherworldly space of mountains and forests, and for this reason they ceremoniously killed these demons or offered prayers to competing deities, such as the "pasture deity" (*makigami*), hoping to protect livestock from wolf predation.

If Japanese killed wolves in highly ceremonial events in the eighteenth century, the nineteenth century witnessed the decidedly unceremonious destruction of Japan's wolves. To contrast the difference between wolf killing in the decades before and after the Meiji Restoration of 1868—Japan's celebrated modern moment—chapters 4 and 5 explore wolf killing on the northern island of Hokkaido. No longer the deities or demons of earlier descriptions, in the nineteenth century wolves became "noxious animals" and the Kaitakushi (Hokkaido Development Agency) mobilized its resources to extirpate wolves from the newly colonized Hokkaido. These chapters reveal that in the nineteenth century, Japanese transformed from a people who invented natural and supernatural realities in an attempt to understand and control wolves, revering them as divine messengers or as deities at shrines, to a people focused on eradicating them with the industrial technologies of strychnine and modern policies of bounty systems. This process did not occur overnight; nor did the advent of modern Japan. Rather, both were gradual processes that occurred over decades. By the late nineteenth century, Japanese no longer imagined control over the natural world but exercised it, and in the post-Darwin world, Japanese, for the first time in their long history, viewed themselves as the alpha species of their islands—the fittest in the struggle of the "survival of the fittest" and no longer subordinate to Japan's myriad animal deities.

Chapter 5 investigates the Kaitakushi's creation of an elaborate bounty system, one modeled after that of the American West and designed to eliminate the Hokkaido wolf and other "noxious animals" such as bears and crows. But the Kaitakushi did not always have to kill wolves to subordinate them. When not killing wolves, the Kaitakushi and other government agencies manipulated wolves in the cultural contexts of zoological gardens and museums and with taxidermy, orchestrating, to the delight of the Japanese public, human superiority over the natural world through the manipulation of animal carcasses and their fabricated settings. Taxidermy, seen in this light, could be used to represent supreme human control over once uncontrollable animals. With taxidermy, wolves can be made to snarl when we want them to, forcing them to conform to our expectations and interpretations, whether these are "accurate" depictions or not. I then contrast these cultural forces with speculation regarding certain natural ones, suggesting that as the Japanese colonized Hokkaido, they confronted the Hokkaido wolf in a war—rooted in both culture and ecology—over deer and territorial dominance, a type of conflict common between different species struggling over the ability to pass down their genes within a limited territory. Consequently, I reserve discussion of wolf behavior for chapter 5, where I make my case that we need to at least consider a wolf's-eye view of Japan's war against the wolf to do justice to our—and their—complexity as a species.

Chapter 6 explores prominent Japanese naturalists and their theories of wolf extinction. Not only did such naturalists offer fascinating ideas regarding why Japan's wolves disappeared so suddenly in the nineteenth century, but their thinking also sheds light on the development of a Japanese discipline of ecology. Such Japanese naturalists, particularly when writing about wolf extinction, presented themselves as "organismic" ecologists (who had a holistic view of animal societies and their environments) and often regarded humans as natural players in ecological and evolutionary change. Japanese ecologists often turned to historical and anthropological sources to craft their theories (much like contemporary environmental historians do), because in these sources lay answers to some of Japan's most vexing ecological questions. It is because of their belief that ecosystems functioned as a kind of holistic organism, and because of their belief in a balance between human and nonhuman forces in generating ecological change, that I situate such scholars as Imanishi Kinji, Yanagita Kunio, Inukai Tetsuo, and Chiba Tokuji within Japan's discipline of ecology.

Together, I hope these chapters paint a compelling picture that illustrates the role of science, culture, and the environment in the extinction of Japan's wolves. In the final analysis, however, I have no doubt that, despite the importance I assign to this topic, the extinction of Japan's wolves will become only a minor blip in an otherwise alarming downward trend tracing the mass extinctions—the Sixth Extinction, as some call it—that originated with the advent of our species. The Paleozoic era (ending about 250 million years ago), the Mesozoic era (about 90 million years ago), and the Cenozoic era (about 65 million years ago) witnessed five mass extinctions, and so there is nothing new about massive disappearances of biota. What makes the Sixth Extinction unique, however, is that humanity, in our unbridled need for total control of the planet and its resources, has caused it.[19]

I know that the cries of wolves that eerily split the crisp winter nights of Hokkaido more than a century ago are only one voice in a chorus of voices warning us of the silent future that surely awaits us.

1

SCIENCE AND THE CREATION
OF THE JAPANESE WOLF

On a warm, Rocky Mountain morning, while in my office preparing to teach a summer class, I received a letter from Inoue Katsuo, one of my teachers and a professor of Japanese history at Hokkaido University in Sapporo. I put down my lecture notes and eagerly ripped opened the envelope. Along with several prominent articles Inoue had recently published, I discovered a newspaper clipping from the *Asahi shinbun* (Asahi newspaper) that featured a color picture of some sort of wild dog. "I thought you might be interested in this," he wrote. As I read the article, I began to realize that when I had ripped open this letter I had opened a metaphoric Pandora's Box, one that would force me to explore the taxonomy of Japan's wolves and, more generally, the cognitive and cultural origins of Japan's natural history.

I learned that on the evening of July 8, 2000, Nishida Satoshi, a high school principal from Kitakyūshū City, in Fukuoka Prefecture, had photographed the medium-sized canine with its tongue hanging from its mouth as it loped within about ten or fifteen feet of him. The photograph, although only of grainy newspaper quality, shows that the canine was mostly gray and black, but that it also had shades of orange on its legs and behind its ears (see fig. 3). As it turns out, Nishida was an animal enthusiast, and he explained to *Asahi* reporters that the wild dog had approached him while he was hiking and then it had just vanished into the mountains. The animal was, quite literally, some sort of mountain dog. Later, the photograph was shown to Imaizumi Yoshinori, former zoologist at the National Science Museum in Tokyo (Kokuritsu Kagaku Hakubutsukan), who said that the

西田智さんが九州中部で撮影したニホンオオカミによく似た動物

明治時代初期に福島県で捕獲されたニホンオオカ
ミのはく製＝東京都台東区の国立科学博物館で

オオカミとみられるとの回

が群れをつくって生活す　増え、オオカミの生息条件

FIG 3. The headline from this *Asahi shinbun* (Asahi newspaper) article on the spotting of a wolflike canine on Kyushu Island reads, "Japanese Wolf? Feral Dog? Debate over an Observation on Kyushu." The top image is the photograph of the wolflike canine taken by Nishida Satoshi. The bottom image is of the Japanese wolf specimen at the National Science Museum, Tokyo, Japan.

animal resembled the extinct Japanese wolf. Needless to say, with the apparent sighting of an animal that now inhabits only the Environmental Ministry's "extinct species" list, the incident caused quite a stir in Japan, and over the next several weeks the photograph was plastered over most major dailies. Controversy surrounded the photograph, however. Maru-yama Naoki, of the Tokyo College of Agriculture and Industry (Tōkyō Nōkō

Daigaku), questioned Imaizumi's suggestion that the animal was actually a wolf. Currently head of the Japan Wolf Association and spearheading an effort to reintroduce wolves to that country, Maruyama explained that, during his studies of Mongolian wolves, even to get within a quarter mile of one was extremely rare, and so he doubted that Nishida could get so close.[1]

In the interview with *Asahi* reporters, Maruyama questioned the existence of wolves in Japan for ecological reasons as well. He explained that a mating pair of wolves can have seven to eight pups per year, and that this family becomes the core of what later develops into a pack. Maruyama thought it unlikely that a lone wolf would be spotted. The dynamics of wolf reproduction, combined with activities in Japan that have caused an explosion of deer numbers, led Maruyama to speculate that if Japanese wolves still existed, there would be significantly more sightings. He added that even more reports could be expected of their distant howls in the mountains. Maruyama admitted that the animal in Nishida's photograph shared some characteristics with the Japanese wolf, but to him, it appeared to be a German shepherd hybrid of some kind. The article went on to explain that Aimi Mitsuru, of Kyoto University's Primate Research Center (Reichōrui Kenkyūjo), had some time ago contacted Leiden (where one of the best-known specimens of the Japanese wolf remains housed) and elsewhere about obtaining samples for DNA testing; such samples might be compared with samples of hair miraculously recovered near Kitakyūshū City. These attempts to obtain DNA from preserved specimens had failed, however, because the harsh chemicals used in the taxidermy process ruined the samples. Even with quality samples, distinguishing dogs from wolves with mitochondrial DNA proves complicated because wolves and dogs became distinct species only about 15,000 years ago.[2] More recently, scraps of dried flesh recovered from a wolf skull discovered in a private home in Kōchi Prefecture, on Shikoku Island, yielded inconclusive results: though wolf DNA to be sure, it nonetheless differed from more-common Eurasian wolves.[3] Basically, the newest marvel of modern taxonomy, mitochondrial DNA testing, failed to distinguish between wolves and dogs with the Kitakyūshū, Leiden, and Kōchi specimens.[4]

One possible genetic reason for the inability of DNA testing to make definitive distinctions between Japanese dogs and wolves might be that Asian spitz-type breeds (which include Japanese breeds such as the Shiba and Akita) are now understood to be the closest genetic relatives to wolves and to the earliest "pariah dogs," the latter of which originated in Asia and migrated alongside nomadic peoples around the globe. Quite simply, not only do Japanese breeds phenotypically resemble wolves, but they also genetically resemble wolves and these early nomadic pariah dogs. As Heidi Parker and her colleagues have noted, this makes genetic testing unusually complicated, particularly if any hybridization between the species occurred in earlier times.

As I stuffed Inoue's letter back in its envelope, I thought that where DNA testing has failed, surely a historical inquiry could solve this vexing mystery of the taxonomic status of Japan's wolves. Even with available sources, however, making any taxonomic determination is complicated because the zoological category of wolf in Japan is a surprisingly recent historical construct, one that did not really solidify in the minds of scientists until the beginning of the twentieth century. The emergence of *Nihon ōkami*, or the Japanese wolf, became possible only with the emergence of many other distinctly "Japanese" things at the turn of the century, such as Japan's unique brand of ethnic nationalism and its imperial ideology.[5] Prior to the early twentieth century, the categories of canine remained diverse and dependent on social situations and ecological contexts; wolves (*ōkami*), sick wolves (*byōrō*), mountain dogs (*yamainu*), honorable dogs (*oinu*), big dogs (*ōinu*), wild dogs (*yaken*), bad dogs (*akuken*), village dogs (*sato inu*), domesticated dogs (*kai inu*), and hunting dogs (*kari inu*) all loped across the boundaries of status and of occupational, religious, and regional understandings of the categories of canine. Only in the early twentieth century, after debates spawned by the introduction of the Linnaean system and after changes brought by the Meiji Restoration, did the stable category of Japanese wolf emerge in the context of the development of Japan's modern national identity. It is this modern Japanese wolf—a product of the twin historical forces of the birth of the modern Japanese nation and the development of its own zoological sciences—that this book explores.

The introduction of the Linnaean system to Japan in the 1820s reconfigured botanical and zoological taxonomies, and so, for this chapter, the Linnaean moment serves as the epicenter of several enduring controversies over Japan's wolves, controversies resurrected by the sighting near Kitakyūshū City. Of course, these controversies tell us something valuable about the way the natural world was classified prior to Japan's Linnaean moment, when the joint legacies of imported Neo-Confucian and, simultaneously, more native Japanese orders ruled discussions regarding the classification of wolves and other animals. But they also tell us something about how the scientific categories inherent in the supposedly "universal" Linnaean system often proved counterintuitive to earlier, more ecologically and regionally valid methods of classifying the natural world.[6]

Scott Atran, an anthropologist of science, has argued that the Linnaean system was actually "universal" in a cognitive sense because it was rooted in earlier folk-biological "life-form" and "generic-specieme" categories shared by most human societies, but ones that Carolus Linnaeus, in the eighteenth century, converted into scientific categories of "species" based on morphotypes and the "rational intuition" of science. Atran insists that with human beings, "there are not *any* intrinsic differences in our cognitive dispositions to classify natural" objects, and Linnaean categories originated from this basic human cognition. That they made sense naturally to early modern Japanese scientists is one reason Linnaean taxonomies so easily transferred and translated into the Japanese context. For Atran, the similarities between European and East Asian folk-biological taxonomies are the result of "a biological conception of the world common to folk everywhere," and this sort of propensity for rational thinking and even science, whether in Europe or in East Asia, "marks the true bounds of our common vision of the world" as humans.[7]

In Japan, with the rise of Western learning, European classifying systems piqued the cognitive interests of many Japanese natural scientists, who, by the first half of the nineteenth century, began uprooting plants from the pots of earlier Neo-Confucian and Japanese taxonomies and planting them in the Western scientific ones introduced by Dutch traders and their German doctors. In this way, the "universal" qualities of the Linnaean system were important in another sense, one less biological than cultural. Like Japan's adoption of the Gregorian calendar and acceptance of world time in the latter half of the nineteenth century, the Linnaean system classified the plant and animal life of Japan according to the same symbolic logic as that used by the Western powers. One suspects that, with the adoption of the Linnaean system, Japan took the first step, in the arena of science, in placing itself on a cultural and cognitive parity with the countries of Europe and the United States.

No matter how progressive this enterprise was to Japan's nineteenth-century modernizers, however, the hasty application of the Linnaean system still failed to address critical taxonomic issues regarding Japan's wolves. This failure opened the door for continued questions (many raised by postwar scholars who harkened back to Japan's pre-Linnaean past) over whether Japan's wolves were really, according to the universally accepted Linnaean system or otherwise, wolves at all. This is one reason the Kitakyūshū City sighting had caused such a stir in Japan's newspapers.

JAPAN'S PRE-LINNAEAN WOLF TAXONOMIES

In the seventeenth and eighteenth centuries, Japan's natural scientists were scholars for whom the semantics of Chinese cosmology, the nomenclature of natural order, the principle inherent in all things, the folklore of the fantastic, not to mention the authenticity of the historical, figured prominently in their categorical logic. Strict morphologies, internal anatomies, the reproductive "seeds" of natural creatures, and other aspects of the "rational intuition" inherent in Linnaean science were far from the minds of these scholars. Of course, they actively ordered their world in a biological and cosmological sense, but they did so with, to use Michel Foucault's words, the "grid created by . . . a language," in their case the elite East Asian written language of Chinese ideographs and pictographs (called *kanji* in Japanese), and with the precedent set by that language in the canons of Neo-Confucian encyclopedic writings from China.[8] Confucius had taught that language provided order in the form of the "rectification of names" (Chinese *chengming*; Japanese *shōmei*). As he explained in the *Lunyu* (Analects), "when the gentleman names something, the name is sure to be usable in speech, and when he says something this is sure to be practicable. The thing about the gentleman is that he is anything but casual where speech is concerned."[9] The learned gentleman strove for order through "speech"—the accurate use of names and language—and so too did early Japanese natural scientists.

Later, because Neo-Confucianism became essentially state supported and intellectually embedded in early modern Japan's educational and scientific landscape, natural scientists emphasized the study of past writings, mostly by certain Chinese and, to a lesser extent, Japanese scholars, and participated in only a limited amount of biological discovery until the late eighteenth and early nineteenth centuries. One historian summed up this climate when he wrote: "Tokugawa Confucianism was not a progressive branch of study, constantly pushing at the frontiers of new knowledge. All that was worth inventing had been invented by the Sage Emperor; all that was worth knowing had been known by Confucius. The task of later generations was simply to absorb this body of knowledge passively and with humility."[10]

For the Chinese scholar Zhu Xi, whom historians credit with synthesizing Confucian humanism with Buddhist and Taoist creeds in the twelfth century to form Neo-Confucianism, this "body of knowledge" and the study of nature did reveal an order, just not the scientific order and natural history advocated by Linnaean taxonomists. Rather, Zhu Xi, much like Confucian scholars before and after him, believed that the "investigation of things" (*kakubutsu* in Japanese), including things in the biological world, exposed the inherent principles of a universal moral order that mirrored and, subsequently, legitimized human social and political hierarchies. Indeed, the study of nature only affirmed the natural place of people within their setting. As Zhu Xi wrote, if scholars "investigate moral principle, everything will naturally fall into place and interconnect with everything else; each thing will have its order. . . . In learning, you should desire nothing more than to understand this moral principle."[11] Most early modern biologists followed this sage advice and investigated the natural world to better understand the relationship between humanity, nature, and social virtue. For this reason, those who became interested in the natural sciences were often part of a broader category of scholars who studied traditional Chinese medicine (*kanpō*) or, later, Dutch medicine, perhaps the two most respected arenas of scientific inquiry in preindustrial Japan because they produced the most social good. Many of the early modern natural scientists discussed in this book, such as Noro Genjō, Ono Ranzan, and Itō Keisuke, came from such medical backgrounds.[12]

Though sophisticated, the taxonomies constructed by Japan's seventeenth- and eighteenth-century natural scientists remained, at least according to Scott Atran's framework, folk-biological in nature for three reasons. First, early modern Japanese taxonomies "represent a holistic appreciation of the local biota" and were centered on immediate human needs. For this reason, the alleged medical and culinary properties of plants and animals remained the two most prominent organizing principles of Japanese taxonomies. Second, the individual taxons, the taxonomic qualities that defined them, and the broader categories that hierarchically ordered them "reflect gross morphological patterns that subsume any number of generic-speciemes," ones recognized as commonsense by most people. But this commonsense understanding of the natural world did have its limitations. That is, relying on human cognition, most peoples around the globe naturally classified wolves and dogs as taxonomically similar species; but "post-rational" symbolic speculations, ones culturally generated, then "go beyond the immediate and manifest limits of common sense" to imbue such animals, even though taxonomically similar, with conflicting meanings. Thus, though their DNA differs only slightly, dogs have coevolved next to humans; wolves, on the other hand, have been persecuted to the brink of extinction. Western countries that practiced animal husbandry policed the boundary between wolf and dog zealously; grain-farming civilizations and hunting peoples such as the Japanese and Ainu did not. Finally, Japanese taxonomists primarily studied local ecologies, even though, with the rise of Western learning, new and exotic plants and animals from around the world became features of some medieval and early modern taxonomies.[13]

Where Japanese science differed from traditional folk-biological taxonomies in Europe and elsewhere, however, and at times exhibited early modern characteristics of full-blown science is in the attempt by some Japanese naturalists to identify distinctions among certain generic-speciemes with simple morphotypes, which means that they attempted to distinguish certain plants and animals by their distinct physiological features—both their external and their internal natures—and not exclusively by their social virtues, life habits, or ecological settings. That is, Neo-Confucianism, with its inherent focus on investigating the innate principle of all objects, in some respects abstracted the study of natural things much as modern science does, allowing objects to be removed from their setting and placed within "scientific" taxons.[14] This is the first task in the creation of a bona fide Linnaean species and in the emergence of scientific thinking. To some degree, this was already under way in Japan even prior to the importation of the European classifying system, apparently bolstering Atran's thesis about the cognitive origins of science.

One reason early modern Japanese taxonomy relating to wolves remains so complicated is that scholars who wrote on the Japanese names for natural things (*honzō wamyō*) referred to two types of wild dog: "wolves" and "mountain dogs." Hiraiwa Yonekichi suggested that references to the mountain dog in Japan prior to the nineteenth century can be explained in three ways: the term was another name for wolf, the animal was a distinct species that resembled a wolf, or mountain dogs were just feral dogs running wild in the mountains.[15] Part of the problem, explained Hiraiwa, stems from the Chinese character for "mountain dog" (Chinese-style reading, *sai*; Japanese-style reading, *yamainu*), which early on was combined with *rō* (or *ōkami*, the Japanese-style reading for "wolf") to form the confusing compound *sairō*, which reads something like "mountain dog-wolf." Hiraiwa insisted that Japanese took this character out of context because in China *sai* often refers to the dhole, or wild Asian hunting dog (genus *Cuon*), an animal distinct from the wolf.[16] However, early Japanese glossed the terms *yamainu* (*sai*)

and *ōkami* (*rō*) as simply referring to the same animal. Tenth-century taxonomies such as Fukane Sukehito's *Honzō wamyō* (Japanese names in natural studies; 922) made this claim, as did Minamoto Shitagau's "Wamyōrui
shūshō" (A collection of Japanese names; 931–37). For classical and medieval
taxonomists (scholars who proved less concerned with the local realism of
scientific discovery than with relaying the "bookish practices" of classical
Chinese categories), the Chinese character *sai* (mountain dog) represented
the same animal that Japanese had traditionally called *ōkami* (wolf).[17]

Scholars of the seventeenth and eighteenth centuries reversed these claims,
however. One assumes that obvious problems within medieval taxons
required that they reject the precedents of Chinese antiquity in favor of local
biological realities. Early forms of "nationalism" are important here as well,
because in the medieval and early modern periods Japanese began to
explore their cultural and political differences, not just their similarities,
vis-à-vis China and elsewhere, and taxonomy came to reflect these identity
shifts.[18] Nonetheless, most early modern taxonomies distinguished wolves
from mountain dogs by identifying the "habits of life" of the animal, their
physiology and behavior, their local ecological settings, and in what capac-
ity they benefited or menaced people. Hitomi Hitsudai, in *Honchō shokkan*
(Culinary mirror of Japan; c. 1695), established eight different botanical and
zoological life-form categories, which were based primarily on ecological
setting and simple morphotype. Moreover, as the title of his "culinary" encyclopedia suggests, Hitomi also focused on the social virtue of biology,
specifically its ability to provide food, medicine, and other useful products
for people, and in this respect his categories remind one of the "gastronomic
taxonomies" written by English natural scientists.[19]

In *Honchō shokkan*, Hitomi provided wolves and mountain dogs with
their own generic-specieme taxons and simply stated that wolves were edible and mountain dogs were not. Interestingly, Hitomi wrote that some
mountain people liked to eat wolves because they believed that the meat
made them "courageous," even though it was reportedly "tough." In terms
of basic life-forms, Hitomi prefaced his multivolume taxonomy with a discussion of water, fire, and earth, the basic building blocks of traditional Neo-
Confucian cosmology, and then discussed cereal grains, vegetables, fruits,
birds, fish (*rinbu;* literally, "creatures with scales"), wild and domesticated
animals (*jūchikubu*), and snakes and insects (*jachūbu*).[20]

Hitomi explained, in vocabulary resembling that used by later Linnaean
taxonomists, that wolves and mountain dogs both belong to the "same class"
of animal (*dōrui*), and so the wolf resembled a large dog and belongs to the

same "genus" (*zoku*) as the mountain dog. But wolves and mountain dogs warranted individual taxons because of their morphotypes and life habits. Hitomi argued that wolves were "more robust" and mountain dogs were "thinner," while the head of the wolf was sharper and the muzzle more tapered than that of the mountain dog.[21] The wolf also had recognizably long, slender legs, and its howl could be heard at greater distances. The front of the wolf was taller than the rear, a characteristic opposite that of the mountain dog. Wolf pelts were "a mixture of yellow and black or a blue ashen color." In terms of life-habit descriptions, Hitomi wrote that "every spring and summer evening, wolves are seen not in the mountain forests but in the towns, where they steal and eat oxen, horses, chickens, dogs, and young girls."[22] Other prominent early modern taxonomies, such as Ono Ranzan's *Honzō kōmoku keimō* (An instructional outline of natural studies; 1803), highlighted similar morphotypes and echoed Hitomi's points about the different physiological characteristics of wolves and mountain dogs, including the long-standing belief that wolves had webbed feet (and hence were strong swimmers) while mountain dogs did not. Ono also reiterated that meat from the mountain dog was inedible (in fact, "poisonous") and that both wolves and mountain dogs lived in the mountains, where they caught and ate game, but when deep snow fell in the winter and spring and food became scarce, they sometimes ventured into villages, where they injured people.[23]

Obviously, as Hitomi's and Ono's descriptions attest, wolves could be dangerous, but Kaibara Ekken, a prominent Neo-Confucian scholar, emphasized the troubling nature of the mountain dog as well. Kaibara believed in the inherent order of the entire natural world and so turned to writing taxonomy.[24] He prefaced his taxonomy, *Yamato honzō* (Natural studies in Japan; 1709), with discussions of Neo-Confucian cosmology and the fecundity of heaven and earth. His taxonomic categories, though original in many respects, were certainly influenced by the Ming scholar Li Shizhen's encyclopedic *Bencao gangmu* (Compendium of materia medica; 1596) and divided the botanical world into (1) cereal grains and other edible foods and drinks and (2) wild plants, which were classified as vegetables and greens, medicinal herbs, useful plants, and several other varieties. He concluded the botanical section with (3) trees. His zoological taxonomies, on the other hand, categorized creatures along the lines of river and ocean fish, aquatic and terrestrial insects, shellfish, and several kinds of birds, and then he commented on animals (*jūrui*) and people (*jinrui*).[25] Kaibara wrote, as did Hitomi and Ono, that anatomically the mountain dog did not resemble the wolf and that, behaviorally speaking, their dispositions differed as well. The moun-

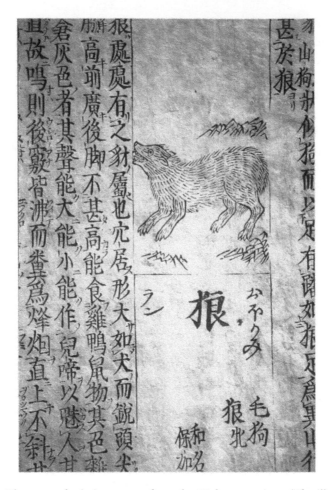

FIG 4. These two depictions come from the *Wakan sansai zue* (The illustrated Japanese-Chinese encyclopedia of the three elements; 1713): (*a*) the *ōkami* (Japanese wolf) (*this page*); (*b*) the *yamainu* (mountain dog) (*facing page*). Courtesy of the Resource Collection for Northern Studies, Hokkaido University Library, Sapporo, Japan.

tain dog, he wrote, actually had a more unpredictable temper than the wolf, which had a "good disposition."[26]

The *Wakan sansai zue* (The illustrated Japanese-Chinese encyclopedia of the three elements; 1713), compiled by Terashima Ryōan, also lists the wolf and mountain dog under separate generic-specieme types, taxonomically distinguishing them much along the life habits and morphotypical lines as

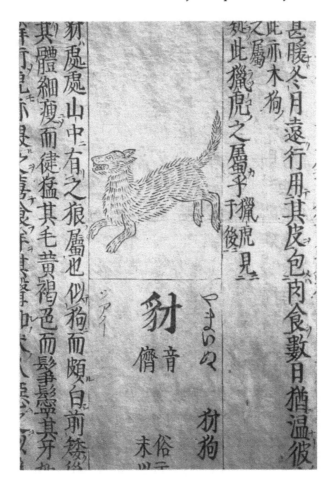

other encyclopedias. As a historical source, the *Wakan sansai zue* stands out as one of the most comprehensive early modern encyclopedias anywhere; the model for the work was Wang Qi's *Sancai tuhui* (Illustrated encyclopedia of the three elements; 1607), another Ming-period work representing the "Chinese classificatory system of knowledge" that so influenced early modern Japanese science.[27] Terashima divided the zoological section into animals and rodents but also included monsters and other mythological creatures. Like Kaibara's description of the dangerous predisposition of mountain dogs, the *Wakan sansai zue* comments that "people traveling in the mountains are much more frightened of mountain dogs than wolves."[28] Illustrations provided in the *Wakan sansai zue*, moreover, depict the mountain dog as exhibiting a behavioral trait of a domesticated dog. Namely, its tail is held high and curled, similar to well-known Japanese breeds. By contrast, the wolf has its tail lowered, as wolves, in the wild, are prone to do for reasons of social communication (see fig. 4).[29]

So far, I have argued that Japan's early modern natural science, though Neo-Confucian in origin and folk-biological in nature, did contain the seeds of modern forms of scientific classification, particularly in its propensity to identify individual generic-specieme taxons by morphotypical principles (not just by the social virtue of animals or their ecological settings) and to understand the basic principles behind the zoological "genus," as when Hitomi described wolves and mountain dogs as being part of the same *zoku*. It is important to point out that, considering the similarities of the descriptions contained in the *Honchō shokkan, Honzō kōmoku keimō, Yamato honzō*, and *Wakan sansai zue*, earlier Chinese and Japanese encyclopedias, not intensive zoological field research, probably provided much of the information for these early modern taxonomies. Nonetheless, the early modern tradition allowed Japanese natural scientists to quickly adopt the Linnaean classifying system in the first half of the nineteenth century, erasing, at least temporarily, Japan's folk-biological past and replacing it with a modern biological present. In this respect, as Atran argues, Japan's folk-biological experience "provided an intuitive underpinning and empirical approximation for the scientific species," which made for a smooth nineteenth-century transition to the Linnaean system.[30] That is, it all appears smooth so long as we disregard the complicated story of Japan's wolves, which, of course, we will not.

NINETEENTH-CENTURY TAXONOMY: JAPAN'S LINNAEAN MOMENT

The Japanese natural scientist whom historians often credit with introducing the Linnaean classification system to Japan is Itō Keisuke.[31] He was born into a celebrated Nagoya family in 1803; Itō's father and elder brother, both accomplished physicians and herbalists, made several important medical and botanical contributions to Owari domain, a region famous for its production of cutting-edge scientists. In 1815, Itō followed in the footsteps of his father and brother and began studying Chinese and Western medicine and botany. In 1828, Itō traveled to Nagasaki, where he met (actually, for the second time; the first had been in April 1826) Philipp Franz von Siebold, who lectured the twenty-five-year-old physician and herbalist on the Linnaean system. Unfortunately, their studies together were cut short by

FIG 5. This depiction of the Japanese wolf appears in volume 5 of Philipp Franz von Siebold's *Fauna Japonica* (1833). Siebold bought specimens of the mountain dog (*jamainu*) and wolf (*okame*) at the Tennōji Temple in Osaka in 1826. Courtesy of the Hokkaido University Library, Sapporo, Japan.

political scandal, but later, in 1861, they continued their discussion of the Linnaean system. The product of the earlier collaboration, and perhaps Itō's best-known work, is the *Taisei honzō meiso* (Western botanical names; 1829), which based its classification of botany on *Flora iaponica* (The flora of Japan; 1784), a monumental encyclopedia published in Latin by the Swedish scientist Karl Pieter Thunberg.[32]

Siebold was from a famous German family of Würzburg physicians. Between 1824 and 1828, while employed by Dutch traders, he maintained a small academy of Western learning just outside the port city of Nagasaki, called the Narutaki School. Two years before Siebold took Itō as a student, the German doctor had acquired a specimen of the Japanese wolf and a specimen of the mountain dog, the latter of which appears to have become the Linnaean standard for the Japanese wolf in Siebold's *Fauna Japonica* (Fauna of Japan; 1833). *Fauna Japonica*, the comprehensive catalog of Siebold's collecting and classifying while in Japan, ignored many (but not all) of the folk-biological debates discussed earlier in this chapter and collapsed

FIG 6. (*a*) *Yamainu* (mountain dog) (*this page*), called *C. hodophilax* in Tanaka Yoshio's "Dōbutsugaku" (Zoology; 1874). (*b*) *Yamainu* (mountain dog) (*facing page*) in Tanaka Yoshio's "Dōbutsu kunmō: Honyūrui" (Instruction on animals: Mammals; 1875). Once more, Tanaka listed the mountain dog as a "wolf." Courtesy of the Resource Collection for Northern Studies, Hokkaido University Library, Sapporo, Japan.

the distinctions between wolves and mountain dogs. In its description, *Fauna Japonica* equates the "new species of wild dog, Jamainu of the Japanese," with the "Japanese wolf." Coenraad Jacob Temminck, the first director of the National Museum of Natural History in Leiden, actually wrote the mammalogical sections of *Fauna Japonica* after Siebold had contacted him from

(WOLF.)
CANIS HODOPHILAX, SIEB.

Japan in 1824. (Hermann Schlegel, also in Leiden, coauthored the sections on birds, reptiles, and fish.)

Siebold had bought the two canine specimens at the Tennōji Temple in Osaka in 1826 while on his way to a meeting in Edo (present-day Tokyo), the capital of the Tokugawa shoguns. He was accompanying the head of the Dutch factory in Japan, Johan Willem de Sturler, as a secretary and had departed Nagasaki on February 15, 1826.[33] Of the two specimens, Siebold labeled one "okame," or wolf, and the other "jamainu," or mountain dog.[34] Whoever sold these specimens to Siebold offered them as distinct generic-speciemes with different folk names. Temminck, however, appears to have received only one whole specimen in Leiden and so ignored the distinction, listing the "jamainu" specimen he received as the "Japanese wolf," which suggests that the proper wolf specimen might never have reached Leiden at all (see fig. 5). *Fauna Japonica*'s designation of the "jamainu" as "Canis hodopylax" thereby established the early Linnaean precedent, followed throughout the late Tokugawa and Meiji years, that mountain dogs were actually Japan's wolves.

Immediately before and after the Meiji Restoration, top Japanese zoologists, because of the Linnaean classification of the mountain dog as "Canis hodopylax" in *Fauna Japonica*, came to believe that Japan's mountain dog was a species of wolf. In 1874, for example, Tanaka Yoshio, a prominent Meiji zoologist, listed together both the wolf (*ōkami*), "which lives in eastern Europe," and the Japanese mountain dog (*yamainu*), which he characterized as a new species of wolf: "*Nihon-san yamainu; C. hodophilax* Siebold" (see fig. 6). Tanaka's precedent, as his citation demonstrates, was the confusing *Fauna Japonica*.[35] The next year, in another zoological text, in a description reminiscent of those found in early modern encyclopedias, Tanaka again described the Japanese mountain dog as a wolf:

> [The *yamainu*] lives in the mountains of the various provinces, where it catches and eats small animals. Its shape resembles that of a dog and it can get exceedingly large. It has a strong disposition. . . . In the winter and spring, when the snow accumulates and food is scarce, it sometimes comes close to houses; and it sometimes injures people or horses. When it gives birth to its young in the mountains and forests, all villagers bring food because it is said that it will not injure people who have fed it.[36]

Tanaka bought the Linnaean categorization of the mountain dog wholesale, but he retained the folk-biological descriptions of the mountain dog from the seventeenth- and eighteenth-century encyclopedias. Tanaka took seriously Japanese stories regarding the mountain dog, but he also took seriously the Linnaean system. For Tanaka and other nineteenth-century zoologists, the mountain dog became Japan's wolf, largely because *Fauna Japonica* had made it so; and with the extinction of the Japanese wolf in 1905 (and serious questions regarding decaying specimens in both Japan and abroad), contesting this Linnaean classification became next to impossible. Nonetheless, the classification of the mountain dog as Japan's wolf not only created disjuncture between Japan's folk-biological past and its nineteenth-century Linnaean moment but later, in the twentieth century, continued to haunt debates among natural scientists regarding the accepted Linnaean status of Japan's wolves, ones raised in part by sightings such as the wild dog spotted near Kitakyūshū City.

IMPERIAL LINNAEAN TAXONOMY

Unlike discussions of wolves and mountain dogs among early modern naturalists and nineteenth-century pioneer Linnaean scientists, early-twentieth-century wolf classification by Japanese scientists became more concerned with the historical process of species creation and was influenced by the rise of nationalism and the fledgling Japanese empire. No doubt the importation of Darwinian thought better acquainted Japanese natural scientists with how, at least according to the logic of natural selection and evolution, species form in the first place. But in at least one instance—the case of the Korean wolf—the geopolitical turmoil of European and Japanese imperialism, not only Darwinian evolution, appears to have partially influenced Linnaean classification. Let us focus our attention in this section on the natural history of Japan's wolves and the formal Linnaean classification of the Hokkaido wolf as told by early-twentieth-century scientists. We will also keep an eye on the role that nationalism and imperialism may have played in the Linnaean classification of the Korean wolf and, more importantly, the disappearance of the name "mountain dog" in favor of *Nihon ōkami* to describe the "Japanese wolf." Remember, the Japanese wolf disappeared, and was subsequently studied and renamed, within the context of Japan's frenzied patriotism after its victory in the Russo-Japanese War and its expansion onto the continent.

Of course, many millennia earlier than these scientific debates, in prehistoric times, the wolf had achieved a Diaspora rivaled among mammals only by the human species and possibly lions, one that saw its dispersion to virtually every corner of the nontropical Northern Hemisphere. Scientists speculate, however, that the species *Canis lupus* was actually born on the Eurasian continent. Sometime about one million years ago, a canine, probably *Canis priscolatrans*, migrated from North America to the Eurasian continent, crossing Beringia, a land bridge spanning the Bering Strait that formed when glaciers lowered water levels. While in Siberia, this canine grew much larger and became the Siberian wolf. Some of its descendants migrated back to North America, recrossing the Bering Strait sometime between 600,000 and 300,000 years ago, in the mid-Pleistocene.[37] Others found their way to the Korean Peninsula and the Japanese Archipelago, which at times in the late Pleistocene were linked by coastal plain.[38] Ron Nowak, a leading wolf taxonomist, has argued that today the lupine Diaspora is represented by fifteen subspecies: six in North America and nine in Eurasia.[39]

Twentieth-century scientists argue that, on this consolidated prehistoric Korea-Japan landmass, three subspecies of wolf evolved. The first was the Hokkaido wolf, which had a historic range of Hokkaido and Sakhalin Islands, the Kamchatka Peninsula, as well as Iturup and Kunashir Islands just east of Hokkaido in the Kuril Archipelago (see fig. 7).[40] The second subspecies, the Japanese wolf, was restricted to the main Japanese islands of Honshu, Shikoku, and Kyushu (see fig. 8). The third subspecies was the Korean wolf (*Canis lupus coreanus* Abe, 1923), which lived on the Korean Peninsula. Some controversy surrounded the Linnaean classification of the Korean wolf, and at least one British scientist maintained that the animal was really a variation of the Chinese (or Mongolian) wolf (*Canis lupus laniger* Hodgson, 1847).

Abe Yoshio, of Hiroshima Bunrika University (Hiroshima Bunrika Daigaku), argued that the Korean wolf was not a variation of the Chinese wolf but in fact a distinct subspecies. Expanding on theories related to evolution and geographic isolation, Abe submitted that in prehistoric times the Korean wolf had separated from the Chinese subspecies and that eventually it developed its own morphotypical traits.[41] However, referring to Abe's skull measurements and pelt-color descriptions, R. I. Pocock, of the British Museum of Natural History, countered that the Korean wolf was not a distinct subspecies but rather a variation of the Chinese wolf.[42] Abe later responded to Pocock's challenge by emphasizing that the canine's slender muzzle, at least when compared to the Chinese wolf, warranted the animal's own Linnaean taxon.[43]

It should not be surprising that the historical backdrop to this scientific debate included such diplomatic developments as the late-nineteenth-century Kanghwa Treaty and Tientsin Convention, not to mention the annexation and colonization of Korea by Japan after 1910, all events that initially pushed for international recognition of Korea as an "independent" country—or at least a country no longer within the orbit of the Chinese tributary order. Eventually, the Japanese empire began exploiting Korea as a "protectorate" without political strings attached to China, which was once the dominant power in the region. Seen in this light, classifying the Korean wolf as a variety of the Chinese kind (as Korea the "satellite" kingdom had once been classified under the Chinese tributary order as a subordinate country) still resonated politically as well as scientifically in the 1920s. Obviously, at a certain level, scientific discussions of the day were being influenced by Japan's colonial presence on the Korean Peninsula.[44]

FIG 7. The only specimens of the Hokkaido wolf. Courtesy of the Hokkaido University Museum of Natural History, Sapporo, Japan.

FIG 8. This male wolf taken from Fukushima Prefecture in the early Meiji period is the best specimen of the Japanese wolf. Courtesy of the National Science Museum, Tokyo, Japan.

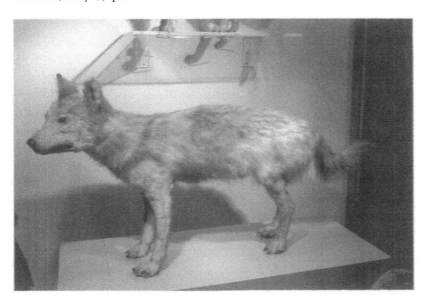

British taxonomy, as described by Harriet Ritvo, generally worked according to the Linnaean system as well, but nationalism and geopolitical dis-
putes also entered the picture, particularly when British scientists were
deciding which species to classify. That is, the British tended to apply the
Linnaean system to animals that were found within the boundaries of their
empire.[45] Similarly, some Japanese scientists, such as Abe, appear to have
erected Linnaean categories in part on the basis of political borders. Indeed,
throughout much of Japanese history, "folk taxonomy," as evidenced by
Kaibara Ekken's *Yamato honzō*, wrestled with extracting individual zoological taxons from the "bookish" precedents of Chinese antiquity and placing them within national boundaries, and so Abe's claim is consistent with
earlier traditions in Japan. Simply, the animal appears to have been a Korean
wolf—*C. l. coreanus* Abe, 1923—because, along with disputable morphotypical traits, it lived within the political geography of Korea.

Unlike the Japanese wolf, the Hokkaido wolf had not been given a Linnaean classification in Siebold's *Fauna Japonica*.[46] This work never listed
the Hokkaido wolf (or other Hokkaido biota for that matter), and so it is
not surprising that some early explorers of the southern Kurils, such as
H. J. Snow, assumed that the wolves spotted on Iturup and Kunashir Islands
were the same species as the Japanese wolf.[47] Earlier, in 1890, George Jackson Mivart, in *Dogs, Jackals, Wolves, and Foxes*, had compared skull specimens of the Japanese and Hokkaido wolves held at the British Museum of
Natural History and concluded that the skull from the Hokkaido wolf was
indeed significantly larger. But Mivart explained away the difference by say-
ing that the size disparity may only represent "local varieties" of the same
subspecies.[48]

In 1913, Hatta Saburō also became impressed by the size of the Hokkaido
wolf and suggested, unlike Mivart, that the animal might be related to the
Siberian wolf. A lack of viable specimens at the time, however, prohibited
Hatta from making a formal Linnaean classification.[49] Later, other Japanese
scientists added their voices to the debate, stating that the Hokkaido wolf
should be classified as distinct from the Japanese wolf.[50] By 1931, Kishida
Kyūkichi, using measurements of a skull taken from a Hokkaido wolf killed
in June 1881 in the Toyohira area, near Sapporo, argued that the canine was
distinct and should be given the name *C. l. hattai* (see fig. 9).[51] This was confirmed only four years later, in 1935, when Pocock measured a specimen
housed at the British Museum of Natural History. He argued that because
of its large size the canine should be named *C. l. rex*. The British Museum
of Natural History had acquired this specimen in November 1886, but little

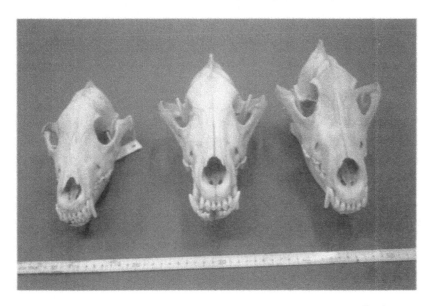

FIG 9. Three of six extant skull specimens of the Hokkaido wolf. Courtesy of the Hokkaido University Museum of Natural History, Sapporo, Japan.

else is known about it. Eventually, the nominal *hattai* became the most commonly used scientific name to describe the extinct Hokkaido subspecies. The Hokkaido wolf, as craniometric data from the British Museum of Natural History illustrate, was indeed larger than its southern cousin (see table 1) and is still considered to be a distinct subspecies.[52]

Interestingly, at the outset of these debates regarding Korean and Japanese wolves, the folksy name "mountain dog," once seen by such nineteenth-century pioneers as Tanaka Yoshio as suitable for Japan's wolf, vanished and was replaced by the more nationally centered and scientifically legitimate name *Nihon ōkami*, or "Japanese wolf."[53] No one person declared that the age of the mountain dog was over, and the name seems to have just slowly disappeared from scholarly writings. Obviously, the rise of ethnonationalism in Japan, geopolitical competition abroad (similar to those forces that influenced the birth of the Korean wolf), empire building, and wolf extinction on the Japanese Archipelago contributed to the disappearance of the "mountain dog." But lingering taxonomic controversies surrounding the Japanese wolf, even after its 1905 extinction, ensured that the mountain dog (in modified form to be sure) would appear in the scientific spotlight once again after the Pacific War.

TABLE 1. Osteological Comparison of *Canis lupus hodophilax* and
Canis lupus hattai Specimens at the British Museum of Natural History

Measurement	*C. l. hattai*	*C. l. hodophilax*
Age/sex	Adult/unknown	Young adult/male
Total length	269.5 mm	186 mm
Basilar length	237 mm	172 mm
Nasal length	105 mm	62 mm
Zygomatic arch width	147 mm	102 mm
Greatest width of brain case	68.8 mm	59 mm
Greatest distance of orbital processes	68 mm	42 mm
Orbital distance (shortest)	49 mm	30 mm
Greatest palatal width	78.5 mm	64 mm
Rostrum width at base of Canini	43.5 mm	32 mm
Width of incisor	32 mm	27.5 mm
Length of the palatinum	48 mm	31 mm
Palatal length	132 mm	97 mm
Diameter of tympanic bulla	29 x 31.2 x 15.5 mm	21.5 x 21 x 11 mm
Length of mandible	201 mm	140 mm
Height of mandible at m1	35.5 mm	24 mm
Height of ramus	81.5 mm	56.5 mm
Width of ramus	51 mm	35 mm
Length of pm4	28.5 mm	20.5 mm
Length of m1 + m2	27 mm	23 mm
Width of m1	23 mm	20 mm
Length of m1	30 mm	26 mm
Length of the sectorial of m1	22 mm	17 mm
Length of the crown of m1 + m2	21 mm	17 mm

SOURCE: Abe Yoshio, "On the Corean and Japanese Wolves," *Journal of Science of the Hiroshima University* Series B, Division 1, 1 (December 1930): 33–36.

POSTWAR DEBATES:
RETURN OF THE MOUNTAIN DOG

Extinction triggered renewed scientific interest in Japan's wolves after the war. In the late twentieth century, two competing postwar theories emerged regarding the evolution and extinction of Japan's wolves: those of the "natural evolutionists" and the "artificial evolutionists."[54] The natural evolutionists posited that migration, isolation, and environmental change on the prehistoric Japanese Archipelago contributed to the evolution of Japan's wolves. These scholars highlight natural forces, such as geographic isolation, dramatic changes in climate, and prey-species size and distribution, as central in shaping the evolution of the two subspecies. For this reason, I call them natural evolutionists. The artificial evolutionists argued that crossbreeding with ancient dogs or a process of semidomestication caused morphological changes in the prehistoric Siberian wolf and led to the emergence of a hybrid species of sorts—the mountain dog—and that these evolutionary changes can be traced through osteological and physiological studies, as well as historical documents. Artificial forces—that is to say, influences emanating from the human-fashioned world, or the "familiar" as in *Canis familiaris*—influenced the coevolution of the two subspecies.[55]

Nakamura Kazue, of the Kanagawa Prefectural Museum, best represents the natural evolutionists. In an important 1998 article defending the Linnaean classification of Japan's wolves as members of *Canis lupus* and employing a "chronocline" model of evolutionary "dwarfing," he argued that Japan's natural history, specifically its becoming a chain of small islands, produced a unique variety of wolf. Nakamura began by stating that the Siberian wolf, once on the prehistoric Korea-Japan landmass, evolved into Japan's two subspecies. Fossil remains of the Siberian wolf have been unearthed throughout much of the Japanese Archipelago, which proves that the larger canine inhabited the islands prior to its extinction just before historic times.[56] Highlighting natural events, Nakamura submitted that its extinction related to spiraling ecological changes, largely caused by climatologic shifts and resulting fluctuations in prey-species size and population.

To explain this, Nakamura focused on the late Pleistocene, about 20,000 years ago, when the flora of Honshu constituted a mixture of subarctic and temperate coniferous growth and when alpine glaciers formed only at the highest elevations. The subarctic coniferous zone primarily characterized northern Honshu, from about present-day Aomori Prefecture to Shizuoka Prefecture. This is where migrant Siberian wolves probably lived (although their fossils have been unearthed as far south as Kyushu). Large ungulates such as moose, steppe bison, and even horses also inhabited the region in the late Pleistocene. The climatologic conditions were perfect for these large herbivores, and so they were also perfect for large predators such as Siberian wolves, brown bears, and, eventually, the early hunting-and-gathering human populations.[57] Even after mainland Japan separated from the continent at the Korea and Tsushima Straits, warm southern currents still failed to enter the Japan Sea, translating into little evaporation and less snowfall than the archipelago has experienced in historic times. Moreover, in the summer a frigid air mass frequently hovered to the north while a torrid air mass usually stayed to the south, meaning that much of the mainland normally received little moisture. These were poor conditions for the development of the thick, lush forests that graced Japan in historic times, but they did create ideal open grasslands mixed with subarctic and temperate growth.[58] Nakamura insisted that such conditions provided excellent habitat for large ungulates and their predators, including the Siberian wolf.

The northern island remained connected to the northern part of the continent via Sakhalin Island, and so more animals continued to migrate south to Hokkaido (and even Honshu) along this bridge. The narrowest sections of Tsugaru Strait are between Cape Shiokubi and Cape Ōma, and between Cape Shirakami and Cape Tappi, where the distance between Hokkaido and Honshu ranges from ten to fifteen miles. By the mid-Holocene epoch, roughly 6,000 years ago, the Tsugaru Strait was open; but it is still possible that as it narrowed with ice, or as chunks of ice broke off and floated across the strait, migrating animals continued to make their way to Honshu. Wolves in search of a mate or new territories, called "dispersers," could have made this journey. Studies in Denali suggest that wolves around three years of age often disperse from their natal territories, traveling an average of 88 miles or, on occasion, over 100 miles.[59] Dispersals as far as 550 miles have been recorded by scientists, situating Hokkaido well within the range of the Siberian wolves of the Amur Estuary and Sakhalin Island.

As the Tsugaru Strait continued to widen throughout the Holocene, it isolated Honshu from Hokkaido, and those wolves on the main Japanese islands became separated from other Siberian wolves. Climatologic warming trends then hit the archipelago and the coniferous forests slowly yielded to primarily deciduous ones, a natural process that together with rising sea levels ate up precious open grasslands and with it large-ungulate habitat. This in turn led to well-documented mass extinctions among many mammal species. By about 13,000 years ago, nearly all the large ungulates that had migrated onto the prehistoric landmass had died off, with the straggler *Sinomegaceros yabei*, a kind of giant deer, taking its last breath on Hokkaido sometime around 7,000 years ago. Today, the largest surviving ungulates south of Hokkaido are the Honshu deer, the Kyushu deer, and the Eurasian wild boar.

Drawing on a range of studies on the relationship between geographic isolation and evolution, Nakamura argued that these extinctions sparked an adaptive process known as dwarfing that altered Japan's wolves, explaining, at least in part, why the Japanese wolf is smaller than all but one other subspecies of *Canis lupus*. According to this hypothesis, the Siberian wolf, under a variety of ecological and climatologic pressures, simply evolved to be smaller over time to meet the changing environmental demands of the Japanese Archipelago.[60]

The Hokkaido wolf, according to Nakamura, evolved into about the same size as the Jōmon wolf (10,000–300 B.C.E.); it experienced the dwarfing process less intensely than its southern cousin.[61] This makes sense, Nakamura explained, because Hokkaido sits at a higher latitude, and throughout most of its natural history, its environmental conditions have resembled those of Siberia. Prey size was also bigger, with the now-extinct red deer (*Cervus elaphus*) and the Hokkaido deer both being significantly larger than the subspecies of deer on the main islands. Scientists speculate that deer, fish, and the occasional beached whale served as the Hokkaido wolf's primary food sources.[62] Finally, and perhaps most importantly, the Hokkaido wolf probably continued to have genetic interaction with dispersing Siberian wolves. The Japanese wolf, now physically isolated much farther to the south, did not.[63]

The picture painted by Nakamura was that of a Linnaean scientist, but one influenced by such ecological theories as dwarfing and chronocline evolution. He highlighted the natural-historical changes that took place in prehistoric Japan but also showed how these changes shaped the distribution and evolution of grasslands, forests, prey species, and, ultimately, predators such as wolves. The prehistoric Siberian wolf that migrated onto the Japanese Archipelago became isolated from the global lupine community and evolved into the Japanese and Hokkaido wolves to meet the changing demands of its small island home. Yet even Nakamura admitted that one question raised by his morphometric and historical analyses of Japan's wolves remained unanswered with his chronocline model. That is, why do Japan's wolves display certain morphotypical traits that resemble those of dogs like the Shiba or Kishū, both ancient Japanese breeds, more than those of other wolves? It is a point, he conceded, that needs further exploration, but he floated the possibility that the isolation of the wolf on the Japanese Archipelago and the restriction of its "range of activity" as a result of human settlement and land clearing over the course of two millennia (Japan's population expanded from about 250,000 some 5,000 years ago to about 5 million by the eighth century[64]) might have sparked evolutionary changes that made the wolf resemble the domesticated dog.[65] For the artificial evolutionists, by contrast, this critical taxonomic question, a question that cuts directly to the morphotypical traits and hence taxonomic indicators of the animal, represented a point of departure for a different theory regarding the evolution and Linnaean classification of Japan's wolves.

The dean of the artificial evolutionists was the late Naora Nobuo (1902–85) of Waseda University. In *Nihon-san ōkami no kenkyū* (Research on Japan's wolves; 1965), Naora boldly went against the grain of most mainstream taxonomy, including Japan's Linnaean bible, Siebold's *Fauna Japonica*. He argued that the morphologic changes that the Japanese wolf underwent related to a historical process of crossbreeding with dogs, implying that Japan's wolves might not be "authentic" wolves at all.[66] Like natural evolutionists such as Nakamura, Naora too drew on ecological models of Japan's natural history, explaining that the breakup of the prehistoric Korea-Japan landmass and the isolation of the Japanese Archipelago from the continent need to be seen as critical events in the evolution of Japan's wolves. "After the early Holocene epoch," wrote Naora, "on the islands that became Japan, without actual fences or chains, a situation similar to a pasture for raising animals emerged." This event, he argued, led to ecological ruin, at least from the perspective of the Siberian wolf. Within a period of about 10,000 years, isolation and a restricted range forced the wolf into what Naora called an "unnatural life ecology" and then disadvantaged it through crossbreeding with dogs within what modern scientists call a "hybrid zone," in this case encompassing the entire Japanese Archipelago. The process drove the Siberian wolf to extinction and assimilated or absorbed their descendants into the physiological world of dogs. In short, little intraspecific homogeneity existed among Japan's wolves, in neither the historical nor the osteological record, making taxonomy unusually complicated.[67]

Of the Hokkaido wolf, Naora argued that this subspecies is also of suspect authenticity as a bona fide member of *Canis lupus*. He explained that in skulls once used in Ainu ceremonies, such as those collected by Sugiyama Sueo, the zygomatic arch (the area of the cheekbone) is rather pronounced, a characteristic some claim to be unusual in wolves.[68] A true Linnaean scientist, Naora insisted that wherever wolves live around the world, whatever their ecological setting or perceived social virtues, they have to be morphotypically similar to one another. Generally, their skulls are characterized by long, slender craniums and less-pronounced zygomatic arches. For Naora, the point was not only that this feature distinguished the Hokkaido wolf from most other wolves but also that it was typical of a dog skull, and so Naora questioned the authenticity of the Hokkaido wolf as an "authentic" member of *Canis lupus* as well.[69]

Nakamura and Naora, in their respective studies, turned to the same natural history to describe either the evolutionary dwarfing of Japan's wolves or their prehistoric demise. Obviously, though looking at the same natural past, they came away with profoundly different futures: the Japanese Archipelago had either become an ecology of chronocline evolution or an enclosure for an "unnatural life ecology" where wolves and dogs hybridized. Of course, both men relied on Linnaean taxonomy and ecology to make their respective cases for and against seeing Japan's wolves as members of *Canis lupus*, but in the process they both, in different ways, unleashed the folk-biological ghost of the mountain dog, much like the photograph of the creature spotted near Kitakyūshū City had. Not surprisingly, both men turned to science to vanquish the ghost of the mountain dog: one sought to eclipse its existence by highlighting the morphotypical consistency of the Japanese wolf; the other sought to accomplish the exact opposite through highlighting inconsistencies. But the critical point remains that, if crossbreeding with dogs led to the emergence of a hybrid species, then the Linnaean classification of these animals might be handled altogether differently. So might discussions of their extinction. Naora's theories raised many questions for me, not the least of which is whether Japan's wolves and native dogs really have created a hybrid species, one that was mistaken for a wolf by nineteenth- and early-twentieth-century Linnaean scientists.

WOLVES AND JAPANESE DOGS

In Japan, some of the oldest fossils of the domesticated dog come from the Natsushima shell mounds in Kanagawa Prefecture, fossils that date to about 9,500 years ago (roughly the same time Japanese wolves became isolated from the continent). Archaeologists have unearthed similarly ancient dog fossils from Ehime and Miyagi Prefectures. At the Miyagi Prefecture site, believed to be from the late Jōmon period, archaeologists exhumed twenty-two such dog remains. Taniguchi Kengo, a historian who has written on dogs, argued that although ancient dog remains exist in Japan, analysis suggests that early human cultures did not domesticate these dogs in Japan. That is, the earliest inhabitants of the Japanese Archipelago did not capture and domesticate Japan's wolves. These dog breeds came to Japan with the earliest human migrants and, genetically speaking, differ from most European breeds.[70] However, once on the Japanese Archipelago, because the Jōmon people used these dogs to hunt, they could have, like Ainu and Native American groups, encouraged crossbreeding between dogs and wolves for convenience and to foster keener hunting instincts in the breeds.[71] On the Japanese Archipelago, the categories of canine never stabilized, and historical ideologies and practices (such as the introduction of the practice of wetland rice farming in about 300 B.C.E.) reshaped interaction between wolves and dogs, altering the coevolutionary trajectory of both species. In this way, as Michael Pollan writes, "Through the process of coevolution human ideas find their way into natural facts," and this process of "artificial selection" may have altered the evolution of Japan's wolves.[72]

Naora argued that the animal that became the Japanese wolf not only was sometimes called "mountain dog" by such nineteenth-century zoologists as Tanaka Yoshio but was, quite literally, a kind of mountain dog. This begs the question: do historical observations of Japanese dogs support the possibility of sustained hybridization? We will take up this topic in earnest in chapter 6, but for now, judging from what we know about Japanese attitudes toward dogs and dog keeping in preindustrial times, early canine crossbreeding was likely, making Naora's hypothesis at least plausible. Of Japanese dog-keeping practices, Tsukamoto Manabu, a historian of Japan's early modern period, suggested that the relationship between people and dogs in the seventeenth and eighteenth centuries was symbiotic at best. At this time, dogs formed packs much like wolves, claimed and defended urban neighborhoods and villages as their territories, and subsisted by scavenging at dumps and receiving handouts from people, for whom they provided only limited companionship. They also hunted and killed game.[73]

As historian Aaron Skabelund points out, nineteenth-century Western observers commented that Japanese dogs looked and acted more like wolves than like the dog breeds the Westerners knew best.[74] Observers such as Edward Morse, an American zoologist at Tokyo University, made it clear that Japanese breeds were of the "wolf variety," animals that "do not bark but howl." In Japan "every village has a pack of dogs, and at night they are very noisy, making sounds like cats, but more infernal; they howl and squeal, but never bark." Evoking early theories of animal evolution, Morse remarked that "Darwin had observed in his work on domesticated animals that when dogs relapse from their cultivated state to a semi-savage one, they lose the bark and take on the howl again." Morse wrote of Japanese dogs and their cousin the wolf that the "[w]ild creatures to which they are related never bark, but howl."[75]

Similarly, George Hilaire Bousquet, a French adviser to the Meiji government, wrote in the 1870s that while traveling in Japan, his caravan had to defend its own dogs from wolflike Japanese breeds. "These creatures," wrote Bousquet, "with their heavy forms, enormous size, long wild hair, elongated snouts, small deep-set eyes and prominent fangs, are far closer to wolves due to their hunter instincts as opposed to the peaceful setter . . . who just came to hide between my legs. Our caravan has been obliged several times since the beginning of our trip to defend our setters by throwing stones at these aggressors."[76] Obviously, Japanese dog keeping contrasted with European practices, which, at least for the nineteenth-century English and French, fostered docile dogs that served as symbols of aristocratic and middle-class power and pedigree.[77] As this handful of descriptions attest, Japanese attitudes toward pets made crossbreeding between wolves and dogs an ecological possibility, even though few examples of lasting wolf-dog hybrid populations exist in the world. Perhaps it was for this reason that Ernest Satow and A. G. S. Hawes, in their nineteenth-century *Handbook for Travellers in Central and Northern Japan*, remarked that "nothing is known to science" of the Japanese wolf and that "[n]o true wolf exists in Japan" other than *C. l. hodophilax*, which "is a sort of lame counterfeit of the European beast."[78]

But in the Linnaean system, dogs are dogs and wolves are wolves, and so let us now return to Nakamura to conclude, at least for the purposes of this chapter, the debate over the classification of Japan's wolves. Nakamura argued that, even if spots of hybridization did occur, careful craniometric analysis, when grounded in ratios and, more generally, allometry, separates wolves from dogs. Nakamura and a colleague, Obara Iwao, wrote on the results of an osteological examination of a "mountain dog" skull taken from the Minami Ashigara Municipal Folklore Museum (Minami Ashigara Kyōdo Shiryōkan). By focusing on measurements of zygomatic breadth, the nasal region, and other craniometries, they concluded that the skull was probably from a male feral dog.[79] The implications are that historically, Japanese attitudes toward dogs did create a situation where hybridization between wolves and dogs occurred, but these hybrid animals can be identified, and, more importantly, their existence does not negate the existence of wolves. Nakamura and Obara appear to have come full circle to agree with their early modern colleagues: perhaps two distinct creatures, mountains dogs and wolves, did exist together in the forests of pre-Meiji Japan.

The Japanese wolf was not the smallest wolf either, and so its doglike size should not be a determining factor for separating wolves from dogs. The full cranial length of the Arab wolf (*C. l. arabs*) is smaller than that of most wolves (averaging about 200.8 mm), whereas a sample group of Japanese wolf skulls from Tanzawa, Kanagawa Prefecture, range from 193.1

to 235.9 mm. (Unfortunately, it is unknown whether all these specimens were adults.) The cranium length for the Hokkaido wolf rivals the large Eurasian continental species, with the British Museum of Natural History specimen coming in at 269.5 mm. Saitō Hirokichi, a canine specialist, surveyed Japanese wolf skulls acquired from Aomori to Ehime Prefectures and came up with an average full cranial length of 216.9 mm and a mandibular carnassial (m1) length of 25.61 mm. These results are within the 99 percent confidence interval of the Tanzawa skulls mentioned above and thus might be seen as representing the mean skull and m1 sizes of the Japanese wolf.[80] (The recent discovery of a large wolf skull at a coal deposit near Kitakyūshū City suggests that the Japanese wolf could have been even larger than previously thought. The wolf skull in question, believed by scholars to be from the Tokugawa period, measures about 240 mm in length, which is slightly larger than any other wolf skull to date. The length of this skull raises questions about possible wolf-dog hybridization and weakens Nakamura's heretofore solid ecological dwarfing model.[81]) The Hokkaido wolf, as mentioned, has a significantly larger size skull. The three skull specimens at the Hokkaido University Museum have cranial lengths that range (according to my rough measurements) from about 240 to 270 mm. But even emphasizing cranium size can be misleading. It is not so much the size of the skull but rather features such as the "inflation of the frontal sinuses" and the "steep angle of the forehead" that demarcate the different species of canine.[82]

CONCLUSION

So what were Japan's wolves? It appears from the evidence provided throughout the ages, both zoopaleontological and textual, that they were indeed wolves, but wolves in the process of sustained, if regionally specific, hybridization with dogs. One fact is sure: the Kitakyūshū City sighting and the lively debates that ensued fit within a longer tradition of Japanese writings on mountain dogs and wolves. The debate has changed little over the ages as well; only the empirical priorities of how scientific knowledge is generated and legitimated have been modified. Where Kaibara and other early modern natural scientists pointed to descriptions of physiology and animal disposition, Nakamura and Obara prioritized the measurement of skulls to suggest that mountain dogs and wolves can be distinguished from one another. The issue became confused in the nineteenth century, when the "mountain dog" became the "Japanese wolf," and ecologists have struggled with this categorization ever since. That being said, however, a more fruitful question raised by this chapter is, why do these creatures need to be distinguished in the first place? Certainly there must have been in-between creatures as well? What do we make of those thick gray lines that separate the individual canine taxons and the evolving qualities that define them?

One reason for distinguishing mountain dogs from wolves is the need for scholars to police the traditional boundaries that separate the inquiry into things "artificial" from the inquiry into things "natural," such as those Linnaean lines separating the categories of dog and wolf in Japan. But what would happen if we heeded the lessons of *both* Nakamura and Naora when talking about Japan's extinct wolves? In the context of Japanese wolf evolution, human artifice obviously contributed to wolf morphology and evolution. The human-fashioned ecotone resculpted the "natural" state of the prehistoric Siberian wolf, through what has been called "artificial selection," into what we identify today as the Japanese wolf. But nonhuman forces, "natural selection," were also at work, such as the chronocline and dwarfing processes described by Nakamura. This raises the question: if artificial and natural selection both played a role in the evolution of Japan's wolves and the emergence of the mountain dog, then what purpose do these terms really serve other than to illustrate some last-ditch effort to distinguish humans, the creators of artifice, from the rest of the natural world? The only thing that separates mountain dogs from wolves is the nature of their apparent connection to us; and the human species is hardly a reliable plumb when it comes to measuring degrees of evolutionary change.

If, as deep-ecologist Bill Devall writes, the "person is not above or outside nature . . . [but] a part of creation ongoing," then that creation extends into the world of zoological morphology and evolution.[83] Rather than see Nakamura and Naora as positing juxtaposed theories on wolf evolution— one focusing on natural selection and the other on artificial selection—they need to be seen as presenting different parts of the same ecological continuum, different stories of the broader "creation ongoing" that was Japan's wolves. Perhaps Japan's early modern naturalists understood this process better than modern scientists. However, *Fauna Japonica* and subsequent works on Japan's wolves up until our times—and even debates regarding the nature of the animal in Nishida Satoshi's newspaper photograph—drew these distinctions to the point of obscuring the complex nuances and slippages that occur in the natural world.

Returning to the photograph sent to me by Inoue Katsuo, from the vantage point of my office it appears that Japan's wolves, like all other creatures on this planet, need to be seen as natural works in progress. Indeed, if there was one concept that all Neo-Confucian natural scientists understood, it was the Taoist notion of eternal change. Creating the wolves of Japan was not solely a scientific endeavor, however. Just as the wolf was in part the product of the powers of categorization inherent in modern science—scientists defined wolves as having certain qualities and then neatly stuffed them into a box labeled the "Japanese wolf"—so too was it the product of the syncretic powers of Japanese cultural and religious traditions, which defined wolves as having certain qualities as well. Shintoism, Buddhism, Confucianism, and the animistic view of the Ainu also created the wolves of Japan, often in ways that contradicted the scientific wolf, and so it is to this part of our story that we must now turn our attention.

2

CULTURE AND THE CREATION
OF JAPAN'S SACRED WOLVES

Deep in the mountains of Saitama Prefecture, some three hours or more by train west of Tokyo, a heavy backpack and lingering traces of winter snow made the two-hour hike up to Mitsumine Shrine from the small village of Ōwa an extremely strenuous one. Less a village than a thin ribbon of roadside residences, Ōwa appeared to be populated mostly by antiquated vending machines: some worked and others did not, but faded advertisements made by youthful celebrities and sports stars I scarcely recognized adorned them all. There were a dozen or more homes of about the same vintage and a small store and gift shop perched dangerously along the winding mountain road. As I walked toward the trailhead, I could barely squeeze between the fronts of these homes and the traffic on the road. I saw only two or three people as I walked through the village; they all stared back at me as they looked up from the open hood of an old Toyota truck. As I crossed a beautifully stylized, brightly colored bridge (typical of shrine architecture in Japan) into the mountains, only the loud compression braking of logging trucks could be heard over the stream below. It was just before my arrival at this village, at an earlier bus stop, that I had learned that the Mitsumine cable car, which normally hauls worshipers and sightseers up to the sacred site several times daily, would be closed for a week for its annual maintenance. As I headed up the trail, the lower cable car terminal stood idle, not a soul in sight anywhere.

Still covered with a foot of snow in some shaded spots, the trail cut along the side of a steep mountain as it climbed just above a tributary of the Ara

River. Massive boulders and fallen trees, deposited by decades of landslides and erosion in the steep valley, forced the stream to tumble boldly and roar loudly with the weight of spring runoff. On the trail, widow-makers, small branches, and eerie tufts of hairlike lichen covered the snow and served as testimony to the harsh winter that had gripped the mountains only months before. It was mostly quiet except for birds, though at one point, as I was recalling some folktale about a woodsman, I thought I heard Japanese macaques. (I may indeed have heard them, as the region has plenty of them.) As I gained altitude, I began weaving through plantation-like stands of Japanese cedar; but I also encountered pockets of red pine, hemlock, and, on the shrine precincts, ancient cypress and cedar, not to mention scattered Japanese beech, oak, chestnut, zelkova, and a thick underbrush of shrub bamboo. After hiking for about half an hour I came across a famous *seijō no taki*, or purification waterfall, located where the stream cascades over a small lichen-stained cliff. *Torii*, or Shinto shrine archways, mark the sacred falls, a mystical site where every year worshipers still come to purify themselves (see fig. 10).

The sacred falls and *torii* contrasted sharply with the rusting vending machines of Ōwa and served as a reminder of the strong ties that Mitsumine once had to traditions of Shugendō, or mountain asceticism. At the front desk of the Kounkaku Inn, where I stayed for several days, I was told that *shugenja*, or mountain ascetics, often stand under the falls and chant in a practice called *misogi*, letting the cold mountain waters fall on their heads and cleanse them of the filthy, vending-machine-like impurities of the human world. That night, as I settled my exhausted body in my room, I heard, rising from the misty shrine precincts, the lonely notes of a *hichiriki*, a flageolet-like instrument, and the low resonance of shrine bells. It occurred to me that, considering the scope of contemporary Japan's industrial clutter and urban filth, it is easy to forget that the people of this country once worshiped at mountain shrines such as Mitsumine, revering their surrounding peaks, forests, and wildlife as divine. The old moss-covered statues of wolves that guarded the Mitsumine trailhead had reminded me of this and had rekindled my belief that wolf worship in Japan needs to be viewed as part of this slowly vanishing culture.

Ichiro Hori, a scholar of Japanese religion, wrote of mountains: "Their height, their vastness, and the strangeness of their terrain often inspire in the human mind an attitude of reverence and adoration."[1] In cultures around the world, explained Hori, people have viewed mountains as pillars that tie the earth to the heavens and that support the heavens, likening such peaks

FIG 10. This *seijō no taki*, or purification waterfall, is located on the trail to Mitsumine Shrine. Several *torii*, or Shinto shrine archways, mark the sacred site, where *shugenja*—mountain ascetics—stand under the falls to purify themselves in a practice called *misogi*.

to the bridges of gods. In Japan, people have historically understood mountains to be sacred places for several reasons, including the frightening
power they wield as volcanoes. When Fuji experienced a major eruption in
865 C.E., for example, many believed they saw a "magnificent divine palace
newly built on the mountaintop in the midst of the flames and smoke." A
second type of worship, one tied to Mitsumine Shrine, celebrated mountains as places of the headwaters of rivers such as the Ara, a waterway that
twists and turns through the Kantō Plain and has for centuries helped support Japan's agrarian society by bringing fertility to the land and its inhabitants. Hori writes that "the sacred mountains in Japan without exception
have sacred waters which proceed from them." At the sacred headwaters of
the Ara River one can discover the otherworldly powers of the *seijō no taki*
because, "whoever wants to possess the divine power of the mountains or
to communicate with the mountain deities must undergo some initiatory
mysteries by these sacred waters." A beautiful *ema*, a small wooden votive
amulet, from Mitsumine pictures two wolves with several pups (see fig. 11).
Today, this *ema* symbolizes human fertility and promises good fortune to
couples hoping to have children; but in the seventeenth and eighteenth centuries, the wolf iconography intersected with the mountain and river
iconography to yield an expanded fertility symbolism. A third type of worship viewed mountains as the home of the spirits of the dead, a form of reverence that might be linked to the development of ancestor worship in Japan.[2]

In all three forms, mountain worship prospered until modern times
because during much of Japan's long history, so much of the country
remained a wild place, an archipelago of rugged peaks surrounded by dense
deciduous and evergreen forests that teemed with wildlife. Gerald Figal, a
historian of modern Japan, points out that in these mountains and forests
the pioneering ethnologist Yanagita Kunio first began his search for the
earliest folk customs of the Japanese, recording the vanishing stories of the
inhabitants of small intermontane villages, the *yamabito*, or "mountain
people." Only after the 1920s did he begin to turn his attention to the wetland rice-farming villages of the plains and their human-centered ancestor
worship, the product of imported East Asian continental culture. That is,
only later did he come to believe that plains villages harbored the earliest
folk customs of the ethnic "Japanese."[3] Nevertheless, in his earlier scholarship, Yanagita explained that mountain people often portrayed mountains
as female, and that these mountains participated in the birth and rebirth of
humans and other animals. Yanagita postulated that three major traditions
of mountain legend could be discerned in Japan: the Kōya tradition, in which

FIG 11. Called an *ubudate ema*, this wooden votive amulet from Mitsumine Shrine pictures two adult wolves with pups. Author's collection.

the mountain goddess or her son grants people permission to build homes or temples in the mountains or forests; the Nikkō tradition, in which the mountain goddess rewards certain hunters for their moral conduct; and the Shiiba tradition, in which the mountain goddess appears as a beautiful maiden who tests the morality of a hunter.[4] Not surprisingly, Yanagita's three traditions of mountain legend revolved around hunting cultures, and one suspects that Yanagita saw them as part of a religious world distinct, at least in part, from the ancestor-worshiping culture of the plains.[5]

In *Tōno monogatari* (Tales of Tōno; 1910), Yanagita presented his findings from a small village in Iwate Prefecture, called Tōno, where he excavated the deep strata of the beliefs of the mountain people, including legends of brave hunters, divine animals, and otherworldly mountains and forests.[6] Yanagita also established a dichotomy between the "mountain people" (*yamabito*) and "plains people" (*heichijin*) of Japan, boldly stating that the mysterious stories of mountain gods would send shivers up the spines of rice-farming lowlanders. Surrounded by the mountains Hayachine, Rokkoushi, and Ishigami, Tōno represented, in Yanagita's mind, an isolated site from which he could retrieve the lost customs of the Japanese, customs untainted, or at least not as tainted, by imported East Asian culture. Tōno harbored spooky stories of wondrous mountains such as Shiromi, where travelers and hunters sometimes heard the shrieks of strange women late at night; of the rude monkeys near Rokkoushi, who stole women for themselves; of fantastic birds whose lonely calls split cold mountain nights; of the powerful *yama no kami*, or mountain deity, who possessed villagers, such as Magotarō, who subsequently became a *shugenja*; and shape-shifting foxes who took human form, often appearing as beautiful women to seduce men. Yanagita also recorded stories about wolves: howling near the mountain Futatsuishi; appearing in packs of hundreds and threatening packhorse drivers between Sakaigi and Wayama passes; and, filled with indignation, killing the horses of the villagers of Iide, who had haplessly raided a den and killed two pups and took one home.[7]

The stories and legends uncovered by Yanagita might be seen as the relics of ancient Japanese beliefs and past customs, though to do so would be to ignore historical invention and the syncretic nature of Japanese religion. Indeed, in the sixth and seventh centuries, Taoist cosmologies, Buddhist notions of compassion, and Confucian humanism became superimposed over more native animistic beliefs when the Yamato rulers of central Japan promoted wetland rice farming and its culture of ancestor worship to strengthen their hold on power in this world and the next. In this context, wolf worship evolved, as over the centuries it was refined to accommodate the canons of East Asian culture.[8] The creation of Japan's wolves continued within this spiritual accommodation, as it did within the scientific process discussed in the previous chapter. Wolves were invented not only by science but by Japanese belief systems and cultural orthodoxies as well. Over time, the wolf found its place among the histories and iconographies of mainstream Shinto shrines and Buddhist temples, not to mention in the more private, spiritual terrain of iconoclastic traditions such as Shugendō.

TRADITIONS OF NATURE
AND ANIMAL WORSHIP IN JAPAN

The importance of mountains and forests as sacred terrain in East Asian culture, including the reverence for such sites in Japan, can be traced to early classical and medieval Buddhist theologians. Between the Nara and Kamakura periods (c. 710–1333 C.E.), after the heavy importation of Buddhism and Taoism from China and Korea, theologians such as Ryōgen, Chūjin, and Saigyō established traditions in which mountains and forests became sacred sites. They recognized that "all sentient beings," including plants, trees, and, of course, nonhuman animals, contained the Buddha-nature (an internal benevolence not corrupted by the filth and materialism of the human world). William LaFleur, a scholar of Japanese culture, argues that an analysis of these theologians' prose and *waka* (thirty-one-syllable poetry) reveals "increasingly great religious significance in nature." Of these thinkers, perhaps Ryōgen is the most interesting for this book because he, as LaFleur explains, created "a nexus between the biological life cycle of a plant and the process of enlightenment as experienced by a human being." LaFleur observes of Ryōgen's writings on the Buddha-nature inherent in plants and trees, "The sprouting forth of a plant is really the mode by which it bursts forth its desire for enlightenment; its residing in one place is really undertaking of disciplines and austerities; its reproduction of itself is its attainment of (the fruits of) enlightenment; and its withering and dying is its entry into the state of nirvana." The discovery of the Buddha-nature, a process that could require so much monastic discipline among humans, was inherent in the life cycle of a plant. Nature came to serve as an expression of spiritual meaning and salvation, and worship of—or within—natural settings became a kind of religion for early Japanese.[9]

The Hōjō-e Ceremony, a rite held annually at the Iwashimizu Hachiman Shrine near Kyoto, and Inari worship, the two largest shrines for which are the Fushimi Inari in Kyoto and the Toyokawa Inari in Tokyo, serve as examples of how animals, not to mention the mountains and forests they inhabited, came to fit within the worldview of early Japanese belief systems. Reflecting medieval Shinto and Buddhist assumptions about animals, a 1280 shogunal edict related to the Hōjō-e Ceremony stated that the "prohibition on the taking of life must be observed in the period just before the Iwashimizu" Ceremony.[10] The Hōjō-e Ceremony itself involved releasing animals, usually birds and fish, into the wild and was rooted in Buddhist canonical sources. The Brahmajāla-sūtra says:

As a [child] of the Buddha, one must with a compassionate heart practice the work of liberating living beings. All men are our fathers. All women are our mothers. All our existences have taken birth from them. Therefore, all the living beings of the *rokudō* [six realms] are our parents, and if we kill them, we kill our parents and also our former bodies; for all earth and water are our former bodies, and all fire and wind are our original substance. Therefore, you must always practice liberation of living beings (since to produce and receive life is the eternal law), and cause others to do so; and if one sees a worldly person kill animals, one must by proper means save and protect them and free them from misery and danger.[11]

The Hōjō-e Ceremony began with dancing, music, and horse riding, and then Buddhist monks released clams and fish into the Hōjō River while others chanted scriptures. Scholars of Japanese religion point out that this popular ritual became an imperial rite when the emperor ordered that the Hōjō-e Ceremony be conducted throughout the country. Records from the Usa Hachiman Shrine and zooarchaeological sites provide additional evidence that monks performed the ceremony at other Buddhist temples and Shinto shrines. As part of state protocol, shogunal officials from the medieval capital of Kamakura abstained from eating fish or meat before visiting shrines.[12]

Even more than the Hōjō-e Ceremony, however, the Inari tradition exposes the intermixing of animistic, Taoist, and Buddhist traditions when revering animals; but Inari also exposes certain tensions regarding the place of animal deities in modern Japanese spiritual life. Anthropologist Karen Smyers explains that the official origins of Inari worship can be traced to about the eighth century, when Inari Mountain, one of the southernmost peaks of the Higashiyama Ridge east of Kyoto, became a sacred site mentioned in Japanese documents (see fig. 12). Archaeological evidence at Fukakusa, at the foot of Inari Mountain, suggests earlier, Jōmon period (10,000–300 B.C.E.) associations of the mountain with the sacred, however, and so Inari, like the Hōjō-e Ceremony, consisted of layers of native and imported beliefs that, over time, overlapped and intertwined with one another to form patterns in the complex tapestry of early Japanese religious life. It is of interest to us that members of the Hata clan, some of whom settled at Inari Mountain, became chief priests of Inari worship sometime after the fifth century when they arrived from the Korean Peninsula: the same Hata clan also became associated with wolf worship. The Hata clan used Inari to protect their finances (Inari served as their *ujigami*, or tutelary deity), and later Hata associations with wolf worship possessed similar economic resonance. Later, in the Heian and Kamakura periods, Inari became associated with the esoteric Shingon sect, as did Shugendō and many of the shrines where people worshiped wolves, which increased the popularity of Inari among the courtiers, not to mention powerful sixteenth-century military figures such as Toyotomi Hideyoshi.

FIG 12. Inari Mountain, just outside Kyoto, is a sacred site marked by hundreds of *torii* that bridge the hiking paths that wander to the top of the mountain.

The introduction of esoteric Buddhism and Taoism to Japan, when combined with native animistic practices, also led to the rise of Shugendō at Inari Mountain. Smyers writes, "Solitary ascetics (*ubasoku, hijiri*) versed in a mixture of Shinto, Buddhist, and Taoist notions spent time practicing austerities in the mountains, developing spiritual powers, which they then used to heal and to perform miracles," and such conduct was often beyond the oversight of official Inari priests. In the seventeenth century, other forms of Inari iconoclasm developed, and Inari became one of the more popular beliefs of urbanites in Edo, who placed Inari talismans in their homes to prevent fires, calling them *hibuse*. As for associations with the fox, Smyers notes that priests at Fushimi and Toyokawa Inari downplayed its role in Inari worship, perhaps not wanting to be portrayed as mountain shamans worshiping animals. Regardless of such modern prejudices, in the syncretic terrain of early Japanese religion, with the importation of East Asian beliefs the fox came to serve as a messenger or assistant to the main Inari deity, even though the fox, as revealed in the *Tōno monogatari*, had long possessed supernatural powers of its own, bringing both good fortune as a guardian of fields (its benign spirit, or *nigimitama*) and hardship as a shape-shifting seducer (its dangerous spirit, or *aramitama*).[13] Wolves too possessed the powers to both help and harm people.

WOLVES IN JAPAN'S SACRED ORDER

As with Inari worship, the Hata clan also became associated with the worship of the Japanese wolf, at least according to the *Nihon shoki* (Chronicles of Japan; 720), a mythologized history of Japan's imperial family. The chronicle describes a divine encounter between a member of the Hata clan and two wolves on a mountain road near Uji, just outside Kyoto. According to this story, Kinmei, before he became the twenty-ninth emperor, was told in a dream that if he employed a Hata merchant named Hata no Ōtsuchi, he would mature to be a successful and benevolent ruler. Kinmei is an important figure in Japan's imperial history, as the *Nihon shoki* credits him with the importation of Buddhism (in either 538 or 552 C.E.), explaining that he received a bronze statue of Sakyamuni (the historical Buddha) and sutras as gifts from King Sŏng-myŏng of Paekche (one of the three Korean kingdoms). After searching throughout the realm, imperial messengers discovered that a man named Hata no Ōtsuchi lived in Yamashiro (at Fukakusa, the site of early Inari worship).

When Hata no Ōtsuchi met Kinmei, the Hata merchant told the emperor an auspicious story. While taking a mountain road home from the Ise Shrine, the head shrine of the imperial family, he had met two battling wolves near Uji, both defiled and drenched in blood. Hata dismounted from his horse, cleansed his mouth and hands in a Shinto practice of purification, and began to pray out loud: "Ye are august deities, and yet ye take delight in violence. If ye were to fall in with a hunter, very speedily ye should be taken." With these words, Hata beseeched the wolves to stop fighting and washed the blood from their fur and let them return to their mountain home. Kinmei believed that Hata had appeared in his dream as a reward for rescuing the divine creatures. Indeed, the story so moved Kinmei that, when he became emperor, he made Hata head of the treasury, and the realm became prosperous.[14] Interestingly, almost a millennium later, Ihara Saikaku, the famous seventeenth-century literary figure, set the conclusion of one of his short stories contained in the *Honchō nijū fukō* (Twenty stories of unfilial behavior in the realm; 1686) in Ōkami-dani—Wolf Valley—near the town of Uji, the same place where Hata had encountered the fighting wolves.[15] The story of the merchant Hata no Ōtsuchi, the emperor Kinmei, and the two battling wolves is important not only because of the intersection with Inari worship and the Hata clan but also because of its early Shinto imagery. Indeed, the imperial family, pilgrimages to Ise Shrine, ritual purification, and notions of defilement and pollution associated with blood continue to be prevalent symbols even in modern Shinto.[16]

Early poetry anthologies also contain references to divine wolves that shared their home with Japan's early emperors; such sacred wolves, in the hands of skilled poets, evoked a strong sense of *wabi*—a powerful Japanese aesthetic that can be likened to a sense of "loneliness" and that elevates the beauty inherent in austerity and simplicity. With the construction of Kiyomihara Palace at Asuka (c. 673), wolves, in poetic verse, brought to the fore memories of the recently pacified wilderness that lay just beyond the gates of the newly built Japanese capital. The *Man'yōshū* (Collection of ten thousand leaves; c. 759) contains at least three poems that refer to the "plains of the Large-Mouthed Pure God," or Ōguchi no Magami no hara, a reference that was probably first used in the *Kofudoki itsubun* (Lost writings on ancient customs; 713) and that conjured the lair of the wolf: a dangerously sacred place just beyond the walls of the courtiers' palaces.[17] Such wild places, much like the otherworldly mountains of Tōno, served as places of transition, portals between this world and the next where only "pure gods" and revered imperial ancestors dwelt.[18] In a poem on snow by Toneri no Otome (Iratsume), for example, the lonesome imagery leaves one feeling uneasy and vulnerable:

> Snow falling on the plains of the Large-Mouthed Pure God,
> O, do not fall so heavily;
> I am without even shelter.[19]

Another poem, this one written from the perspective of a court woman, reads:

> From Mount Kamunabi of
> Mimoru
> Clouds overshadow the sky,
> Bringing heavy rain;

> The rain is swept in spray,
> And the storm gathers.
> Has he reached home,
> He who went back
> Across the plains of the Large-Mouthed Pure God,
> Deep in thoughts of me?[20]

This poem conjures the lonesome image of a courtier safe within the confines of her palace, thinking of a single traveler braving a storm and navigating the dangerous windswept plains of the Large-Mouthed Pure God. In ancient Japan, people imagined the wolf as a deity in a sacred landscape associated with the uncertainty and dangers of the world outside the palace, and this image was contrasted with the ritualized certainty of court life. The late-seventh-century poet Kakinomoto Hitomaro, on the occasion of the temporary enshrinement of Prince Takechi, used the plains of the Pure God of Asuka masterfully to evoke a space inhabited by "pure gods" and revered emperors and princes: a moor hosting sacred animals, noble souls, and austere moss-covered tombs—only wolves and imperial ancestors shared such divine terrain. One also gets the sense, however, that being able to build an imperial palace on such a barren plain that was home to the large-mouthed beast was one contribution Prince Takechi had made in bringing courtly civility to the realm.

> Awesome beyond speech,
> O dread theme for my profane tongue!
> That illustrious Sovereign, our mighty lord,
> Who reared his imperial palace
> On the plains of the Pure God of Asuka,
> Now keeps his divine state,
> Sepulchred in stone.[21]

It should not be surprising that the wolf, at least in the minds of those who rarely ventured far from the palace gates, was seen as symbolic of loneliness and impermanence, of a windswept plain inhabited by beasts, and of decaying stone memorials to ancestors past. In the wolf, wrote naturalist Barry Lopez, "we have not so much an animal that we have always known as one that we have consistently imagined."[22]

Early Japanese projected the wolf through the lens of Shinto and Buddhist traditions as well as imperial legends, placing it in the simultaneously sacred and dangerous terrain outside the capital. The Shinto tradition, in which deified natural phenomena, or *kami* (the Japanese wolf was known as *ōkami*, which phonetically can mean "great deity"), inhabited the natural landscape, cast the natural world as a divine space teeming with spiritual life of a kind preserved in the stories and relics of Tōno. Coming from a different perspective, Buddhism blurred the line between humans and the rest of nature, viewing various natural phenomena (depending on the creed) as sentient because of their inherent Buddha-nature. All living things were part of a shared "continuum of life." For this reason, mountains—also connoting wilderness and similar in their spiritual quality to the plains of the Large-Mouthed Pure God and Wolf Valley, where wolves once lived— remained important sites for discovering the Buddha-nature. In this sense, the wilderness teemed with spiritual life, becoming a "zone of transition," suggest Pamela Asquith and Arne Kalland, "a mediator, between the 'outside' (in relation to society) and 'inside' (in relation to the other world). Being both outside and 'betwixt-and-between,' mountains and forests are potentially dangerous areas inhabited by a multitude of spirits, and they therefore make ideal sites for pilgrimages and other religious exercises."[23] In this way, wolves became associated with Shinto shrines and Buddhist temples as well, and so it is to these institutions, where wolves often served as divine messengers to other deities, that we must now turn our attention.

SACRED SITES OF WOLF WORSHIP

In *Ōō hitsugo* (Stories of Ōō; 1842), after briefly retelling the story of Hata no Ōtsuchi and the plains of the Large-Mouthed Pure God, Nonoguchi Takamasa mentioned the Ōkawa and Mitsumine Shrines as important sites for wolf worship.[24] While in Japan doing research for this project, I decided to use (as much as possible) Nonoguchi's work as a guidebook and travel to sites mentioned in the *Ōō hitsugo* and elsewhere. On the one hand, it was a journey filled with the sort of immense satisfaction I get from, as Wallace Stegner characterized it in *Wolf Willow*, taking in completely the "colors and shapes" of the places I study. Oddly, even faded colors and industrial shapes, such as the vending machines and rusted sheets of corrugated metal roofing at Ōwa, when contrasted against the earthy colors and spontaneous shapes of their mountain backdrop, felt familiar to me, as the same colors and shapes can be found in many old Montana towns. On the other hand, the journey was filled with a sad irony created by the void left by wolf extinction in Japan. Thus, there was always a tension inherent in my trips to these shrines and other wolf sites in Japan, a tension between my contemporary experiences and the images left by my research into the past.

I first traveled north of Kyoto near where the Yura River runs into the Japan Sea. Just outside Nishi Maizuru in Kyoto Prefecture, a port city visited daily by massive ships from Russia, stands the humble Ōkawa Shrine. The head priest and caretaker at Ōkawa, a charming elderly man who frequently smiled as we drank tea together, told me that Nonoguchi had correctly associated the shrine with the Daimyōjin, a Shinto deity, and that, according to tradition, the Daimyōjin used the wolf as a divine messenger (*tsukawashime*), much as the Inari deity used the fox as one. Nonoguchi had explained that although many wolves reportedly lived in the mountains around Ōkawa Shrine, they rarely caused injury to people, and that whenever deer damaged local crops, farmers beseeched the wolf messenger through prayer to guard their crops against hungry ungulates.[25] Interestingly, when I asked the head priest about such prayers, he said that they once did take place and that even though wolves became extinct, their power still lingers at Ōkawa, as damage caused by wild boars and deer has never been much of a problem in the area around the shrine.

Hiraiwa Yonekichi, in his brief survey of the Ōkawa Shrine's history, explained that the site was founded before 500 C.E. during the reign of the twenty-third emperor, Kenzō, and that it was connected to the worship of the Hōkura deity. In fact, according to Hiraiwa, the name "Ōkawa" was once read "Ōkami," the Japanese word for "wolf," and some Japanese once worshiped the wolf as the Daimyōjin deity itself at the shrine, not simply as a messenger.[26] On the precincts, tall Japanese cedars stand near the main shrine buildings; the surrounding forest, however, like many forests in Japan, consists of stands of younger commercial cedar (see fig. 13). To this day, the shrine distributes *ema*, small wooden votive amulets, with images of the divine wolf messenger; the amulets promote *mubyō sokusai*, "perfect health," and *kanai anzen*, "domestic safety" (see fig. 14).

Far to the east, in those rugged mountains of Saitama Prefecture described earlier, worshipers also associated Mitsumine Shrine with wolves. The main deities worshiped at Mitsumine Shrine are Izanami and Izanagi, two powerful figures who appear in the Japanese creation myth. Legend explains that Yamato Takeru no Mikoto founded the shrine. During his pacification campaigns in eastern Japan, this princely unifier wandered astray on the Karisaka mountain road in the province of Shinano. According to the *Nihon shoki*, Yamato Takeru no Mikoto had lost his way until a white wolf god led him out of the mountains, serving as his divine guide.[27] Later, the prince's father, the celebrated twelfth emperor, Keikō, retraced his son's route through the mountains during an imperial tour. Legend explains that after climbing the mountains, the views of Shiraiwa, Kumotori, and Myōhōgatake, the three peaks of Mitsumine, so stunned Keikō that he bestowed on them the name Mitsumine-gū, or the "shrine at the three peaks." Over time these three peaks became objects of worship among commoners as well. Spring thaws caused swift and clean streams, such as the one mentioned earlier that fed the purification waterfall on the trail to Mitsumine. Such streams flowed from Mitsumine and ultimately converged with the Ara River, and so, not surprisingly, local farmers revered the mountains in their agrarian traditions. After the Heian period the mountains became a favorite spot of Japan's mountain ascetics. The peaks stretch about five miles south to north (Kumotori, 6,727 feet; Shiraiwa, 6,403 feet; and Myōhōgatake, 4,607 feet). Today, the Mitsumine Shrine stands at about 3,600 feet, on the northwest slope of Myōhōgatake, on a smaller peak now called Mitsumine.[28]

Mitsumine's history reads much like that of other ancient shrines. Worship there shifted between early Shinto and many popular Buddhist creeds, responding to the tides of religious popularity as new theologians and theologies came from Korea and China. Mitsumine was eventually incorporated into the esoteric Tendai and then Shingon sects, much like other Shugendō sites such as Kumano, and became associated with the Honzanha branch of Shugendō. (The other major branch of Shugendō is the Tōzanha, which was associated with the Sanbō'in Temple.) In the twelfth century, Mitsumine became a favorite place of worship for powerful samurai such as Hatakeyama Shigetada, eventually coming under his family's protection. In the seventeenth century, the shrine became more closely tied to the Shingon temple Shōgo'in and mountain asceticism, and historians believe that at about this time it became associated with wolf worship. Eventually, the shrine became a kind of seminary for the Shōgo'in.[29]

Because Mitsumine incorporated elements from both the Shinto and the Buddhist traditions, following the Meiji Restoration officials targeted it during the implementation of the *shinbutsu bunri rei*, the "orders to separate Buddhist and Shinto deities" at temples and shrines. As a result, it became an official Shinto shrine under the Meiji system.[30] The Meiji government also outlawed Shugendō there. Nonetheless, Hiraiwa notes that for the inhabitants of the Kantō, the plain around Tokyo, the Mitsumine Shrine constituted the sectarian center of wolf worship and retained its strong Shugendō heritage.[31] At the shrine gates (both the Seidō torii and the Zuishin hon), instead of the traditional Yadaijin and Sadaijin guardians, two wolves watch

FIG 13. Ōkawa Shrine (once Ōkami jinja, or Wolf Shrine) is located on the Yura River near Nishi Maizuru City in Kyoto Prefecture. Legend has it that the twenty-third emperor, Kenzō, discovered the shrine site and that the wolf served as a divine messenger for the Daimyōjin deity.

over the precincts; inside the sacred Haiden and inner Honden sections of the precincts, wolf guardians watch over the most sacred relics of the shrine. Reflecting Mitsumine's syncretic heritage, these statues strike a pose similar to that of the Guardian Neva Kings, or Niō in Japanese, two fierce Buddhist figures who protected temple precincts.

Whether drawn on *ofuda* (talismans) or carved out of stone and wood, the wolves at Mitsumine are often depicted as facing each other (see fig. 15). The wolf on the right usually has its mouth open, symbolizing *a*, or the sound of an open mouth and also the ideographic representation of the first syllable of the Siddhamatrika syllabary, an ancient script used to write Sanskrit. The wolf on the left, by contrast, has its mouth closed, symbolizing *un*, or the sound of a closed mouth. Together, *a* and *un* form the word *om*, which is the eternal word for, as the *Upanishads* explains, "all [of] what was, what is and what shall be." The three sounds contained in the sound *om*, or *aun* in Japanese, refer to the three states of consciousness: *a* being the first state, of waking consciousness; *u* being the second state, of dreaming consciousness; and *m* being the third state, of sleeping consciousness. When combined, *om* is the fourth state, of supreme consciousness. "It is beyond the senses and is the end of evolution. It is non-duality and love."[32]

What made my journey to sites of wolf worship eerie at times was best reflected at the town of Misakubo in Shizuoka Prefecture, the train stop closest to the Yamazumi Shrine (see fig. 16). At Misakubo, continual explosions, like the blasts one might hear on a battlefield, rang out in the valley. As it turned out, I had indeed stumbled onto a kind of battlefield, because the sounds were that of wild boar, deer, and macaque deterrents. These devices, called *bakuonki* in Japanese, ignite a natural-gas explosion set on a timer, so that five or six such explosions might be heard in a single hour. An elderly man at the Misakubo train station, eager to assist with my research, explained that other than green tea, the most important local crop is shiitake mushrooms, and that the boars, deer, and macaques routinely ravage these crops, forcing the people of Misakubo to resort to such disruptive tactics.[33] At a location where Japanese once worshiped wolves, beseeching the canines to protect their crops, farmers have resorted to mechanized sentries to guard their mushrooms.

FIG 14. (*facing page*) Amulets distributed by Ōkawa Shrine read *Ōkawa jinja shugo*, or "protection by Ōkawa Shrine of Tango Province." As the inscriptions on either side of this amulet reveal, it promotes *mubyō sokusai* (perfect health) and *kanai anzen* (domestic safety). Author's collection.

FIG 15. (*a–b*) *above*
These two wolf statues
stand near the Seidō torii,
or bronze archway, of the
Mitsumine Shrine. Like
the Yadaijin and Sadaijin
of traditional Shinto
shrines or the Guardian
Neva Kings of Buddhist
temples, these two wolves
protect Mitsumine's
Haiden and Honden
inner precincts. (*c*) (*left*)
The hanging-scroll print
is a more stylized version
of the *ofuda*, or talismans,
distributed by Mitsumine
Shrine. Author's collection.

FIG 16. The Yamazumi Shrine, located outside Misakubo in Shizuoka Prefecture, also used the wolf as a divine messenger. As at Mitsumine Shrine, wolf guardians watch over the shrine's inner precincts.

As I walked around this charming mountain town, I noticed that several residents had *ofuda* (talismans) from Yamazumi, which they displayed on their *kamidana*, a shelf for Shinto and Buddhist icons, or in the entryways to their homes (see fig. 17). Just as Inari talismans protected Edo urbanites from fire, talismans associated with wolves acted as charms that protected people from theft (*tōnan yoke*) and infectious disease (*ekibyō yoke*).[34]

To return to Mitsumine, the Edo physician Negishi Yasumori, in his eighteenth-century *Mimibukuro* (literally, "Ear bag"; a collection of fantastic tales and intriguing rumors from early modern Japan), wrote of the power of such wolf talismans to reveal the divine Providence of the Mitsumine avatar (in Japanese "Gongen," or an incarnation of the Buddha):

> People have a habit of saying that, in the creed of the Mitsumine avatar, when
> you receive a protective talisman to guard against the calamities of burglary
> and fire, you have "borrowed a dog." People believe that when you "borrow

FIG 17. *Ofuda*, or talismans, from Yamazumi Shrine, can be found pasted to the doorways of residences and businesses in Misakubo in Shizuoka Prefecture (*below*). Such talismans, like those distributed by the Inari shrines (*left*), are said to guard against theft, fire, and disease.

a dog" you are never exposed to the calamities of burglary and fire. A certain person earnestly requested to a shrine attendant, "You say you lend a dog, but after all you only hand out a protective talisman. How about lending people a real dog? Somehow make clear the deity's Providence [*myōkan*]." The attendant, deep in thought, gave the man a talisman. But when the man was getting ready to descend from the mountain, he had the distinct feeling that a lone wolf was loitering around, at first behind him and then in front of him. The man, for the first time, was truly enlightened to the deity's Providence and, trembling with fear at the thought of taking a wolf back home, turned back toward the shrine. The man told the attendant the story and apologized for being suspicious of the deities; he humbly requested another protective talisman. The attendant prayed for a second time and gave the man one. It is said that, on the homeward journey, the man did not see the figure of the wolf again.[35]

Once a person brought home a Mitsumine talisman, he needed to make the customary offerings of food to the divine wolf messenger. Nativist (*kokugaku*) scholar Hirata Atsutane cautioned that failing to make daily offerings to the "borrowed wolves" of Mitsumine made them weak, emaciated, and vengeful. Hirata claimed to have known a man who had neglected the wolf messenger for four or five days and, to the man's surprise, he was visited by misfortune.[36]

Given that wolf legends in Japan were deeply intertwined with ancient Japanese emperors and poetic anthologies, as well as notions of religious purity and famous Shinto shrines, it is not surprising that Hirata—who, as a nativist scholar, celebrated Japan's language, legends, religions, stories, and history—told other tales about wolves. He told the story of one woman who had dutifully fed a famished wolf she had encountered while traveling in the mountains to visit a lover who had been rejected by her parents. One day, the wolf saved the woman from another suitor, who tried to force himself on her; later, she was rewarded for her fidelity by gaining permission to marry her lover.[37] People who treated wolves kindly—people such as Hata no Ōtsuchi and the filial woman of Hirata's tale—were smiled upon by the gods through the agency of divine wolves and often experienced good fortune as a result.

In some cases, peasants used talismans to prevent their crops from being ravaged by deer and wild boars and called them *shishi yoke*, or "boar deterrents."[38] As Sutō Isao has observed in his study of the contemporary war waged between wild boars and rural communities throughout Japan, mountain people still use such "boar deterrents" with images of wolves to rid their crops of ungulates.[39]

OKURIŌKAMI: THE SENDING WOLF

The deification of the Japanese wolf was not confined to shrines, temples, and Japan's agrarian culture, however. When the Nara and Heian imperiums built Chinese-style capital cities, they also laid out travel circuits that ran through Japan's countryside and connected provincial headquarters with the capital. Seven circuits in all, they extended outward from the capital like spokes from a hub and were heavily traveled near Kyoto; but the farther one progressed from the city, the more primitive the circuits became until they devolved into mere trails. On these circuits, travelers, such as the irreverent man mentioned by Negishi, reported encounters with the Japanese wolf. By the seventeenth century and after the rampant roadside banditry of the medieval years had passed, travelers often spotted wolves on these circuits and, subsequently, the canine became associated with travel lore. For reasons of commerce, employment, and religious pilgrimage, travel in Japan underwent a boom in the late seventeenth and eighteenth centuries.[40] With the proliferation of roads, and an increase in the number of people using them, wolf encounters became more common, and tales from the roads began to tell of the Okuriōkami, or the "sending wolf," a deity who guarded but also sometimes menaced travelers.[41]

Legends from the travel circuit depict the Okuriōkami as unpredictable and elusive, but it could also be helpful, reminding one of the duality of other Japanese deities such as Inari or even Marebito. Marebito was a "visiting god" that came to villages from the other side of the sea, from a world inhabited by immortals. Once it visited a village, it displayed both *nigimitama*, or benign power, and *aramitama*, or dangerous power, behavior similar in some respects to the duality of the Inari deity, the Okuriōkami, and even the Christian deity.[42] It should be noted that even the Latin nomial for the Japanese wolf, *hodophilax*, refers to Okuriōkami lore. In Greek *hodo* means "way" or "path," while *philax* means "guard."[43] Here, Japanese folkbiological descriptions determined the Linnaean name given to the Japanese wolf. The sending wolf, like Marebito, possessed a dual personality, however. The seventeenth-century *Honchō shokkan*, mentioned in the last chapter, explains that if a traveler should be followed by Okuriōkami on the road, and if that traveler tried to run or looked back and stumbled, the wolf would pursue and attack. To ward off attacks by the cagey Okuriōkami, the *Wakan sansai zue* recommends bringing along musket fuse (*tsutsuhinawa*) because the smell reportedly frightened wolves.[44] Negishi, the Edo physician mentioned earlier, wrote that some of the fresh and salted fish deliverers who worked the mountain roads between Nagoya and Minō, in Owari Province, knew to give bits of fish to wolves encountered at night on the roadsides. This, wrote Negishi, delights the wolves; and the deliverers generally pass without being harmed by the wild canines.[45]

One suspects that Okuriōkami lore was on the mind of Furukawa Ko-shōken, an eighteenth-century geographer and extensive traveler, as he made his way to Japan's far northeast.[46] In *Tōyū zakki* (Miscellaneous notes on travels in the east; 1788), he noted forms of wolf worship and the use of boar deterrents during his travels to Morioka domain. He reported that north of Sendai in Oinokawara—where the name "Oino" was actually spelled out with the Chinese character for wolf—farmers worshiped wolves, beseeching them to protect their fields from hungry ungulates. Oinokawara is in the valley of the Futamata River in Miyagi Prefecture. It was reportedly founded when a dispute erupted during a game of Japanese chess and resulted in the death of a samurai from Ichinoseki named Yonekawa Gaiki. His older siblings, eager for revenge, confronted their brother's murderer at this location in Morioka domain. During the confrontation, the Mitsu-mine avatar appeared in the form of a white wolf and helped them avenge the murder. Following this auspicious event, the village name was subsequently changed from Kawara Village to Oinokawara, in honor of the divine wolf avatar.[47]

That people in Morioka referred to the wolf as *oinu,* attaching the honorific *o* to *inu,* "dog" in Japanese and thus rendering it "honorable dog," illustrates the celebrated place of the wolf in this region. "Even if you say the word *ōkami* [wolf]," wrote Furukawa, "they do not understand." He reported that people were grateful for the wolves that lived in Oinokawara, because deer were numerous and frequently decimated crops. He said that the locals insisted that they kindly greeted wolves they met at night because they knew that the canines dutifully chased deer from grain fields.[48] When people of Oinokawara encountered wolves in the wild, explained Furukawa, it was customary to say, "O lord wolf, what do you say? How about chasing the deer from our fields?"[49] Nishimura Hakū made similar observations in the *Enka kidan* (Strange stories from the mist; 1773), writing that mountain people did not fear wolves and that when these people crossed paths with them (as they frequently did), even after the spread of rabies, they did not panic and showed respect and so were seldom injured.[50]

TSUNAYOSHI: WOLF SHOGUN

Not only among the mountain people and the grain farmers of the Japanese countryside but even in the grand halls of the palaces of the Tokugawa shoguns and their councilors, forms of wolf reverence fell in and out of fashion. Starting in 1687, shogun Tokugawa Tsunayoshi issued the infamous *shōrui awaremi no rei*, or "clemency for living creatures orders," which, with some exceptions, forbade killing animals throughout the realm. Tsunayoshi's critics believed (as do many contemporary historians) that these orders sprouted from the seeds planted by soothsayers and priests in the shogun's confidence and grew out of his increasingly intense Buddhist anxieties that during his past life he had been responsible for great bloodshed. As Karmic retribution, his only son, the dear Tokumatsu, had died at the age of four in 1683. To atone for his past bloodshed, Tsunayoshi decided to make it a crime to take the life not only of human beings but also of other animals, because in the Buddhist tradition, animals too participated in the grand cycle of reincarnation and the "continuum of life." However, other historians point out that such clemency orders, like the Hōjō-e Ceremony of the medieval era, were common in Japan, often taking the form of state-sponsored rites, and need to be seen as less the product of manipulative soothsayers and priests than of deep-rooted Buddhist convictions concerning the inherent Buddha-nature of all sentient beings.[51]

Under Tsunayoshi, the presence of soothsayers and such Buddhist beliefs led to what historian Harold Bolitho calls a "torrent of legislation" making it a crime to "kill, abandon, or torment animals of any sort, or to eat their flesh; even insects were included in this protection of all living things." In particular, however, because Tsunayoshi and two leading councilors were born in the year of the dog according to the Chinese calendar, he became known as the Inu Kubō, or Dog Shogun. Tsunayoshi went so far as to establish special dog inspectors (*inu metsuke*), whom the shogun instructed to compile elaborate registers of dogs throughout the realm and to construct massive kennels for housing the thousands of dogs in Edo that might otherwise have been dispatched as strays.[52] At least since the Kamakura period (1185–1333), samurai had targeted dogs during archery exhibitions and confined hunts, and so Tsunayoshi's clemency orders even went against the grain of traditional samurai practice.

Such Buddhist anxieties, Chinese cosmological associations, and intense feelings for dogs extended to other canines, including wolves. For Tsunayoshi, as the *Tokugawa jikki* (True chronicle of the Tokugawa; 1809–43) demonstrates, these clemency orders extended to all wild animals, even those considered by some to be dangerous. Tsunayoshi insisted that when dealing with animals, a compassionate heart should govern people. If people did not act compassionately but wantonly shot at wild boars and deer running in the fields or killed wolves chasing horses, they would be held responsible and punished under the new orders. When necessary, farmers were allowed to drive animals away, but in doing so they were to show caution. Only under extreme circumstances, if a problem animal could not be stopped, a pledge could be written at the local government office, and then access to firearms could be arranged. If peasants killed animals for no reason, however, not only did the Edo shogunate promise to hold the guilty party responsible, but the chief magistrate and the local lord (the *daimyō*) would be held accountable as well.[53]

In many instances, Japanese farmers and samurai alike obeyed these orders. Examples of people following the letter of Tsunayoshi's law litter the *Tokugawa jikki*, and some of these examples involved the treatment of wolves. In the fourth month of 1690, a hunter named Inoue Masatomo, in Sakura of Shimōsa Province (Chiba Prefecture), shot a pack of problem wolves; but, showing compassion, he spared three pups, taking them alive from their den and relocating them to a distant area.[54] In the fourth month of 1693, moreover, hunters frightened away problem wolves by firing blanks; when this failed to stop them, the hunters used fowling guns instead, which presumably only frightened the animals. These hunters later briefed an inspector on the situation.[55] Such restrictions, boasts the *Tokugawa jikki*, fostered compassion among the people of the realm. In the fifth month of 1695, when locals reported that bears, wild boars, and wolves ran out of control in the countryside and posed a threat to domesticated animals, officials carefully drove them away with canes; even when dogs and cats attacked birds or when "comrades" fought with each other, people pulled them apart carefully so as to not hurt any of them.[56] On hearing such reports, one suspects that Shogun Tsunayoshi beamed with delight at the new tenor of compassion engulfing his realm.[57]

However, not all people followed the clemency orders, and Tsunayoshi often punished such uncompassionate offenders with surprising severity. In the seventh month of 1690, Nagahara Tanshichirō, a vassal of Kaga, a powerful domain on the Japan Sea coast, reportedly killed a wolf. When the incident was reported to the Edo shogunate, it punished Nagahara by confiscating his fief, and thus his lifeline and genealogical entitlement, which the shogunate placed in the custody of relatives. But Nagahara's problems with canines had only begun. Seven years later, in the second month of 1697, officials reported that he had struck again, this time killing a dog that had allegedly attacked his son, Sōjirō. Consequently, both he and his son were banished. While at Shima Village, Sōjirō became ill and died; the shogunate pardoned his father, Nagahara, in the eighth month of 1709, following the death of Tsunayoshi. For killing a wolf and a dog, Nagahara had spent twelve years in exile. In another instance, in the second month of 1692, a samurai named Teranishi Magokyūrō was exiled from his domain after he stabbed to death a wolf that reportedly had entered his home. His son, Jūrōzaemon, who was also there, voluntarily cut off his hair, proclaimed penance, and entered monastic life; his crimes, we are told, were graciously overlooked by the Edo shogunate.[58] Following Tsunayoshi's death, the clemency orders were repealed. But the reign of the Dog Shogun, as bizarre as it might have been, does illustrate how deeply Buddhist sentiments toward animals ran in traditional Japanese culture. Or, more cynically, his reign illustrates how well-known Buddhists anxieties could be manipulated to legitimize the autocratic power of the shoguns.

That Tsunayoshi, the Dog Shogun, protected wolves, that for people of the far northeast the word *oinu*, "honorable dog," referred to wolves, and that in ancient times the Hata clan associated itself with both foxes and wolves should remind us of the complexities of the categories of canine discussed in the last chapter. But wolf associations with early Japanese religion should also remind us of the way in which people can, without actually knowing wolves all that well, imagine them as divine, imbue them with cosmological meaning, and thereby envision them as powerful Shinto messengers, as loyal Buddhist guardian kings, and as faithful Confucian protectors of grain fields.

In creating the wolves of Japan, the recognition of commonsense forms similar to zoological taxons always coexisted with cultural values attached to those same animals. And cultural traditions nearly always trump the "universal" cognitive abilities of our species, as the case of the Japanese wolf attests. Of course, as humans, we recognize the obvious similarities of dogs and wolves, yet we let dogs sleep in our beds at night while we relentlessly poison wolves with strychnine, shoot them from airplanes, or run them over with snowmobiles. In Japan's case, the wolf became a sacred category of animal, not just a zoological one, in the context of courtly literary traditions, shrine and folk Shintoism, Buddhist doctrines, and Confucian agrarianism, in a process that was, in retrospect, anything but commonsense or universal to our species. The Linnaean scientist might say wolves are wolves; but to most Japanese the animal meant something entirely different than it did to other peoples around the planet who lived in the company of wolves. Though there might be a universal wolf taxon, with recognizable morphotypical qualities, the wolf has no universal meaning: even the Ainu, who shared the archipelago with the Japanese, viewed wolves differently from their southern neighbors.

WOLF WORSHIP AND THE AINU

Wolf worship was not confined to ethnic Japanese but occurred in certain Ainu communities as well.[59] The Ainu, the native inhabitants of northeastern Honshu and Hokkaido, knew the wolf as the high-ranking god Horkew Kamuy (literally, the "howling god"). No doubt it was as part of their ethnobiological understanding of the world that Ainu realized that their hunting habits resembled those of the Hokkaido wolf, and such recognition fostered reverence for the animal. Over time, Ainu chiefdoms came to feature the Hokkaido wolf in regional creation myths, place-names, epic poetry, folklore, and selected ceremonial events. Ainu hailed the wolf as a deity, or *kamuy* (much as Japanese hailed the wolf as a *kami* or a divine messenger of the Daimyōjin), and Ainu sacrificed wolves, as they did bears and owls, in "sending-away" ceremonies called *iomante*.[60]

In certain Ainu communities, particularly those in the Tokachi and Hidaka regions, there flourished alternative versions of an origin myth about a white wolf that mated with a goddess; the offspring from this union became the ancestors of the Ainu people. Sarashina Genzō and Sarashina Kō, who collected stories of the plants and animals connected to Ainu village life, pointed out that several regional versions of this origin myth exist and that some feature a white dog rather than a white wolf. This difference appears to have been unimportant to the Ainu because both dogs and wolves inhabited much the same space in their classifying imagination. One version of this myth, from Shizunai, in the Hidaka region of southeastern Hokkaido, explains that Retaruseta Kamuy (White Wolf God), the god of Poroshiridake Mountain, could not find a suitable mate, even though he searched the entire island. So Retaruseta Kamuy summoned his divine powers to peer all the way to lands across the seas, and in time he spotted a mate in a distant country. Again drawing on his divine powers, he coerced the woman to get into a small boat and cross the sea; once on the island, she became his wife. From this union the Ainu people were born.[61]

In a version of this story from the Tokachi region, in eastern Hokkaido, the White Wolf God appears as Yukkoiki Kamuy, the God Who Takes Deer, or sometimes as Horkew Kamuy-dono, Lord Wolf God, while versions from Iburi used the name Horkew Retara Kamuy, White Wolf God. In Ainu folklore, not only do male wolves take human brides, but sometimes female wolves become the wives or concubines of Ainu chiefs. In Shizunai, for example, Ainu told a story of a white female wolf who transformed into a woman and then married a chief, a story reminiscent of the shape-shifting fox tales of *Tōno monogatari* and other collections. In Abuta, in southern Hokkaido, stories tell of female wolves marrying Ainu men, while in nearby Oshamanbe, stories tell of a white wolf that became the concubine of a local notable.[62]

With so many wolf gods inhabiting the mountains, it is not surprising that legend had it that if Ainu hunters left behind portions of their kill, wolves came and ate the leftovers; and if Ainu hunters cleared their throats near a wolf kill, the wolf god generously made room for them. Because of its divine status, Ainu believed that the Hokkaido wolf should not be killed with poison arrows or firearms, because then it would not be resurrected as a god; and if a hunter wasted the wolf's meat and pelt after properly killing one, the remainder of the pack would come to kill the guilty hunter.

Interestingly, in Ainu lore, wolves usually were friendly toward people, as in one story from Tokachi in which a wolf saved an elderly Ainu woman from an evil bear god while she picked wild plants. Other myths explained that if a person tied the sinew of a wolf around his wrist, he would acquire great strength and could even grapple with large bears.[63] Such descriptions share similarities with Native American lore about the wolf, which frequently celebrated the canine as a friend of people. To cite just one example of such lore, "The Legend of the Friendly Medicine Wolf," as told by medicine man Brings-Down-the-Sun to Walter McClintock in Montana in 1905, relates the story of a captured Blackfoot woman who, on escaping a Crow village, received aid from a wolf when she was on the brink of starvation. The wolf brought her a bison calf, an act that saved her life, and when she returned to her Blackfoot village, she asked people to be kind to the wolf because it was a friend.[64]

Further exposing traces of a landscape inhabited by wolves, Ainu also associated certain places with wolves, and many of these places were mountains. Ainu in the Hidaka region gave the name Setaushinupur, "divine mountain inhabited by wolves," to a mountain located between Mitsuishi and Shizunai. Ainu in Tokachi once called a mountain at Lake Shikari-betsu, Setamashinupur, the "mountain where wolves descended from the heavens." Legend also claims that wolves dropped Usu Mountain from the heavens. The name for wolf that appears in these words, *seta*, actually means "dog" in the Ainu language. But as these cases illustrate, the Ainu classifying imagination made little, if any, distinction between the two canines. Rarely did Ainu prioritize distinctions between wolves and dogs when making natural identifications or classifications; nor did they prioritize distinctions between artifice and nature when crafting identities for themselves. Just as earlier in this chapter Negishi Yasumori spoke of "borrowing a dog" at Mitsumine Shrine but then seamlessly slipped into a folksy story about a lone wolf menacing a shrine visitor, so too did Ainu see the two kinds of canine as similar and their distinction, when one was needed, as largely situational. When in the village aiding people the canines were dogs; when in the mountains hunting deer they were wolves. This same point was made around 1890 by Nagata Hōsei, who studied Ainu place-names. He pointed out that Ainu once called a location on the Biei River, in Ishikari, Setaushinay, or a "place where wolf pups are born" or sometimes a "swamp where wolves live." The important point is that *seta* is rendered as "wolf." Nagata also explained that in Monbetsu, Ōseushi referred to a location that once had many wolves.[65] The story goes that deer and wolves once thrived together at this place, but that eventually the wolves chased away all the deer.

In an earlier Ainu world, a world yet to be disrupted by the Japanese intrusion from the south, the landscape was alive with wolves, busily hunting deer, raising their young, and, at magical times, aiding people and descending from the heavens to inhabit sacred mountains and forests, much as wolves did in the traditions of some Japanese villages.

HOKKAIDO DOGS AND WOLVES

One reason that Ainu drew such a fine line—if any line at all—between dogs and wolves in their classifying imagination was because Ainu appreciated the hunting skills of wolves and relied on those of their dogs, and so, not surprisingly, Ainu tried to reproduce wolf traits in their own dogs (a breed once called the Ainu-ken, now called the Hokkaido-ken) through both accidental and intentional breeding. Accounts from Kushiro tell of Ainu domesticating wolves and then using them to hunt deer; sometimes Ainu simply brought dogs into the mountains while wolves were in heat and allowed them to breed. One suspects that though adult wolves became domesticated only reluctantly, pups proved fairly easy to train and made fine hunting companions.[66] Probably the most reliable account of the intentional interbreeding of wolves and dogs comes from an 1882 survey conducted by Kaitakushi (Hokkaido Development Agency) officials Uchida Kiyoshi, Tauchi Suteroku, and Fujita Kusaburō. At one Ainu village at the mouth of the Biboro River, in Kitami (northeastern Hokkaido), they reported that Ainu there tried domesticating wolves. The survey team also reported that in Tokoro Village, also in Kitami, Ainu kept wolf pups at the village for about two years, and once they had become accustomed to people, Ainu allowed them to venture into the mountains on their own to hunt and kill deer, after which the wolves returned to the village.[67] Earlier, in 1792, Kushihara Seihō recalled that during his travels an Ainu dog at the trading post at Sōya, on the northernmost tip of Hokkaido, went into the mountains while in heat and mated with a wolf. She then returned to the post and whelped three pups. The wolf father was more elusive, however. He continued to show up periodically at the village only to quickly disappear. In time, the wolf and dog took the pups deep into the mountains, but the dog continued to come and go from the post. After a while she stopped coming, however, and hunters discovered her half-eaten carcass in the mountains.[68]

Kushihara portrayed dogs as pervasive features of Ainu domestic life and hunting practices. He wrote that throughout Hokkaido, dogs sat in front of every Ainu household, and that when the inhabitants went hunting, they took these dogs into the mountains with them. When hunters came across a bear, they released several arrows that, not surprisingly, caused the bear to charge at them. But then the hunting dogs took over, barking and howling, chasing the bear and biting at its hindquarters. The bear eventually fell back, and after hunters lodged several more arrows in its body, they dispatched it. He wrote that Ainu kept at least five or six such hunting dogs at the Sōya trading post.[69] Matsuura Takeshirō, who left vivid descriptions of early-nineteenth-century Ainu life, explained that dogs were used not only to hunt but also to guard millet plots from destruction caused by deer. At night deer quietly tried to get into the millet plots, but dogs sounded the alarm and chased them away.[70] This is precisely what Japanese in Morioka expected of wolves. Finally, as Tessa Morris-Suzuki points out, some Ainu communities (particularly on Sakhalin Island) also "farmed" dogs, eating their flesh and using their skins for clothing.[71]

In the nineteenth century, with the Japanese settlement of Hokkaido, the Hokkaido wolf began to loiter near Ainu villages, particularly when local dogs were in heat. In interviews, Kawamura Muisashimatsu, of Asahikawa in central Hokkaido, remembered being told as a child not to wear dog-skin clothing outdoors at certain times of the year or wolves might attack her. She remembered wolves howling and recalled that it frightened the dogs in the village, which all fled into nearby homes. Another elder from Kussharo, Deshi Kanji, claimed to have watched a wolf kill a deer. In a natural drama similar to one I witnessed once with Montana State University seminar students while in Yellowstone, Deshi observed a deer (in our case it was a cow elk) nervously looking over its shoulder and then noticed that a wolf was following it. When the deer rested, so too did the patient wolf. But ever so slowly the wolf closed the gap, until it finally killed the fatigued deer with little difficulty.[72]

So neither preindustrial Japanese nor Ainu clarified the distinction between wolves and dogs in the same way that Linnaean science has tried to. Where Japanese scientists tried to separate them according to the logic of zoological taxons, Ainu hardly recognized the difference between the two species at all. It appears that these animals were wolves when they behaved as wolves in the wild setting of the mountains, but when in villages, and in the company of human beings, they became dogs. Behavior and setting, not abstracted morphotypical traits, separated the two species of canine. In Ainu culture, perhaps no story makes this point as vividly as the "Horkew kotan kor kur" (The wolf god and the village man; 1967).

THE HOKKAIDO WOLF IN AINU POETRY AND FOLKLORE

Ainu *kamuy yukar*, or divine epic poetry, also exposes traces of the wolf in Ainu culture. One such example is the "Horkew kotan kor kur," as recited by *ekashi* (storyteller) Hiraga Satano and recorded in October 1967. The narrator is no other than Horkew Kamuy, the wolf god. Horkew Kamuy begins by describing its lavish treatment by its Ainu master. Once more exposing the blurry line between wolves and dogs, we learn that Horkew Kamuy (treated as a dog when in the village and the proximity of people) was fed in lacquered wooden bowls while sitting on stylized mats. Horkew Kamuy and its master also hunted together in the mountains. As Horkew Kamuy recalled:

> My master
> when he goes hunting
> I, too, go with him to the mountains.

> We take deer,
> we take bear.
> When I do this
> my master,
> his work is made easy . . . [73]

However, Horkew Kamuy got into trouble when other wolves (from outside the village) talked it into going hunting with them. In a sequence that celebrates the effortless hunting skills of wolves, characteristics obviously admired by Ainu, Horkew Kamuy said:

> Before my master came
> I ran ahead,
> not so much running as flying.
> When I went to the hunting grounds,
> my friends,
> the other wolves,
> came in large numbers.
> "Today,
> for just one day,
> let's run together,
> let's play together."
> "Ohhh," Horkew Kamuy howls.[74]

The wolf pack then went to a second and a third hunting ground; until well into the evening they hunted and played together. When Horkew Kamuy finally returned home, the master beat the hapless animal to death with a piece of firewood from the hearth because it had not hunted with him. The Ainu master ranted:

> "This bad dog,
> this rotten dog,
> today, for the entire day,
> what has it been doing?
> Not even one deer
> has it brought to this house."[75]

After having killed Horkew Kamuy—now a beaten "dog" in the village—with firewood, the master threw the body onto a heap of trash. When Horkew

Kamuy woke up, it found itself floating downstream in an unknown river as if in a dream. Horkew Kamuy explained:

> I realized
> that at this point my
> heart was standing tall in my chest,
> my feelings were standing in my chest;
> this is what happened to me.
> From this point I flowed downstream
> like I was flying,
> like I was running,
> forever downstream.[76]

Midway down this dreamlike river, Horkew Kamuy found itself at the home of an even wealthier Ainu master. Unlike the earlier master, this one acknowledged Horkew Kamuy as a high-ranking god. He performed appropriate acts of reverence, such as whittling ceremonial *inaw* (wooden fetishes) and adorning the site with important *ikor* (treasures). Horkew Kamuy even dined at the *iyoykir* (treasure altar). Horkew Kamuy said:

> Whittling *inaw*
> he fastened the *inaw* to a sword guard [*seppa*].
> He then took this and came outside:
> "Upstream
> I heard gossip
> of lord wolf god [*horkew tono*],
> a high-ranking god."[77]

This new Ainu master then invited Horkew Kamuy into his house and to sit near the sacred treasure altar so that it could be worshiped accordingly. Eventually, the previous master, armed with gifts of apology, came to retrieve the wolf god. Horkew Kamuy consented to go back, where the exact same scenario occurred again: pack members invited the wolf to go hunting, it neglected to hunt for its master, and the master beat Horkew Kamuy to death. This time, however, Horkew Kamuy decided not to stop at the house of the previous wealthy master but rather continued downstream even farther to the house of another Ainu. When this Ainu came out with *inaw* and *ikor*, eager to worship Horkew Kamuy, the wolf god fled across the ocean, where it encountered the wolf boss.[78]

When the wolf boss heard the story of the evil Ainu who beat Horkew Kamuy, he was outraged and pledged "to go and kill all the evil Ainu."[79] The wolf boss assembled many boats, and the narrator, Horkew Kamuy, served as guide. When they arrived at the place of the Ainu, they were told by a frail-looking youth, who appears on the scene in another boat and whose knees wobbled when he walked, that the Ainu village was in the midst of a terrible famine.[80] Although Horkew Kamuy was moved into not returning to the village to seek revenge, the other wolf gods were not. Horkew Kamuy explained:

> They arrived at that village,
> the village upstream
> where I was raised.
> Without saying a word,
> and without leaving behind a single insect,
> all the people of the village
> were killed.[81]

This epic poem then concludes with a fairly straightforward message that highlights the interconnected place and interrelated nature of wolves and dogs in the Ainu imagination:

> Therefore,
> simply put,
> a dog,
> even if you kill one,
> should not be sent in the direction
> of the ocean.
> Its ancestors are wolves.
> It should be sent in the direction of the mountains.
> That's the lesson of this story.[82]

Other examples of wolf-related folklore, such as "Ōkami o tasuketa shiro kitsune" (The white fox who saved the wolves), recorded in 1929 by Kinjō Matsu, also celebrate the wolf as a "high-ranking god" and as a "skilled hunter." In this story, after famine struck the land of the Ainu, a white fox god came across from *kamuy moshir*, or the "land of the gods," and saved both the gods (animals) and humans (Ainu). The wolf gods, in particular, although excellent hunters and higher-ranking gods than the fox, had proved unable to catch game because of the presence of two terrible monsters. The first monster, Kinaposoinkara, or Looking through Grass, had spooked deer before the wolves could approach. The monster had big eyes that shone with the clarity of "dewdrops being lit by the morning sun." Peposoinkara, or Looking through Water, had scared the salmon. The white fox god, portrayed as a trickster, slew these two monsters with *yomogi* arrows, a type of arrow fashioned from the first grasses that had grown in the land of the Ainu and hence having sacred powers. Ainu crafted these arrows when the tall stocks of the grass hardened in the fall, creating solid arrow shafts. When struck by these magical shafts, both monsters turned to skeletons and crumbled under the trampling herds of deer and swimming schools of salmon. Some of these deer ran wild and fell off a cliff, while the salmon similarly made themselves available to the villagers and wolves, bringing enough food to feed both the humans and the gods. The Ainu showed their appreciation by making offerings of delicious saké in *tuki* (sacred lacquer bowls), carving sacred *inaw*, and making prayers of gratitude, which the white fox god acknowledged, eventually marrying into the wolf family and siring several children. Certainly, this story casts the white fox god and Horkew Kamuy in a favorable light, and the animal gods themselves, as an immediate part of the natural community Ainu lived in, often took on anthropomorphic characteristics: a shared emotional terrain of sorts, such as that explored in the Introduction. The story described the wolf sister, for example, whom the white fox god married, as having long flowing hair that reached the ground. In this story, Ainu remained bystanders to the pageantry of the animal world—*kamuy moshir*, the divine world—a vantage point indicative of the prominence animals held for most Ainu.[83] Again, artifice and nature seldom served as meaningful categories; at times neither did the categories of human and nonhuman.

One final tale worth mentioning is "Kibori no ōkami" (The carved wooden wolf), published in 1998 by Kayano Shigeru as a children's story. The narrator, not surprisingly, was a young hunter from the lower reaches of the Ishikari River who, led by unexplainable impulses, was drawn to the distant upper reaches of that famous river. There he found a village, at the center of which stood a large house. He noticed, however, that in front of the house, at a sacred site adorned with old bear skulls and *nusa* (groups of carved fetishes), no new bear skulls were present, which suggested that the *iomante* (sending-away ceremony) had not been performed recently. He speculated that a strong hunter, one strong enough to have caught those bears, must live in the house; but the fact that no new bear skulls adorned the sacred site meant that recent hunts had proved unsuccessful. Later, he met the occupants, whom he described as sad and miserable. After sleeping in their house, he woke up early the next morning and (again drawn by unexplainable impulses) found himself running on distant mountains.

He discovered a small house surrounded by what he thought were dog tracks in the snow. A lone woman with a child inhabited the house. The woman explained to him that her normally kind father-in-law had abandoned her, and that, periodically, a large dangerous bear god came and tried to attack her. She explained that the attacks had been foiled by what she first thought was a dog but then later realized was a wolf. She said that the wolf was in fact a small wooden wolf that her brother had made for her as a charm. She knew this because when the wolf fought the bear, the wooden one vanished from her pocket, where she always kept it. That night, when the hunter was at the house, the angry bear god returned, and it fought a brutal battle with the wolf until morning, when the hunter shot an arrow into the bear and killed it. At that moment the wolf disappeared, and later the hunter found the wooden wolf just outside the house on top of the snow. The next night, the bear god appeared to the narrator in a dream, explaining that it had loved the woman and so had cast a spell on the father-in-law, prompting him to abandon her. Ultimately, the narrator brought the woman back to the village, which led to much celebration.[84]

The themes that resonate through these Ainu tales are much the same themes that resonate through the tales of Japanese villages such as Tōno. In these stories, one consistently encounters the sacred space of mountains and forests, as well as the divine animals that inhabit them. But Ainu saw themselves as part of this animal community, unlike many Japanese, who, at least in the eighteenth and nineteenth centuries, created the deities *ta no kami* (paddy deity) and *yama no kami* (mountain deity) to differentiate between the "this world" of settled agricultural villages and the "other world" of mountains, a wild space inhabited by powerful animals and gods.

SENDING AWAY THE DIVINE WOLF

As mentioned, Ainu used the wolf in the *iomante*, or sending-away ceremonies, in which Ainu chiefs sacrificed certain animals, liberating their *kamuy*, or godlike essences, from their earthly bodies. During these ceremonies, Ainu chiefs wore headgear called *sapaunpe*, often adorned with carved images of wild animals; a *sapaunpe* recovered from the Usu area has the carved image of a wolf. Most descriptions of *iomante* tell of bear cubs

FIG 18. Ainu sometimes sacrificed the Hokkaido wolf in a ceremony called the *iomante*, or the "sending away." This specimen, housed in Shizunai, has a hole bored into the left side of the cranium, suggesting that the animal was a male. Courtesy of the Shakushain Memorial Museum, Shizunai, Japan.

or owls being sacrificed by Ainu; but some evidence, such as that from Chikabumi and Biboro Villages in Kitami, suggest that Ainu chiefs also sacrificed wolves (which might also explain why Ainu raised pups in Kitami).[85] This should not be altogether surprising, because Horkew Kamuy, as the above epic poetry and folklore reveals, ranked among the most revered gods in the Ainu pantheon.

Probably the most compelling evidence of the use of wolves in the *io-mante* comes from the Shakushain Memorial Museum in Shizunai. Housed at this site is a wolf skull, complete with decaying fragments of flesh and fur, and a hole in the left side of its cranium, anatomical evidence that Ainu used the wolf in an *iomante* ceremony (see fig. 18). In the ritual use of skulls, Ainu believed that the spirit of the animal, its very *kamuy* essence, resided in the cranium between the two ears, and so during the *iomante* they bored a hole there so that the spirit might be released to *kamuy moshir*, the "land of the gods." Ainu ritualists knew this part of the *iomante* ceremony as *unmemke*, to "make up" or to "arrange" the skull, and ritualists considered it to be among the most important parts of the ceremony. In the *unmemke*, if the animal sacrificed was a male, ritualists bored the hole in the left side of the cranium, while in the case of a female the hole was bored in the right side. The specimen at Shizunai is of a six- or seven-year-old wolf killed in the early Meiji years; the *unmemke* hole on the left side of the skull suggests that it was a male.[86] Today the skull is adorned with wood shavings, a presentation designed to show what it might have looked like in its original ritual setting.

CONCLUSION

In this chapter we focused on wolf symbolism in Japanese and Ainu culture. By exploring the sacred life imposed on wolves by both societies—a life superimposed over the forms of "scientific" classifications discussed in the previous chapter—we have exposed rare common ground between these two cultures in the rugged mountains of mainland Japan and Hokkaido. In these mountains, the hunters of Japan, whose culture Yanagita Kunio brought to life in *Tōno monogatari*, and the Ainu of Hokkaido, whose divine epic poetry and folktales preserve a now extinct way of life, celebrated wolves as deities, skilled hunters, and divine messengers. But important differences in the two traditions have been exposed as well. Perhaps the most fundamental difference was that Japanese viewed mountains as distinctly otherworldly, different from the "this-worldly" agricultural villages of the lowlands. One suspects that the underlying reason behind the difference between Japanese and Ainu perceptions of wolves stemmed from the fact that Japanese, with technological advances from Korea and China, had established large-scale paddy and upland agriculture, while Ainu, though farmers of select grains in small plots, had not. Farming villages, and the paddy fields they serviced, became the "this world" of most Japanese; and "this world" became a source of Japanese notions of civility and even their distinct ethnic identity. Religious traditions imported from Korea and China (most notably Buddhism) influenced Japanese attitudes toward wolves as well, offering iconographical templates, such as those found on talismans, amulets, and stone statues, for sacred wolves to inhabit. Like the establishment of the "this world" of agricultural villages, then, Buddhist theologies further relegated wolves to a different plane of existence.

By contrast, the only otherworldly place for the Ainu was *kamuy moshir*, "land of the gods," which was a metaphysical place that people too could inhabit. In short, the Ainu were more a people of one world than the Japanese, a realm that they shared with all other creatures. For Ainu, animals and humans shared a common existential space as members of the same earthly community of being, even more organic than the abstract "continuum of life" emphasized in Buddhist doctrines. For Ainu, the "continuum of life" was the here and now of life on earth. Hence, animals in Ainu stories share many of the same values as the Ainu themselves; and Ainu and these animals always lived in close proximity to one another in the forests and coastal areas of Hokkaido. In Ainu stories, a sense of admiration for animals is revealed; in Japanese stories, a sense of anxiety.

In Japan, then, wolves represented otherworldly creatures that, much like the otherworldly mountains they inhabited, were unpredictable. Consequently, the activity of hunting wolves, at least for some, carried with it deep cultural anxieties. In fact, as we shall see in the next chapter, wolf hunts in Japan often took on a highly ceremonial character. And sometimes, these otherworldly creatures suddenly went "mad," lashing out with sharp teeth at travelers or the inhabitants of small villages; such behavior led to a major reconfiguration of the place of wolves in the Japanese imagination. The appearance of rabies in the eighteenth century led to just such a reconfiguration, transforming wolves from benign, if highly unpredictable, creatures to dangerous man-killers.

3

THE CONFLICTS BETWEEN WOLF
HUNTERS AND RABID MAN-KILLERS
IN EARLY MODERN JAPAN

Shiga Naoya wrote some of Japan's most celebrated modern fiction during his long and productive life, and "Takibi" (Night fires; 1920) ranks among his finest. Filled with lucid descriptions of colorful sunsets, rainbows that stretch between towering mountain peaks, crisp reflections on calm alpine lakes, and the lapping flames of bonfires that illuminate coal-black nights, "Takibi" is a literary feast of color, light, shapes, and natural imagery. Similar to the tales told in Yanagita Kunio's *Tōno monogatari,* moreover, "Takibi" takes place in the once otherworldly space of the mountains and so the story strikes a familiar cord, one related to changing Japanese attitudes toward mountains and of interest to our discussion of the extinction of Japan's wolves.

Midway through the story, while the characters gather birch to start a bonfire, Mr. K, a young mountain innkeeper, comes running out of the woods after having been startled by a strange worm with a glowing tail. The story's other characters—the artist Mr. S and the narrator and his wife—have a hearty laugh and then begin to talk with Mr. K more seriously about his experiences living in the mountains. In a question that only the people of modern societies can even conjure—a question that would have baffled the legend creators and storytellers of Tōno and even the epic poets of Ainu villages— the narrator's wife, somewhat anxiously, asks, "Is there anything really frightening in these mountains?" For the Japanese of preindustrial mountain communities, dark forests roused fears of animals, both imagined and real, and of scarcely understood deities and monsters that always lurked dangerously in the night. *Tōno monogatari* is full of such frightful imagery as shape-shifting foxes and vengeful wolves. Mountain villagers understood themselves to live on the boundary between the realm of the *ta no kami*, the paddy deity, and the *yama no kami*, the mountain deity. In the winter, the *ta no kami* crossed over to the other side to become the *yama no kami*. In preindustrial Japan, animals possessed strong associations with this mountain deity, and so one suspects that stories of such creatures had the ability to humble or even frighten those inhabiting this border space. But as "Takibi" reveals, such was not the case with modern societies like Japan in the Meiji period (1868–1912), the newly modernized world of Shiga. Mr. K could confidently reply, "Nothing at all," and in the ensuing conversation, the innkeeper even dismisses traditional supernatural monsters, specifically the Ōnyūdō, as mere superstitions. Now all things, animals and monsters alike, could be explained by science and controlled, if need be, by its technologies.

The dense forests of Shiga's imagination contained no dangerous monsters, but he chose to resurrect the memory of wolves in "Takibi." As Mr. K recalled, his dark face partially lit by the flickering bonfire, "When I was a child, often we just heard their distant cries. Hearing their cries in the middle of the night, I remember that a lonely, uneasy feeling came over me." As everyone huddled around the bonfire, Mr. K began to talk about wolves (*yamainu*). He told how his deceased father, who liked night fishing, had one evening become encircled by wolves. Being close to the lakeshore, he was able to return home by wading through the water. Mr. K also recounted how, the same year that they cleared land for horse pastures, wolves ate about half the horses. "That year we put dynamite in the horse meat and killed them," he continued. "In one week we wiped them out."[1] Once creatures that haunted the otherworldly realm of mountains, and that occasionally ate horses after the denuding of their forest home to make pastures, in "Takibi" wolves became the stuff of childhood memories. Seen from a literary perspective, the technology of modern society, in the form of dynamite, had blown the Japanese wolf to bits. By Shiga's day, the myth of the mountains and the divine animals who called this otherworld their home had all but vanished: wild places became simply resources to exploit, or pretty places to recreate at, within Japan's new order.

This chapter has begun with a brief discussion of Shiga's "Takibi" because the story anticipates Japanese attitudes toward wolves in the late nineteenth and early twentieth centuries: an extinct animal but one still part of childhood memories and local storytelling practices. How did wolves come to be seen this way? Once viewed as the sometimes troublesome but largely

beneficial Large-Mouthed Pure God, wolves were worshiped by the peas-
antry and mountain villagers of preindustrial Japan at grain fields and at
certain shrines. For about two decades in the late seventeenth century,
Tokugawa Tsunayoshi, the Dog Shogun, protected wolves from the halls of
the shogunal palace in Edo. However, early in the nineteenth century, the
wolf was targeted as a dangerous pest and thereafter vigorously hunted. At
this historical juncture, Japanese began creating a new kind of wolf, one that
could be justifiably killed en masse with traps, guns, poisons, and even dynamite with no divine repercussions.

Two related developments—one cultural, one ecological—facilitated this
change in earlier attitudes toward the wolf. Culturally, even before Japan's
Meiji modernizers targeted wolves for destruction (labeling wolves "noxious
animals," or *yūgai dōbutsu*), the transformation of the wolf from a revered
to a reviled animal was well under way. One reason for this transformation,
referred to briefly in Shiga's "Takibi," was that wolves preyed on horses in
early modern domains from Satsuma in the south to Morioka in the far
northeast, causing economic losses for already financially strapped domain
lords. *Ōkamigari*, or "wolf hunts," were in part responses to horse depredation; but starting in the early eighteenth century, Japanese documents
began referring to an epizootic called *inuwazurai*, almost certainly rabies,
which transformed the wolf from a benign creature to a deadly one. Wolf
hunts became standard features of life in Japan's far northeast, evolving into
forms of ceremonial bloodletting that served as responses to wolf attacks
that, by the late eighteenth century, had become commonplace in many of
Japan's most remote areas.

In this chapter, ecology and culture intersect, this time, however, to create one of the most dangerous kinds of animals: a "mad" wolf. After the outbreak of rabies, stories of wolves attacking travelers became numerous,
leading to one of the strangest episodes in global experiences with wolves.
In the late eighteenth century, the Japanese wolf became—and came to be
viewed as—a man-killer. To set the stage, however, we need to explore some
of the earliest Japanese encounters with wolves before rabies landed on the
Japanese Archipelago.

ANCIENT WOLF ENCOUNTERS

Most of the earliest reports of wolf encounters can be traced to the beginning of the eighth century, about the time when Japan constructed its first
capital cities at Nara and Kyoto.[2] Drawing on imported East Asian culture,

Japan's elites saw these cities as geomantic representations of celestial order and imperial power, and they modeled cities like Kyoto (called Heian-kyō at the time) after the great Tang capital of Changan. As Hiraiwa Yonekichi points out, from the first reports of wolves to about the 1030s (about 330 years), sources reveal some twenty documented human encounters with wolves. These wolf encounters are of interest because, among other reasons, they cast Kyoto in a new light.

Historians commonly portray Kyoto as a cultural space: the site of imperial government and high courtly culture. Kyoto was in fact a fairly large city housing a royal regime complete with palaces and ministries and served as the site of poem-writing competitions, calligraphy contests, scent-judging exhibitions, and other events designed to show off the refined skills and sensibilities of the courtiers. It was also the home of Genji, Murasaki Shikibu's fictional prince of the mid-Heian-period *Genji monogatari* (Tale of Genji). In this self-indulgent dominion of the Heian elite, in an otherwise mountainous and rugged Japanese Archipelago, powerful men, such as Fujiwara no Michinaga, came to epitomize the political savvy of their world, carving out illustrious careers by becoming synonymous with the culture of the day and by manipulating it to their own advantage. Given the importance of Kyoto as a site for such cultural activity, it is, according to one scholar, "small wonder that to be away from the capital, as an exile or even *en poste*, was regarded as a form of living death."[3] To leave the city was to be banished to the mountains and forests, remote places inhabited by large-mouthed beasts such as wolves. Such is the image that we commonly convey and teach in the classroom. However, early wolf encounters near Kyoto cast the ancient capital in a more rustic light. Kyoto emerges more as a frontier city carved out of only recently felled woodlands than as simply the splendid arena of the imperial family and its playful courtiers.

The fiftieth emperor, Kanmu, established Kyoto in 794 when he ordered that the capital, then at Nagaoka-kyō, southwest of Kyoto, be moved north to the new location along the Kamo River. Traditional explanations for moving the capital range from religious issues concerning pollution and Shinto, wherein the death of the emperor necessitated relocation to an unpolluted site, to the growing political power of Buddhist temples and their clever monks, a situation highlighted by the Dōkyō Incident, when a monk of that name attempted to expand his position by manipulating an empress and seizing power. Kanmu thought it prudent to relocate the capital away from these increasingly powerful Buddhist institutions, so as to reduce their influence in courtly political life and to cement his own political power.[4]

Historian Conrad Totman, however, offers a third, and equally intriguing, explanation for why Kanmu relocated the capital to Kyoto. He argues that it also had to do with the accessibility of natural resources, particularly timber supplies. These great capitals, with their splendid wooden palaces, required abundant supplies of special timber, such as clear planks of *hinoki*, or Japanese cypress, lumber that courtiers adored for its beauty and rich scent: the move to Kyoto ensured that stands near areas such as Tanba and Yamaguni could be harvested and brought to the capital.[5] Seen from this perspective, along with serving as the seat of imperial authority and the center of Heian culture, Kyoto was also a city situated amid only recently cut forests and lands still inhabited by wildlife, and so it should not be surprising that Japanese courtiers occasionally encountered wolves.

Ecological factors other than denuded hillsides also shaped life in the ancient capital and attracted wolves. Sadly, Kyoto and vicinity—called the Kinai—witnessed several severe harvest failures, famines, and epidemics, particularly in the twelfth century, the worst of which Kamo no Chōmei lamented in *Hōjōki* (An account of my hut; 1212). Of a famine that broke out in 1180, for example, he wrote, "The number of those who died of starvation outside the gates or along the roads may not be reckoned. There being no one even to dispose of the bodies, a stench filled the whole world, and there were many sights of decomposing bodies too horrible to behold."[6] The ripe "stench" of "decomposing bodies" (carrion in the eyes of wolves and feral dogs, as well as crows and other carnivores; many medieval paintings depict hellish images of corpses being devoured by dogs) surely attracted wildlife from near and far, as it certainly did in nineteenth-century Kaga domain during the Great Tenpō Famine. In the medieval years, even when not plagued by famines and epidemics, the residents of Kyoto often discarded human corpses along the banks of the Kamo River, a practice which established an outcaste class of *hinin* (i.e., "nonhumans") who handled the cadavers that, once discarded, attracted scavengers near the city's environs.[7] Given Japan's early history of forestry, famine, warfare, and burial practices, it should not be surprising that people sometimes encountered wolves near the ancient capital. Cultural practices such as burials reconfigured the ecology of Kyoto and also that of the animals who lived nearby.

When corpses were available and wolves were hungry, the canines fed on human carrion. Stories tell of wolves eating corpses in places other than Kyoto. Take two examples from Shinano Province and one from Kakunodate City, site of a wolf hunt to be discussed later in this chapter. Sometime before 1335, Go-Daigo's son, Prince Moriyoshi (Munenaga), stayed in Shinano while fighting the rebellious Ashikaga Takauji. One night, when throwing a net into the Koshibu River (a tributary of the Tenryū River in Nagano Prefecture) to catch fish, the prince spotted a famished wolf ready to eat the corpse of a drowned man. He bravely drove the wolf away and buried the poor fellow. In an even grimmer tale, in the aftermath of an earthquake that struck Zenkōji Temple in present-day Nagano City in 1847, victims sought shelter from the rain and cold in relief huts. With the smell of the dead wafting into the air, "mountain dogs" soon appeared looking for corpses to eat and terrifying people. And northward, in Akita, reports tell of villagers who defended a newly dug grave from hungry wolves with sickles.[8]

The *Kofudoki itsubun* (Lost writings on ancient customs; 713) offers the first documented trace of wolves, explaining that in ancient times at Asuka in Nara Prefecture, not far from where Hata no Ōtsuchi encountered wolves near Uji and the same place mentioned in the *Man'yōshū* as the plains of the Large-Mouthed Pure God, there lived an elderly wolf that was called the Large-Mouthed God (Ōguchi no kami) by the terrified locals because it had allegedly eaten so many people.[9] But, unfortunately, that is all we are told. Later chronicles mention that in 802 a wolf was killed while running down the capital's main boulevard, the Suzaku Ōji, and nine years later, during the reign of Emperor Saga, a wolf entered a military station near Kyoto, where it was subsequently killed.[10] In 826, a white wolf (four early encounters with white wolves have been cataloged in Japan) entered Emperor Junna's court in Kyoto, and chroniclers interpreted the event as an auspicious sign from heaven.[11] Such early Japanese interpretations are of interest because cultures around the world have seen white wolves as auspicious. Legends told by the Crow nation, for example, explain that White Wolf once was a servant to the Sun and saved the Crow nation from starvation; but for doing so, the Sun eventually cast out White Wolf to be a wanderer. "'Go to the Crow people,' the Sun said to White Wolf, 'and tell them that I shall no longer work to destroy them. You yourself will forevermore be a vagabond, an outcast among the animals of the world.'"[12]

Other early encounters with these vagabond wolves at court were more grisly. On a "violently rainy night" in 851, during the reign of Emperor Montoku, a wolf entered the house of a Shinto priest of the Hōjuku sect and attacked and ate a thirteen-year-old acolyte. Eerily, the next morning, only the acolyte's skull and two leg bones remained in front of the cooking stove.[13] In 855, Fujiwara Mototsune recorded that a wolf attacked a man one evening but that the next morning an archer shot the animal.[14] Two other encounters, recorded by Fujiwara Tokihira, include reports of wolves howling near the Dajōkan (council of state) office in 881 and a deadly wolf attack in 886 at the Kamo Shrine. This latter wolf was found and stabbed to death.[15] In 957, a wolf reportedly killed three women at the Gakkan'in Hokuchō (a branch of a Heian-style university founded in 845 C.E.) during the reign of Emperor Murakami, and in 968, a wolf entered the west gate of the compound of the crown prince.[16] The two final reports discovered by Hiraiwa relate how a wolf unsuccessfully pursued a deer into Emperor Ichijō's court in 998 and how one pursued a deer into the Kamo Shrine in 1034, where it captured and killed the deer.[17]

These early accounts, though lacking in details, are important because they contrast with later, more descriptive reports from the seventeenth and eighteenth centuries. Classical and medieval chroniclers portrayed wolves as only sometimes dangerous, as when one allegedly ate the young Hōjuku acolyte or when one killed three women near the university: incidents involving wolves were just sightings, sometimes auspicious ones, or observations of wolves carrying out presumably normal behavior, such as chasing and killing deer near shrines or eating carrion. These chronicles portray a canine neither despised nor celebrated, simply one of many animals that inhabited the mountains and forests surrounding the frontier capitals of classical Japan. As such wolf sightings illustrate, Kyoto was, in some respects, only a small island of East Asian civility carved out of a still otherwise wild archipelago, and so it should not be surprising that Japanese encountered wolves near the city. By the late eighteenth and nineteenth centuries, however, wolves had become either predators at horse pastures or dangerous when rabid, and hence Japanese designed techniques to kill problem wolves.

WOLF-KILLING TECHNIQUES

Carving out cities such as Kyoto from an otherwise wild archipelago was not the only activity that placed wolves in closer proximity to human populations. In regions ranging from northeastern Honshu to the southern tip of Kyushu, places where horse breeding remained an important part of the culture and economy, wolves often entered pastures and killed and ate prized horses. It should be mentioned that pastures in Japan were mostly small parcels of land, often tucked among steep, densely forested hillsides, giving wolves a degree of accessibility to livestock and cover from guards that would have sent chills up the spine of even the most seasoned Montana rancher. As early as 967, a document reports, "On account of wild boars and wolves, twenty-seven horses have fallen dead."[18] Centuries later, in the southern domain of Satsuma, in 1681, officials conducted prayers (*kitō*) to prevent wolf attacks. "Many bad wolves have appeared," explains one chronicle, "and they have repeatedly attacked and injured horses at the stables. Official Machida Jirōzaemon made an official trip to the stables to announce preventive measures, and then on the twenty-eighth day of the twelfth month, for a total of seventeen days, he offered prayers to the pasture deity [*makigami*] to stop damage caused by wolves."[19]

It was in the far northeast, however, more than any other region, where horse breeding developed into an important part of domain economies in the early modern period and where wolf predation became a problem. In Morioka domain, where poor soil and a short growing season made raising lucrative cash crops difficult, and where large-scale wetland rice farming was nearly impossible, officials turned to horse breeding as a way to supplement revenues.[20] When horse depredation occurred, domain lords turned to punitive measures to stop wolves. Hiraiwa counted some fifty-one incidents of wolf predation between 1644 and 1672 in Morioka domain alone, resulting in about 270 injured or dead horses. In the early eighteenth century, wolf predation became even more severe. Typical of wolf predation around the world, most horses attacked by wolves appear to have been older mares grazing in the pastures.

Wild horses also appear to have constituted a source of food for wolves. Tsumura Masayuki wrote that the great number of wild horses (*nouma*) inhabiting some areas of the northeast necessitated that domain authorities, on the twenty-sixth day of the third month of every year, sponsor ceremonial (*sairei*) hunts and roundups, conducted by samurai dressed in full armor and divided into military squads, to help deal with wild horses that damaged crops.[21] Other than wolves killing wild horses or horses out to pasture, documents contain only two examples of their appearing in villages and harming horses or people. The first occurred in the sixth month of 1740 in Yanagizawa (Iwate and parts of Aomori Prefectures), when wolves injured horses, and the second occurred in 1763 nearby in Hiromiyazawa, when a wolf attacked a man named Mansuke.[22]

To suppress these problem wolves, horse ranchers and hunters requested firearms at government offices (because the Edo shogunate had restricted access to firearms).[23] Records of such requests exist from throughout the northeast. In Morioka domain, according to one such request from 1724:

> Wolves have gathered in large numbers, attacking and eating five horses in the pasture, while wounding another three. Therefore, we respectfully request

that honorable officials see to the organization of a hunt using firearms. For our impoverished villages, sponsoring this kind of hunt is difficult, and so we ask to borrow four or five guns, some shot, and four or five soldiers, for the purpose of a wolf hunt for a day or two. We need hunters who specialize in killing wolves.[24]

Such specialized wolf hunters (or "wolfers," as they were called in the American West) did in fact exist in northeastern domains. Morioka domain included among its rank and file a man known as Ōkamidori Shōjurō, or Wolf Hunter Shōjurō. Beginning in 1742, Shōjurō reportedly killed many wolves and turned in their skins and canine teeth to domain officials for reward. On one occasion in 1743, authorities ordered Shōjurō to kill a wolf that had attacked a fawn-colored packhorse at Sumiyano (Aomori Prefecture). He reported the results of the hunt to an inspector, who forwarded the report to the Miyauchi magistracy.[25] Wolf depredation was not confined to horses. In 1697, during Shogun Tsunayoshi's reign of compassion, wolves killed four dogs as well as forty-seven horses in Takamatsu, Akata, and Kuwanokita.[26] Often, wolves are attracted rather than repelled by guard dogs, no matter how fierce the latter might be. I have heard stories from ranchers in the Tom Miner Basin near Yellowstone about wolves making a beeline from the mountains to guard dogs in order to kill them, avoiding the "guarded" livestock altogether. They left tracks in the snow that suggested as much. "It's almost as if they woke up one morning and said to themselves, 'Let's go kill those stupid dogs,'" a friend and rancher told me.

Wolf hunters killed wild canines with poison as well as firearms, but the practice appears to have been quite limited until the early Meiji years, when strychnine was used to kill wolves in Hokkaido. In 1759, for example, in Morioka domain, hunters inserted poison into a packhorse killed by wolves. The poison was weak, however: although a male and female wolf fed on the carcass, and hunters followed them into a nearby field, they never recovered the bodies.[27] Negishi Yasumori, the Edo physician mentioned in the previous chapter, reported a "strange way" to kill wolves in the Mineoka pastures and Kogane region of today's Chiba Prefecture. Wolves relish the taste of salt, he explained, and so ranchers, to protect their horses, conceal poison in raw salt and place it in areas where wolves frequently loiter. Interestingly, the poison mentioned by Negishi was *machin*, an extract from the plant *Strychnos nux-vomica*, which is native to India, the Malay Archipelago, and parts of China and contains high concentrations of alkaloid strychnine. Strychnine does possess some medicinal qualities, and so

Negishi, as a physician, was probably familiar with the drug; but its use as a poison in predator control did not become widespread in Japan until after the Meiji Restoration.[28]

In the second month of 1822, according to another report: "From winters past, wolves run wild in the fields and chase pasture horses. Preventive measures were taken with poison and firearms, and hunters killed two or three wolves with the poison. However, afterward, the wolves stopped eating the poison and continued to harass horses throughout the night."[29] Such descriptions remind one of Ernest Thompson Seton's observation in *Life-Histories of Northern Animals* that wolves transmit knowledge and learn how to avoid traps and poison. In the American West, wrote Seton, some wolves, over time, "learned how to detect and defy traps and poison, and in some way the knowledge was passed from one to another, till the Wolves were fully possessed of the information."[30] Scientists know that many animals are capable of passing down learned, adjustable behavior, often in the form of what we recognize as local cultures or traditions, and so Seton's observations should not be all that surprising. But sometimes wolves were slow to learn. In 1835, in the villages of Onizawa and Dōgazawa (Aomori Prefecture), reports claim that a pack of ten wolves appeared and harassed the village inhabitants. Kuraemon, from the village of Taruzawa, knew the recipe for a poison used by wolfers and smeared the deadly concoction on the carcass of a horse, and three wolves turned up dead.[31]

Hunters also used game calls to lure wolves to traps or within range of guns. In his collection of stories entitled *Tankai* (Sea of conversations; 1795), Tsumura Masayuki described hunters killing wolves in this way. "For killing wolves," he explained, "hunters make a blind out of stones and then hide in it. If they then start howling like a dog, the wolves will hear this and come out of hiding. The hunters place bait, a piece of meat from a wild animal or something, in front of the blind, and when the wolf begins to eat the bait the hunters shoot it from behind the blind."[32] Shooting wolves from blinds might be combined with pit traps and other devices. In 1799, a brief note from Morioka domain explains that along with poisoning the carcass of an ox or horse, pit traps (*otoshi ana*) also served to prevent horse depredation by wolves; other popular traps, called *oshi wana*, crushed wolves and foxes. A twentieth-century observer who reported such trapping techniques found that hunters used pit traps near villages in Akita Prefecture. Such pits were oval, about three feet across and six feet deep. Earlier, Tanamori Fusaaki, a late medieval Shinto priest, mentioned traps that crushed wolves: "Even in Bōshū there is wolf hunting. One or two were taken by crushing them;

later, five or six were also taken." The pioneering educator Watanabe Hayashi wrote that in the 1870s some young people, while walking with snowshoes in deep drifts, killed a wolf with a lance when they ran across the animal at a place where horse carcasses were dumped (*uma sutedokoro*) in Niigata Prefecture. As wolves came under increasing ecological and human-generated pressure in the early modern and early Meiji periods, such sites became frequent haunts for wolves, as did human gravesites.[33]

EARLY MODERN BOUNTY SYSTEMS

We hear of some of the earliest wolf bounties (*shōkin*) from Morioka domain, where horse depredation remained a problem throughout the seventeenth and eighteenth centuries. In 1701, documents indicate that the killing of three wolves from Yūto and two from Sumiyano produced a one *ryō* bounty in gold. Other eighteenth-century sources suggest that for wolf hunters, on average, domain lords paid 700 *mon* for male wolves; female wolves, because they could produce pups, brought in 900 *mon*. If peasants or field guards (*nomori*) killed wolves, however, they received considerably less. In Takashima domain, Takayama Zen'emon and Kubōjima Zenzaemon noted that on the third day of the sixth month of 1702, a hunter named Yoshizaemon killed two wolves for a reward of two *ryō*, while on the fourth day of that same month, Shōsuke from Fukuzawa killed a wolf for one *ryō* in gold.[34]

The "Satake-ke gonikki" (Chronicle of the Satake family; 1673–1854, often called the "Kita-ke nikki") contains some of the best information on wolf hunting in northeastern Japan. This chronicle suggests that often domain lords paid bounties by allowing wolf hunters to keep some of their game (presumably to sell on their own) and even by allowing hunters to drink saké with the Satake lord himself (a practice sometimes called *sakazuki*, and a signal privilege in early modern Japan). In 1715, for example, hunters spotted eleven wolves during a wolf hunt and managed to kill six. Of these six, the Satake lord kept three in return for lending the hunters weapons, carrying-pole boxes (*hasamibako*), and other tools; the hunters, for having killed the wolves, kept the remaining three. In 1812, moreover, three hunters took one wolf each, and, as reward, the Satake lord allowed them to keep their kill as well as "drink saké in the lord's private chambers."[35] Such bounty systems or payments in kind continued into the Meiji years and contributed to the ultimate extinction of the Japanese wolf. Unlike Meiji bounties, however, early modern bounties were never designed to eliminate wolves entirely: wolf extermination (*ōkami no kujo*) was a product of the post-Meiji mind-set, the same one depicted by Shiga in "Takibi."

In perhaps the most symbolically important example of such wolf bounties, in 1876 the Meiji emperor led an imperial procession into the far northeast. These imperial processions, as historian T. Fujitani points out, strengthened the national authority of the refashioned monarch, and they became common in Meiji's day.[36] In July of that year, while in Morioka, where horse depredation was traditionally high because of the proximity of wolves to pastures, the emperor reportedly inspected a wolf pup at a museum (*hakubutsujo*) that had been kept alive by a hunter. It was a symbolically potent moment in the history of the Japanese wolf.

First appearing on June 9, 1876, Kishida Ginkō's "Tōhoku gojunkōki" (Northeastern imperial procession chronicle) was published in the *Naniwa shinbun* (Naniwa newspaper) and later in the *Tōkyō nichinichi shinbun* (Tokyo daily newspaper). Kishida wrote that prior to the northeastern imperial procession, officials had undertaken mass wolf hunts because of an increase in horse depredation in the area. In recent years wolves had been killing up to forty or fifty horses per year. Starting in August 1875, the newly established Iwate prefectural government offered bounties (*shōreikin*) for wolves: just over ¥7 for a male and over ¥8 for a female. This represented a substantial amount of money at the time, and the pup that the Meiji emperor inspected had been taken from its den and left alive despite the fact that even it could have been exchanged for a ¥3 bounty. If wolf hunters came from more distant areas, Iwate Prefecture often paid their travel expenses; and reports suggest that this campaign alone produced some forty dead wolves.[37] The moment is a salient one because when the emperor went to Morioka and viewed the small pup, he inspected a vanishing, and once sacred, part of Japan's far northeastern landscape, one that would be sacrificed, along with centuries of culture, in the name of creating a modern nation.

While at this important historical intersection, let us briefly review our story so far. The initial front in Japan's war against wolves was opened at the horse pastures of southern and northeastern Japan. Domain lords offered wolfers, such as Wolf Hunter Shōjurō, lucrative bounties to kill wolves that harassed, killed, and ate valuable livestock, but these hunters, though wolf killers to be sure, were not wolf exterminators. Nor did they demonize wolves. By the seventeenth and eighteenth centuries, the Japanese wolf had found its niche in the landscapes and cultures of early modern Japan, a fluid topography more diverse (as the story of the wolf suggests) than the more homogenized one of modern Japan. Some Japanese worshiped wolves while others, out of economic necessity, killed them. Importantly though, at this point in our story, we have arrived at the intersection of the ecological and cultural forces that created and killed the wolves of Japan, and yet no monolithic attitude toward wolves—whether regarding their classification, the religious significance attached to them, or their nature as a pesky predator—had crystallized in Japan. Ultimately, what pushed Japanese toward creating a more monolithic, and decidedly pejorative, image of wolves had little to do with Neo-Confucian science, wolf worship, or horse farming but rather was a force that came in the form of a deadly virus that, for a time before it killed them, made wolves go completely mad and turned them into man-killers.

REPORTS OF MAD WOLVES

Beginning in the early eighteenth century, the behavior of some wolves became as violent as that of the men who pursued them, and they acted mad, as if possessed by some unspeakable demon, becoming man-killers and no longer the benign guardians of agricultural crops or pesky horse killers. Consequently, the people most affected by these changes in wolf behavior—upland villagers, travelers on isolated mountain passes, and the horse ranchers of the far northeast—began transforming perceptions of wolves in Japan through storytelling, mythmaking, culture creating, and, ultimately, by killing them. Importantly, the increase in conflicts that started in the late eighteenth century allows us to trace not only the degree of change in Japanese society during Japan's critical nineteenth century but the degree of behavioral change in wolf society as well.

In the 1830s, Tadano Makuzu, the daughter of a Western Learning scholar, wrote fantastic stories while living with her husband in Sendai in northeastern Japan. Itself a kind of geographic intersection in wolf history, Sendai is near the Oinokawara area where farmers once revered wolves, but it is also near where later, as Kishida Ginkō reported, bounty hunters killed them in preparation for a visit by the Meiji emperor. Tadano retold a great deal of animal lore from the northeast, including the story of a wolf attack in late-eighteenth-century Rikuzen Province (Miyagi Prefecture). In the *Ōshū banashi* (Tales from Ōshū; 1832), she described the harrowing tale of a farmer named Yoshirō from Yokura Village. She prefaced her story by explaining that, throughout the eleventh month of 1789, reports claimed that "sick wolves" (*byōrō*) had been roaming the area, and that they had even attacked several villagers.

Tadano's story goes something like this. While walking home at night after talking late with friends, Yoshirō was suddenly attacked from behind. As Yoshirō turned, his assailant lunged at him and bit him, ripping his flesh and drawing blood. He tried to run away, but again his attacker bit him near the ribs. It was only at this point that Yoshirō realized that his attacker was a wolf. He screamed for help as loud as he could: "Hey, it's Yoshirō, I'm being attacked by a wolf. Somebody, please come out and help!" But nobody came, and the wolf continued its violent onslaught. With deadly swiftness, the wolf flew at Yoshirō from front and behind, from left to right, slashing with its sharp, powerful teeth. Without even a stick with which to defend himself, he tried to push and kick the wolf away; but the crazed canine only bit him on the hand and then sank its teeth deep into his leg. In a matter of seconds, wolf bites covered Yoshirō's entire body. Realizing that he was in a fight for his life, Yoshirō finally got the upper hand by breaking one of its legs, and then, in an extraordinary scene, he tore apart its tired, gnashing jaws with his bare hands. Yoshirō then exposed his own teeth and bit the wolf on the throat, killing it.

In the aftermath of this violent attack, Yoshirō was thoroughly drenched in human and wolf blood. Mustering his strength, he eventually made his way to a nearby home, where the inhabitants looked after him until a physician arrived. The physician, a wolf-bite specialist it turns out, examined Yoshirō and said that the wolf had bitten him in forty-eight places. The physician (although a specialist) had never seen a person so badly mauled by a wolf. Tadano explained that in order to treat the "canine poisons" (*kendoku*) the physician applied *kuri nuki*, a chestnut extract, to the wounds and then cauterized them with moxa treatment (the burning of a cone of the therapeutic herb mugwort on acupuncture points on a body). The physician also restricted Yoshirō's diet, forbidding him to eat oily foods like trout, pheasant, and even his favorite, red-bean rice cakes. Although Yoshirō had some minor dietary indiscretions one New Year's Day, he lived a long and healthy life. In 1812, we learn, he turned fifty-two years old.[38]

Tadano's colorful story, though no doubt embellished, is nonetheless a frightening one. In global experiences with wolves, reliable accounts of healthy wolves attacking people, although not unheard of, are exceedingly rare, with some documented cases in Europe and Russia, not to mention other cases in the largely unrecorded oral traditions of Eskimos and Native Americans. In prehistoric times, human and wolf communities, as groups of social hunters, often lived alongside one another in proximity to prey, and so there can be little doubt that conflicts broke out that resulted in human deaths. As Barry Lopez speculates, "I see no reason why under the right cicumstances—a desperately hungry wolf and an unarmed man, for instance—the wolf wouldn't kill." The explorer Vilhjalmer Stefansson and Canadian government biologist C. H. D. Clarke spent years trying to track down accounts of wolf attacks in North America, but neither man substantiated any of the tales they encountered. Clarke concluded that most accounts of human depredation by wolves could be attributed to rabies or wolf-dog hybridization, both the result of increasing human proximity to wolves. When speculating why European wolves frequently bit people, Clarke wrote that these wolves, much like the sick wolves described in the *Ōshū banashi*, were "frequently rabid."[39]

On surveying French documents related to wolf attacks, Clarke also learned that most wolf bites actually killed their victims, further evidence that the canines probably carried rabies. Of the twenty-eight people bitten by a wolf in Puy-de-Dôme, for example, twelve died; three people in Indre and seventy people in Auvergne died from rabid wolf bites. Even more horrific, 161 people died of rabid wolf bites in Russia in 1875. Clarke suggested that the "demoniacal possession of a single animal," often celebrated in France through the telling of stories about the Beast of Carmarthen, the Beast of Orléans, the Beast of Ardennes, or the beast of something or other, may have actually been stories of widespread rabies epizootics in southern Europe's canine population. The fact that French bounties offered more money for "wolves rabid or having attacked man" demonstrates that French authorities understood rabid wolves to be more dangerous and a menace. Perhaps the most celebrated French example of wolves attacking people occurred only decades before the attack on Yoshirō in Rikuzen. In the wild Cévennes Mountains, in south-central France, the Beast of Gévaudan reportedly attacked, killed, and often ate at least sixty-four people, perhaps as many as one hundred people, mostly children, between July 1764 and June 1767.

With the Beast of Gévaudan, a steady stream of victims, both children and adults, followed the first one, the fourteen-year-old Jeanne Boulet, and locals quickly notified French authorities. But even as large hunting parties combed the complex "network of ravines and gullies" of the Cévennes, the Beast continued to devour children. Of the many stories surrounding the Beast of Gévaudan, the tale of Portefaix rivals even the story of Yoshirō in the *Ōshū banashi*. With the help of six other children, Portefaix defended the smallest child in his group from the Beast as it grabbed the young boy by the face and tried to pull him away. Finally, Antoine de Bauterne shot the first Beast of Gévaudan, a large male canine, in 1766; hunters shot its mate nine months later. Clarke, in his analysis of the autopsy performed on the first Beast of Gévaudan, concluded that the canine was not rabid but probably a wolf-dog hybrid. Indeed, both were exceedingly large, were strangely colored, and had completely lost their fear of humans.[40] The scenario of wolf-dog hybridization on the Japanese Archipelago is one explanation proposed by ecologists for both the evolution and the disappearance of Japan's wolves.

In contrast to Europe, Clarke concluded that documented examples of rabid wolf attacks in North America are quite rare. On occasion, however, such attacks did occur. In July 1833, several decades after the Gévaudan and Rikuzen incidents, fur traders and their Native American guides in the Green River area of Wyoming, near Yellowstone, encountered a "mad wolf." Charles Larpenteur, Warren Ferris, and Washington Irving all told the story of the "mad wolf" that entered the trapping camp and bit several people, some in the face much as the Beast of Gévaudan did. The fur trappers had "heard of the like before," when wolves attacked and bit people; they had heard of how sometimes their victims had "gone mad" too. Larpenteur wrote that one trapper, George Holmes, was "badly bitten on the right ear and face," while other versions of the story tell of a wolf biting Native American guides. The next night, after howling near the camp, the mad wolf bit a large bull. After the trapping party broke camp and before it reached the Yellowstone River, the bull started showing signs of rabies, "bellowing at a great rate, and pawing the ground." Holmes, who also had become infected by rabies, became delirious and reached the point where he refused to cross small streams, holding back the entire party of traders. Eventually, they covered him with blankets and left him behind. Later, however, they "found only his clothes, which he had torn off his back. He had run away quite naked, and never was found."[41]

John Richardson told the story of a wolf attack that occurred at about the same time. "I was told of a poor Indian woman who was strangled by a Wolf," he recounted, "while her husband, who saw the attack, was hastening to her assistance; but this was the only instance of their [wolves] destroying human life that came to my knowledge."[42] Later, in the 1870s, Richard Irving Dodge, an officer in the U.S. Army, told of a rabid wolf that entered the hospital at Fort Larned, along the Arkansas River, and bit several people, who later died. The rabid wolf "charged round furiously," biting people on the hands and legs and over the rest of their bodies. The wolf "moved with great rapidity," wrote Dodge, "snapping at everything within his reach, tearing tents, window curtains, bed clothing." In an attempt to stop the virus, soldiers who received bites from the rabid wolf at Fort Larned were "thoroughly cauterized with nitrate of silver," something like the moxa treatment administered by Yoshirō's physician. Later, on interviewing nearby Native Americans, army officials learned that rabid wolves frequently entered their camps in February and March. The report cryptically added that "in no instance have any of them ever known a person to recover after having received the smallest scratch from the rabid animal's teeth."[43] Rabies could be absolutely deadly to both wolves and humans.

RABIES: ITS ETIOLOGIES AND ECOLOGIES

To understand what struck these "mad wolves" and turned them into man-killers we must briefly explore the life cycle of the rabies virus and the pathology of canine disease; the rabies virus, too, created and killed the Japanese wolf. Among canines, foxes are actually most susceptible to rabies; but scientists still consider dogs and wolves to demonstrate "intermediate susceptibility" to the virus. Cruel experiments conducted on laboratory animals reveal that pups between the ages of two and four months contracted rabies 100 percent of the time but that the rate gradually decreased as the pups grew older, meaning that younger canines, at least in the hands of scientists, prove more vulnerable to the rabies virus. Natural transmission of rabies occurs almost exclusively through bite wounds, when virus-infected saliva enters the body, something like the "canine poisons" described in the *Ōshū banashi*. Once inoculated, the rabies virus acclimates to its new environment by attaching itself to the plasma membrane. It then enters the cell, replicates, and spreads into the intercellular environment of the body. When cell infection occurs, the virus then makes its way to the central nervous system within only a few hours. Once in the central nervous system, the rabies virus incubates in the victim for an average of three to eight weeks, although great variation exists, with some viruses incubating for as long as six months. Lengthy incubation occurs as forms of viral harborage in the cell, both because of the long retention of the virus in myocytes at the bite site and because so few cells in the central nervous system become inoculated in the first place. For this reason, forms of cauterization or moxa treatment, such as those performed at Fort Larned and in Rikuzen, were only partially successful at best.

Symptomatologically, rabies runs a course of three life stages: prodromal, excitative, and paralytic. The prodromal stage may last only two or three days and is difficult to detect in canines. A slight change of temperament usually accompanies a rise in temperature, the dilation of the pupils, and a slow corneal reflex. The excitative stage can last as long as seven days. This stage occurs as the virus begins to spread within the central nervous system. This form of viral amplification begins in the Nissl granules and then spreads to the motor neurons in the brain stem or the spinal cord, to the primary sensory neurons in the cerebrospinal ganglia, and to the neurons in the autonomic ganglia. The rabies virus can spread rapidly in the central nervous system, although it may take four to five days for the canine to actually exhibit symptoms. At first, the canine may only hide in dark places. Later, however, it becomes irritable and nervous, with a heightened reaction to stimulation. Eventually, by the later part of the excitative stage, the canine may start to "wander aimlessly," becoming increasingly irritable and violent, with a "tendency to bite anything that it encounters, be it human, animal, or an inanimate object." Such excitative behavior accompanied the Rikuzen, Green River, and Fort Larned incidents. The slow paralysis of the laryngeal muscles and repeated spasms cause the canine to drool infectious saliva, frothing at the mouth as it gasps for air. At this point the canine may also experience convulsive seizures and a loss of muscular coordination, and a "far-off" look—again, as if the animal is possessed by a demon—takes over its eyes. Such behavior expresses impaired neuronal function in the central nervous system. Finally, in the paralytic stage, the canine loses muscular coordination over its entire body and eventually falls into a coma and dies.[44]

RABIES IN EIGHTEENTH-CENTURY JAPAN

In the 1730s, this strange new disease began spreading in the canine population of Japan, leading to unpredictable behavior among dogs, badgers, wolves, and foxes. Noro Genjō, a physician and Western Learning scholar of some acclaim, wrote in 1736 that in the old days people had never heard of "mad dogs" biting people in eastern Japan. He speculated, however, that the disease had spread from the western districts to afflict the canine populations in central and, eventually, eastern Japan. By locating the origins of the disease in the west, he implied that it probably reached Japan from Korea and China. (Similarly, continental-origin theories continue to be offered for the arrival of smallpox and syphilis in Japan as well.[45]) Importantly, the epidemiology of the disease described by Noro is that of *kyōkenbyō*, or rabies—the "mad dog disease."[46]

After its introduction, the rabies virus spread and impaired the neuronal functions of wolves, rendering them prone to intense delirium and violence. Subsequently, incidents of wolf attacks increased throughout the eighteenth century. Kanzawa Teikan, in *Okinagusa* (Pasqueflower; 1851), tied wolf attacks to the spread of "canine poisons." Although published posthumously, Kanzawa's work contains many reflections from his early life until about 1795. In 1732, Kanzawa traced the spread of rabies among dogs in the regions of Kinai (Kyoto and vicinity), Nangai (Shikoku Island), Yamage (Miyazaki Prefecture), Sanyō (Yamaguchi Prefecture), and Nishikaidō (Kumamoto Prefecture), regions for the most part in western Japan, where dogs stricken by madness bit and killed people. He explained that once people came into contact with the "poison" (*dokuke*, literally meaning "noxious air" or "virulence") carried by these "mad dogs," it sometimes meant instant death or death after a short period of time. This dog disease, called *inuwazurai*, spread eastward to the Nangai and Kinai in the spring of 1736, and by the summer of 1737 to the Tōkaidō, or Eastern Sea Circuit, a major travel route during the Tokugawa years. Dogs were not the only victims, and wolves, foxes, and badgers also died in large numbers. People, oxen, and horses that received bites, he continued, later developed strong fevers and sometimes experienced severe pain for thirty to fifty days, or even a year, before they died. They might also experience a loss of appetite before they went mad; but eventually they died just like the dogs, he added.[47]

Kanzawa's description of the spread of *inuwazurai*, the "mad dog disease," matches the symptomatology and epidemiology of the rabies virus. In the wild, rabies spreads through virus-infected saliva introduced by bites. The incubation period for rabies is three to eight weeks, about the thirty to fifty days described by Kanzawa. Nishimura Hakū, whom you might recall from chapter 2, also noted the spread of the "mad dog disease." He explained that as early as the 1730s, Japanese reported dogs stricken by a disease that rendered them "mad." When they bit people, the "poison" (also described as *dokuke*) spread to the human victim. Much like Kanzawa, Nishimura wrote that these infected people ran around and acted "mad," just as the dogs did, until they died. He reported that between the regions of Suruga and Tōtōmi (Shizuoka Prefecture) the disease had spread to wolves and that occasionally these wolves bit people. "The sick wolves fly at you just like birds," wrote Nishimura. "When they see people they just come at them and bite." He noted that infected wolves had appeared in about ten villages.[48]

Ansō Mutaka (probably a pen name), a samurai from Shōnai domain in the northeast, explained that in midsummer of 1742, dogs had started biting people in that domain. In some instances, the wounds took thirty to forty days to heal. Sometimes, he added, the victims relapsed and died, and it was unknown how many people had died in Sakata Village (Yamagata Prefecture) from the "mad dog disease." By 1743, in Yamamoto Kawachi (Akita Prefecture), wolves had started to bite people as well, with two or three people having been killed by wolf bites in a single morning. Clearly, such overwhelming evidence suggests that the rabies virus had infected Japan's canine population by the 1740s, even far from ports in western Japan.[49] Eight years later the situation became acute in Shōnai. One collection of historical documents contains a story from 1750 about two "mad wolves" that went on a rampage for three days in the area of Atsumi (Yamagata Prefecture). In Yuatsumi Village alone, these mad wolves attacked nineteen people, six horses, and one ox; in Hitokasumi Village, they attacked five people. Because of the undivided attention of a local physician, many of the victims temporarily recovered. However, thirty days later the disease reoccurred in eight people and they all died.[50]

Yet another account, this one in Tachibana Nankei's *Hokusō sadan* (Brief conversations through a northern window; 1825), reports that in the 1770s, between Higo (Kumamoto Prefecture), Bungo (Ōita Prefecture), and Hyuga (Miyazaki Prefecture), at Nasu Village, a number of "sick wolves" (*byōrō*) appeared and injured people. Tachibana, a Kyoto physician, claimed that hundreds of people had sustained injuries. The local lord dispatched a physician to tend to the injuries, and the physician remarked that not even the children felt any pain.[51] As for the symptoms exhibited by "mad canines," Noro Genjō, the Western Learning scholar mentioned above, reported that signs ranged from the tip of the nose becoming dry, red eyes, and a loss of appetite to madness, excessive slobbering, and the tail hanging lifelessly between the legs. Many of Noro's veterinary observations correspond exactly with the most visible symptoms of rabies. It was such rabid canines, their brains infected by disease, that lost their fear of humans and began attacking people and farm animals throughout Japan.[52]

By the 1750s, the high number of canines afflicted with rabies translated into a frightening number of wolf attacks. The chronicle of the Kumatani family, a clan from the mountainous region near Kamihara Village (Nagano Prefecture), tells of a farmer by the name of Shōzaemon whom a wolf bit in the second month of 1762. However, because Shōzaemon was such a strong man, he grabbed the animal by the neck and squeezed it so hard that its eyes bulged out. The Kumatani chronicle says that nearly two months later, in the fourth month, Shōzaemon, despite having killed the wolf, became ill and died. Once more, consistent with our understanding of rabies epidemiology, the two-month period between the initial bite, Shōzaemon's inoculation with the virus, and his death probably represents the incubation period for the virus. Finally, the Kumatani chronicle mentions that in the third month of that same year, wolves continued to appear, and a local man named Hiraishi Sakusuke was bitten; he also ended up beating the wolf to death, this time with a wooden sword. At this time others had also received bites from the rabid wolves.[53]

Tachibana Nankei, the Kyoto physician, evoked Japan's growing anxiety about wolf attacks with his almost humorous story "Ōshū no oni" (The demon of Ōshū), from the *Tōyūki* (Lyrical record of a journey to the east; 1795). Tachibana offered a snapshot of real late-eighteenth-century anxieties regarding wolves, canines that many in the northeast increasingly viewed as more a violent menace lurking in the mountains than a benign guardian of their grain fields. As a creature that killed people ever so slowly and horrifically by injecting its hideous "canine poisons," the wolf no longer fulfilled its role as an *ekijū*, or "beneficial animal," in either the agrarian or the hunting mentalities of early modern Japan. In this respect, the Japanese wolf, afflicted with the rabies virus, crossed and recrossed the line between the otherworldly mountains and the world of settled villages (as well as the travel routes that linked them), doing so with the whimsical insanity of something gone mad.

Tachibana remembered that it was already past the hour of the monkey, about 4:00 P.M., when he arrived in the pouring rain at the Osa River in Ugo Province (Akita Prefecture). He asked an old-timer at the roadside village if it was too late to travel on the next leg of the road that wound between Tachibana and his destination. Frowning, the old-timer replied, "If you hurry and don't do anything else you can probably make it, but a demon [*oni*] appears in this area and kills people and eats them. At first, the demon only appeared at night, but now it also appears during the day. People or horses, it doesn't discriminate, it just eats anything. You're lucky. Though the demon often appears on the road you've just passed, you made it without being eaten." The owner of the next home then chimed in, "The road you just traveled without incident has many demons. Yesterday, a man named Hachitarō, from this village, was eaten, and today, from the neighboring village, Kyūrōsuke was killed. It is terribly frightening." The owners of every house Tachibana inquired at spoke of the demon. Tachibana just laughed at first. But then it occurred to him that, separated as he was from his hometown by nearly 100 miles, he should seek lodging for the night. When he did finally find lodging, Tachibana learned the real identity of the demon of Ōshū. It was a mad wolf.

Tachibana eventually discovered that at several places, including near the Osa River, a wolf (or several wolves) had killed as many as six or seven people. In one attack, the wolf even leapt over a gate and bit a man. As Tachibana explained, however, because the man was particularly strong, he managed to hold the wolf down with one hand and smash in its head with a rock held by the other. Because the man had been wounded in several places during the brawl, however, on returning home he died. "This was because of the sickness of the wolf," Tachibana explained. "Even in broad daylight, some ten wolves might appear and attack and injure people." The next day Tachibana inspected the jaw of the wolf that had been killed with the stone, but its body was nowhere to be found.

Interestingly, Tachibana thought that the villagers referred to a demon, or *oni*, because he had misunderstood the local northeastern dialect term for "wolf," *oinu*, or, literally, the "honorable dog."[54] Furukawa Koshōken, as seen in chapter 2, also had explained that people in the northeast rarely used the name *ōkami* for "wolf," instead preferring *oinu*, which sounds something like *oni*.[55]

In 1808, further south in the Kantō, locals dissected a sick wolf at Momichi Village in Ibaraki Prefecture. Komiyama Masahide, a Confucian scholar from Mito domain, wrote that villagers had killed the wolf after it had attacked an elderly woman. Eventually, they cut the belly of the wolf open and discovered worms that resembled *hariganemushi* (literally, a "hairworm" or "wireworm"). The villagers, erroneously it appears, believed that the worms caused the wolf to be sick.[56] These worms were likely intestinal parasites. While villagers from Momichi thought that the root of the sickness resided in the belly, one story in the lore related to Tōdō Takasawa explains that the "canine poisons" transmitted by wolves passed through their *dokuga*, or "poison fangs." Nagashichi, an Iga (Mie Prefecture) farmer who in 1822 rescued his mother by killing a "mad wolf" that had come bolting out of the woods, later died of wounds he sustained from the animal's "poison fangs," despite the efforts of a physician.[57]

Finally, an old family chronicle, discovered by Komatsu Shin'ichi in 1932 in Ojiya Village, in Niigata Prefecture, contains the story "Ōkami ni kamikorosaretaru hanashi" (The account of being bitten to death by a wolf).

> One night, during a heavy snowfall, Jiemon, the village head, because his dog was barking, let it into the inner garden to sleep. Then a wolf broke through

the inner gate, invaded the inner garden, and attacked the dog and killed it. Not realizing it was a wolf, Jiemon and his wife tried to drive it away, and the wolf killed them as well. The people of about seven or eight neighborhood homes heard the commotion, and in full force, with clubs in hand, they gave chase to the wolf. They caught up to the wolf in a small stream in a local ravine and killed it. However, at this point, Tarōemon was bitten, and because of the poison, in the end he died.[58]

These numerous stories of rabid wolves and dogs, as dramatic as they are, may belabor the point, but one theory posited by ecologists and historians for the extinction of the Japanese wolf emphasizes the role of rabies. As the rabies virus impaired the neuronal functions of wolves, skewing their judgment, the disease also transformed their behavior. The rabies virus too struggled to live out its life cycle in the canine environment of wolf bodies; and as it spread through the central nervous systems of Japan's wolves, it induced the canines to hide in dark places and lash out at human faces and bodies. Such behavior necessitated, in the minds of many Japanese, wolf hunts. With the demon rabies burning red hot in the brain of the Large-Mouthed Pure God, the wolf became a monster of sorts, and ceremonial hunts, events that might be seen as forms of otherworldly exorcisms, focused on cleansing the mountains of the demon wolf. Blowing loudly into conch shells and marching through the forests of northern Honshu armed with spears and firearms, hunters coordinated such wolf hunts with the rhythms of traditional festivals such as the celebration of New Year's Day. In doing so they began the process of erasing the wolf from its once sacred mountain home.

RABIES AND HOKKAIDO'S CANINES

Before we turn to northeastern wolf hunts, one more piece in the puzzle of the history of rabies among Japan's wolves needs to be put in place: rabies and the Hokkaido wolf. In the fifth month of 1861, the Hakodate magistracy, the local Japanese authority on the northern island, reported an outbreak of rabies in that port town. The virus had jumped across the Tsugaru Strait, from Japan's far northeast, onto Hokkaido. Later, rabies outbreaks, as documented in Hokkaido police histories, erupted in the spring of 1873, and rabid dogs reportedly bit both people and livestock in Sapporo (the capital of Hokkaido). That same year, the Kaitakushi (Hokkaido Development Agency) ordered the killing of dogs afflicted with rabies. In time, such orders became part of standard police manuals. The 1872–73 "Rules for Patrolmen" and the 1878 "Kaitakushi Police Administration Rules" both explicitly state that rabid dogs should be killed and reported to local officials.[59] Much like the nineteenth-century rabies epizootics in London described by historian Harriet Ritvo, Kaitakushi police received orders to protect the public by killing dogs and disciplining their owners in the urban settings of Meiji Japan.[60]

This police commitment to controlling and killing mad dogs strengthened in 1877 when the Hakodate branch office issued strict new rules for domesticated dogs. It appears that even without the rabies virus, dogs ran out of control in agricultural lands and caused damage. One year later, in 1878, both the Sapporo and Nemuro branch offices adopted the Hakodate plan. Probably the most important element of the new rules was the requirement that all dogs have licenses, and any dog found without a license (similar to the situation in London) would be killed.[61] At the Nemuro police station, to enforce these rules, officials offered bounties for dogs found without licenses, leading to an open season on dogs and a spate of dog killing. Between 1876 and 1881, bounty hunters killed 219 dogs in Sapporo and 165 in Nemuro. Licensing dogs was one strategy for keeping rabies under control, and the Kaitakushi executed the rules with ruthless efficiency.[62]

No archival evidence that I found confirmed definitively that rabies had spread from dogs to wolves on Hokkaido; but considering the sheer number of documents that speak to the problem of "epidemics among dogs" (*inu no ryūkōbyō*), "wild dogs" (*yaken*), "bad dogs" (*akuken*), and "mad dogs" (*kyōken*) running wild throughout Hokkaido, it is highly probable that rabies did spread to some of Hokkaido's wolves.[63] One assumes that it is also highly probable that rabies contributed to their extinction in the late 1880s.

THE RHYTHMS OF WOLF KILLING

In *Suzumigusa* (Cool grass; 1771), the bohemian-like author Takebe Ayatari related how he participated in a wolf hunt in Michinoku (Akita Prefecture) that probably took place in the 1740s. Takebe's description reminds one of Ohio's "Great Hinckley Hunt" of 1818, when six hundred hunters, coordinated by loud bursts of horns and conch shells, created an encircling human line around wolves and other wildlife in the "Western Reserve" of the American Midwest. In what can only be described as a slaughter, Ohio hunters killed every living thing that tried to escape the human fence.[64] One notable difference between the hunt in northeastern Ohio and that in

Michinoku was that Takebe and his party were not out to kill everything that moved; rather, they tried to capture wolves alive. In an astonishing story, Takebe writes:

> When I went down to Michinoku and stayed over the New Year, I saw an extremely rare event. While it was already the second month and spring rains were indeed falling, the snow had accumulated to a depth of nearly ten feet, and the surface had frozen, so, even if you treaded on it, it would not collapse. Marshes and pools had not yet broken up the snow, so a vast expanse of field seemed to extend over the land. A decaying carcass of a horse or such beast was left there, so that wolves would approach it. Choosing an opportune time, many hunters [*sachibito*] spread out over a distance of about twelve miles and encircled the area overnight. They stood in the snow through the night, and when dawn broke, they began to tighten the circle. The wolves suddenly started up and ran west and east in search of an opening. However, since they had made a human fence, the wolves eventually tired and were easily captured alive. That night I became one of the hunters and stayed up the night—the cold was beyond compare.[65]

What Takebe and these hunters did with live wolves remains unknown. Indeed, Japanese hunters designed most wolf hunts to kill wolves, not capture them. Probably the best records relating to these northeastern wolf hunts come from a chronicle of the Satake family of Akita domain.[66] The "Satake-ke gonikki" offers a vivid description of a wolf hunt held in 1810, one that I want to explore at some length in the remainder of this chapter. But to do so, I first need to take you to the part of Japan once known as Michinoku (a region that encompassed the ancient provinces of Iwaki, Iwashiro, Rikuzen, Rikuchū, and Mutsu), where Japanese hunted wolves in events that resembled otherworldly exorcisms.

Some two and a half centuries after Takebe made his trip, I traveled by bullet train to the city of Kakunodate, in the interior of Akita Prefecture, to investigate a wolf hunt described in the pages of the "Satake-ke gonikki." Tracking down original copies of the Satake chronicle (a source that had hitherto proved elusive) was my primary purpose in traveling to the northeast, but I had others as well. As in my jaunts to Yellowstone and elsewhere, I hoped to make a personal connection to the region, its people and wildlife, and its landscape by mapping out and then retracing the wolf hunt I came to call the Kakunodate hunt. Of course, settlement, hunters, and rabies had long since destroyed the wolves of Kakunodate, but still I needed to see the wild terrain of ancient Michinoku. I hoped to "read" the landscape as one more historical source, just as archaeologists study artifacts in museums or anthropologists observe villages in the hope of better understanding the complexities of past cultures. After a brief conversation with the front-desk clerk at my hotel near the Kakunodate train station, I made contact with members of the Kakunodate Historical Information Association (Kakunodate Rekishi Annainin Kumiai), a study group made up of volunteers who teach history lessons to tourists who visit the city. The study group held their lessons in the Kakunodate historical village. Specifically, the lesson I attended took place at the Iwabashi home, a samurai-style residence built during the late Tokugawa period.

When I entered the main room of the old, smoke-stained Iwabashi home, I discovered that several members of the Kakunodate study group had already gathered around an open hearth to roast chestnuts over red-hot charcoal. Above them, dried daikon radishes hung over the hearth, a curing technique that gives them a smoky flavor, one reputedly particular to the pickled radishes of Kakunodate (called *iburigakko*). In the Tokugawa period, these pickled radishes proved a critical source of sustenance during the infamously harsh winters of Akita. Finding a place next to the hearth, I sat down with the study group and, together with several Japanese tourists from Kyoto, listened to stories of old Kakunodate as we gingerly peeled and ate warm chestnuts. Transported back in time, I listened to the tale of how the village of Kakunodate became a feudal castle town under the northern branch of the Satake family; how officials ordered the creation of a controversial firebreak (*hiyoke*) to protect the town from fire, and where traces of it could still be found in the modern city; the history of the now famous cherry trees of Kakunodate that line the eastern bank of the Hinokinai River; and how deep snow forced the inhabitants of Kakunodate to wear straw wrappings on their feet (called *fumi dawara*) as makeshift snowshoes. For a historian always fascinated by Japan's hinterland, the session transformed the Iwabashi home into a time capsule for a memorable journey into Kakunodate's rich past.

While at the Iwabashi home, I met the leader of the Kakunodate study group, Tozawa Shirō. Familiar with the Satake chronicle, Tozawa had transcribed several sections of this important source on Kakunodate history, and he explained that he was familiar with the sections involving Kakunodate wolf hunts. Later, he brought me copies of the Satake chronicle, and together, in an old samurai home rich with the smoky aroma of roasted chestnuts, we began to piece together what had happened on the nineteenth day of the first month of 1810. As it turns out, between 1674 and 1838, a period of 164 years, the Satake chronicle lists eleven such wolf hunts, often in the form of *makigari*, grand hunts, or executed as military training exercises sponsored by the domain lord. It is also worth mentioning that the great majority of the hunters returned from the mountains empty-handed.[67] Some of these hunts stand out, almost like eerie scenes from Mary Shelley's *Frankenstein*, because of their immense size, the frenzied mass participation of all ranks of Tokugawa society (not just elite samurai), and their strong ceremonial nuances.

The Satake chronicle explains that participants of the 1810 Kakunodate hunt first met that year to deliberate (or to conduct a *ōkamigari hyōgi*) for two or three days on how to kill an extremely crafty animal. An official announcement was then made on a village placard, where officials wrote the date and time of the impending hunt. On that day, at the hour of the tiger (about 4:00 A.M.), the first party of six samurai, probably seasoned hunters, all armed with swords, lances, firearms, and travel boxes and accompanied by firearm standard-bearers (*matoi mochi*), headed out into the nearby mountain forests of Akamochi with forty villagers and about sixty others from neighboring rural districts (probably just west of Kakunodate, see map 2).

At the hour of the hare (about 6:00 A.M.), another one hundred townspeople, all armed with lances and carrying travel boxes, formed two groups. One party made its way from what is now the Tama River, while the other party started pushing its way through nearby fields. All parties, including recruits from villages along the way, such as peasants from the mountain pass at Kaifuki, planned to converge on Takanosu Mountain. (After studying maps together, Tozawa and I came to believe that Takanosu Mountain is downstream of Kakunodate in Nakasenmachi. On today's maps, it is a 743-foot unnamed hill near Matsugura Mountain, about six or seven miles southwest from Kakunodate.) The Satake chronicle records that by the time the party reached its destination, some twenty-one groups boasted a total membership of an astonishing 930 people. In the end, however, the Satake chronicle reports that "no wolves were met; the hunt was without results." Not to be denied, however, the massive hunting party then made its way to Furushiro Mountain, which overlooks Kakunodate and is the site of an Inari shrine (see figs. 19 and 20). As before, hunters killed no wolves and the party was forced to be content with two foxes shot near the Saimyōji Temple (located in Nishikimura, just north of Furushiro and Kakunodate). The group disbanded at about 4:00 P.M.

Unfortunately, the Satake chronicle does not mention what precipitated

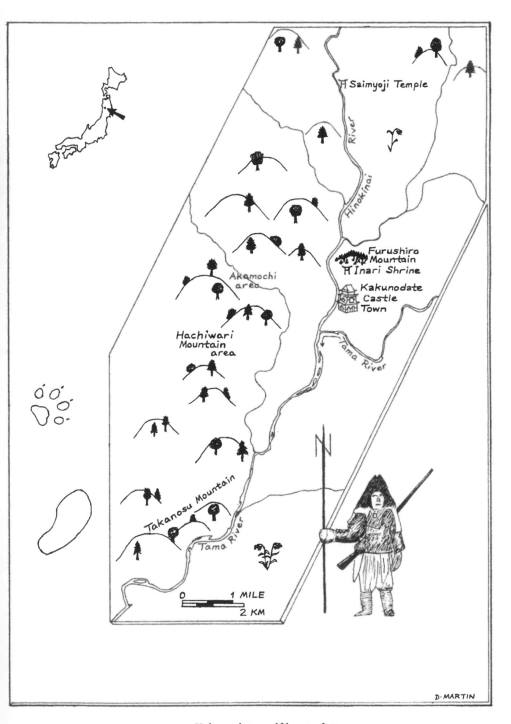

Saimyoji Temple

River

Hinokinai

Furushiro
Mountain
Inari Shrine

Kakunodate
Castle
Town

Akamochi
area

Hachiwari
Mountain
area

Tama River

Takanosu Mountain

Tama River

N

0   1 MILE
           2 KM

D·MARTIN

MAP 2. Kakunodate wolf hunt of 1810.

FIG 19. Documents mention Furushiro Mountain as one site where hunters sought wolves in the 1810 Kakunodate hunt.

the 1810 Kakunodate hunt. But other sources reveal that horse depredation led to wolf hunts in 1678 and 1690, and attacks against humans in mountainous areas led to hunts in 1836. Later, though, in 1853, Santō Kyōzan (older brother of Santō Kyōden, the famous author of stories involving the Yoshiwara pleasure quarter) wrote about an announcement for a wolf hunt posted on a placard at Arigano Village and on the Kisokaidō Circuit in central Japan. On the placard, the village head declared, "Attention! Attention! Has it eaten again? Hunters assemble for a wolf hunt!" suggesting that wolf attacks on travel circuits also prompted hunts. Moreover, instigating such wolf attacks was the spread of rabies.[68] When the village head posted such inflammatory announcements on the Kisokaidō Circuit, hundreds of people could be assembled, much as in the case of the 1810 Kakunodate hunt. Of course, local townspeople or farmers had to obtain official sanction for such hunts. One suspects that the thought of nearly a thousand villagers and peasants armed with rakes and spears and marauding through domain forests sent shivers down the spines of even the most hardened lords. Indeed, records tell of some thirty to forty youths who set out for a wolf hunt at Hachiwari Mountain (about two miles west of present-day Kakunodate) and who were punished for hunting wolves without permission.[69]

In general, as Hiraiwa Yonekichi documented, volunteers in these hunts

FIG 20. Today, on Furushiro Mountain, no sign of wolves or wolf hunts exists, but an Inari Shrine graces the side of the mountain.

gathered at designated meeting places, carrying large drums, torches, and banners, as well as *bentō* lunches. With the loud burst of a conch shell—also a standard feature of military assaults, some Shugendō rituals, and Ohio's Great Hinckley Hunt of 1818—they set out for the surrounding hills. The results of these hunts varied, but the Satake chronicle lists some dates and kills. Of the thirteen records from the early nineteenth century, most hunts yielded only modest success, with six wolves killed. However, wolves were not the only animals taken: Japanese serows, rabbits, and some game birds fell into the sights of these northeastern hunters. One other element stands out with these hunts. With the exception of one, which was held in the second month of the lunar calendar, all hunts, like Takebe's New Year's

Michinoku hunt and even the Christmas Eve Great Hinckley Hunt, took place in the context of the rituals of the second and third weeks of the first month.

The apparently ceremonial nature and regularity of these wolf hunts (and the fact that participants killed animals other than wolves) suggest that domain lords, samurai, and villagers had integrated wolf killing into the rhythms of daily life and yearly ceremonies in the northeastern region of Michinoku. Traditionally, the coming of the New Year and the passing of the winter solstice represented a time when Japanese cleansed their homes and renewed community relations through ceremony, and perhaps they projected these rituals into the natural world. Of course, it is not hard to imagine mustering hundreds of people, gaining state sanction, and arming volunteers with flaming torches and sharpened lances to march in the dark, early-morning hours through snow-covered forests and remote roads in search of a mad wolf that had reportedly killed a traveler. But that hunters killed all animals, including the cagey, shape-shifting fox, meant something different altogether. An event more meaningful was being played out in Michinoku, part of a slow, multifaceted, and complex transformation in Japanese attitudes toward the natural world in general and the wolf in particular that led to a mind-set that could countenance the use of dynamite to eradicate wolves, as told by Shiga Naoya.

CONCLUSION

Speeding away from Michinoku by bullet train, I watched out my window as we sliced through rail crossings dividing small towns and strangely shaped plots of farmland, through tunnels cut into snowy mountains, and across bridges spanning half-frozen rivers and streams. From my seat, these communities and mountains flew by like film moving through an old clicking movie projector. Traveling with such dizzying speed through these places reminded me of the troubles historians face when they attempt to narrate so much time and space and interpret so much history. As our train approached Tokyo, we had covered hundreds of miles in a matter of hours; but in that time I had managed to glance at most of my field notes, covering centuries in minutes, and a picture began to emerge of the ecological and cultural processes that worked in concert to create the wolves of Japan in the early modern period.

This picture consisted of splashes of historical color; of Neo-Confucian and, later, Linnaean classification at the hands of Japanese and Western scientists; of Confucian definitions of wolves as "beneficial animals" because they protected grain farms from wild boars and deer, an image so different from the one fostered by beef-hungry America; of Shinto and Buddhist symbolism at mountain shrines and temples; of predator instincts that led wolves, quite naturally, to the horse pastures of the far northeast, where they found deadly confrontation with hunters as well as horses; and of "mad" wolves, animals whose mental faculties rabies had blurred, possessing them as a demon and turning once benign creatures into man-killers. Simply, in preindustrial Japan, wolves were all these things and more; and wolf imagery was as diverse as the ecologies and cultures of Japan itself before the Meiji Restoration of 1868. The picture that emerged reminded me that, though historians often search for meaning in the patterns, causal relationships, and consistencies discovered in the human past, diversity and inconsistencies are just as meaningful. Whether in Japan or elsewhere, it is the modern nation—which shapes our story from the next chapter forward—that struggles to define how we view the past, and it celebrates our discovery of patterns, relationships, and consistencies, as in doing so we often make its case for a legitimate national order. Patterns mean order; order facilitates power. But it was the very absence of patterns and consistencies in Japan's preindustrial world, like those odd shapes flashing by my train window, that had allowed wolves to survive for centuries. Indeed, one imagines that for every wolf killed by an angry northeastern lord, another was offered a ceremonial dish of red beans and rice by grateful mountain villagers.

Overlapping this early modern act of creating wolves was the act of killing them. When wolves killed and ate horses in northeastern pastures, they competed with human populations, which led to increased confrontations between hunters and wolves. Killing wolves, at least before the Meiji Restoration, was neither uniform nor designed to exterminate them and instead looked like isolated flashpoints of confrontation. As wolves slowly encroached on pastures or as rabies made them into man-killers, Japanese reacted with hunts, undertaken mostly by specialists and designed to rid farmland of wolf predation; but at times such hunts projected different meaning into the forests and mountains of preindustrial Japan. That most hunts corresponded to New Year's ceremonies suggests that they, along with other activities, solidified community relations, cleansed nearby forests, and reconfigured human relationships to the land. But all this changed in the decades surrounding the Meiji Restoration. The relentless patterns and consistencies of national policy replaced the ceremonial synchronicity of preindustrial hunts, and wolves, in one generation of Meiji Japanese, were erased from the Japanese Archipelago. To understand how the need to kill wolves accelerated in the middle of the nineteenth century with the birth of Japan's modern nation, we must turn our attention to the experience of the Hokkaido wolf in the Meiji era.

4

MEIJI MODERNIZATION, SCIENTIFIC
AGRICULTURE, AND DESTROYING
THE HOKKAIDO WOLF

The Meiji Restoration of 1868 ranks among the most important events in Japanese history. After over two and a half centuries of samurai rule, the Tokugawa shogunate fell to what historians call the Satchō alliance—essentially, a political and military alliance between Satsuma, Chōshū, and a handful of other disgruntled feudal domains—and, in the course of the next several decades, the alliance replaced Japan's decentralized early modern polity with a more centralized modern one. Within months of the 1868 transfer of power, the emperor issued the Charter Oath (Gokajō no seimon), a short document that outlined the priorities of the new Meiji government. Most importantly, the Meiji government pledged to end centuries of carefully constructed isolation from most Western countries.[1] Instead, the Charter Oath proclaimed that "knowledge shall be sought throughout the world so as to invigorate the foundations of imperial rule." For a time, knowledge was indeed actively sought throughout the entire world: Japanese wrote a Prussian-style constitution, built an English-style navy, and established an American-style agricultural college on Hokkaido, among many other noteworthy achievements. Of course, Japan did not simply mimic other countries and their institutions but rather refashioned the knowledge and expertise garnered from foreign advisers and returning Japanese embassies to fit its emerging vision of modernity.[2] One such foreign adviser who assisted with Japan's modernization in the Meiji period, a man who offered expertise in the arena of modern agriculture and scientific ranching, was an Ohio rancher named Edwin Dun (1848–1931).

Dun came to Japan as a foreign adviser in 1873 (see fig. 21) at the suggestion of Albert Capron, a cattle broker. Earlier, officials with the Meiji government had asked Capron to find a qualified rancher in the United States who could oversee the establishment of a modern livestock industry in Japan. Officials planned to focus the new industry on the recently acquired island of Hokkaido, where Capron's father, Horace Capron, a former commissioner of the U.S. Department of Agriculture during the administration of Ulysses S. Grant, served as chief adviser to the Kaitakushi (Hokkaido Development Agency).[3] As part of the colonization of Hokkaido, the Meiji government promoted ranching, largely in the form of state-run experimental farms. In the eyes of Meiji officials and their Western counterparts, ranching was progressive and scientific, and it produced the primary cuisine of modern nations—beef. Most officials on Hokkaido, moreover, believed that ranching represented the agricultural future of northernmost Japan. Meiji officials and public intellectuals, many of whom had visited North America during the 1871 Iwakura mission, also knew that the United States had settled the American West through the expansion of ranching and other forms of agriculture.[4] Indeed, the development of ranching and scientific agriculture suited Kaitakushi needs perfectly on underdeveloped Hokkaido.

Correspondence between the Kaitakushi's branch office in Hakodate, in southern Hokkaido, and superiors in Sapporo, home of the Kaitakushi's central offices, illustrates the place that ranching was to hold in the economic future of northernmost Japan. While debating whether to raise bounties on predators to quell increasing losses at the experimental farms, one Kaitakushi official emphasized the importance of the new commitment to ranching. Harkening back to the agronomic system of the Tokugawa period, wherein the wealth of feudal domains was measured in bushels of rice (the so-called *kokudaka* system, in which one *koku* equaled 5.2 bushels), he remarked that until now grain farming had served as the cornerstone of Japanese agriculture. He submitted, however, that Hokkaido's cold springs and early fall frosts made grain farming risky. What made sense for Hokkaido was raising livestock such as cattle, horses, and sheep. The same official pointed out, nonetheless, that despite the best efforts of Japanese settlers and the Kaitakushi, livestock numbers, particularly horses, had not increased. The reason was that year after year, bears, wolves, and wild dogs killed and ate free-ranging horses on the farms, devoured all the foals in the pastures, and once even attacked a ranch hand. This official suggested that predators even caused hardships among the Ainu, the native people of Hokkaido, who hunted deer for the Kaitakushi after their forced deculturation and assim-

FIG 21. Edwin Dun, "the father of Hokkaido agriculture," drew on his experiences in the American Midwest and used strychnine-laced bait to exterminate wolves in southern Hokkaido. Courtesy of the Resource Collection for Northern Studies, Hokkaido University Library, Sapporo, Japan.

ilation in the early nineteenth century.[5] According to this writer, wolves and wild dogs, not the native Ainu or even the menacing Russians in the North Pacific, most immediately stood in the way of the Japanese settlement of Hokkaido. The Kaitakushi must act, and act decisively, or hungry carnivores would devour Hokkaido's future.[6]

Dun oversaw the task of eliminating wolves and wild dogs from southeastern Hokkaido. Through the services of such "live machines," meaning Dun and the many other late-nineteenth-century foreign advisers hired to assist with Japan's transformation into a modern nation, Meiji officials adopted foreign attitudes about wolves that combined with, or in some instances superseded, existing ones formed in the early modern period.[7] In early modern Japan, Neo-Confucianism had basically served as the official ideology of the Tokugawa shogunate, and it generally viewed grain farming as the most noble and productive of all enterprises (with the possible exception of governance). "The treasure of the people is grain," wrote the scholar Kumazawa Banzan. "Gold, silver, copper and so forth are the servants of grain."[8] By the late eighteenth century, such Neo-Confucian ideologies had combined with more native Japanese ones to form prejudices that associated raising and eating livestock with foreign "barbarians" and elevated farming and eating Japanese rice to a borderline religious experience.[9] Hence, many Japanese came to see grain farming as a noble endeavor; they viewed animal husbandry and meat eating with a certain amount of disdain.[10] Not surprisingly, Japanese farmers principally used oxen and other farm animals for draft work and not as sources of protein. As the Christian Socialist Katayama Sen wrote nostalgically of the pre-Meiji years, "I was born in a farm house, and I worked as a farmer. The family ox was absolutely necessary for plowing, and we loved him as one of ourselves. I followed behind him working, and I made money on his labor. I had so many memories of the animal that I would never have wanted to eat meat."[11] Given such attitudes, those wild animals that threatened Japanese grain farmers, such as deer and wild boars, came to be seen as truly "noxious" in the context of Japan's early modern agronomy, while wolves, which chased these ungulates from grain fields, became friends of the noble Confucian farmer.

We have also seen that the geographer Furukawa Koshōken had noted at the close of the eighteenth century that wolves in Morioka domain graciously chased deer from the grain fields of farmers. For this reason, when peasants there encountered wolves in the wild, it was customary to say, "O lord wolf, what do you say? How about chasing the deer from our fields?"[12] By the 1870s and 1880s, however, after Japan's appropriation and refashioning of newly imported Western assumptions about predators through the agency of Dun and others, a new greeting for wolves was created: shoot them, sever their ears or legs, bring these to a municipal office, and collect a lucrative bounty. Whereas in the seventeenth and eighteenth centuries peasants revered wolves at Shinto shrines as guardians of Tokugawa agriculture or reviled them as "mad" demons, the new Meiji regime ruled out both these "primitive" interpretations in favor of modern Western ones and categorized wolves as "noxious animals" (*yūgai dōbutsu*). The Meiji government, much like governments in Australia, South Africa, and the United States, then mobilized its resources to exterminate them with the same chemicals labeled "strange" by the early modern observer Negishi Yasumori. As a rancher born in the American Midwest, Dun was comfortable with the task of killing wolves at the experimental farms of Hokkaido. A product of mid-nineteenth-century American beliefs in the power of scientific agriculture and hostilities toward the wolf, Dun came to Japan quite prepared to expand ranching and eliminate a predator that needed to be wiped out to facilitate the march of civilization.[13]

A historical lesson in the story of Dun and wolf extermination on Hokkaido transcends even the troubling question of why our species is capable of so hastily destroying another. Changing perceptions of wolves in Japan after the introduction of industrial ranching highlight the significance of new attitudes toward the natural world in early Meiji definitions of modernity. With striking ease, Meiji officials brushed aside centuries of reverence for and fear of wolves—and, to a lesser degree, the entire East Asian order that supported such traditions—replacing them, brick by brick, with the edifice of a modern order. Even without knowing stories such as "Little Red Riding Hood" or worshiping at Christian churches, traditions often targeted as sources of wolf hatred, Meiji officials saw the wolf as a "noxious animal" because the bottom line of industrial ranching demanded that they do so.[14] In the new order, Hokkaido's wolves were to be sacrificed, their dead faces grimacing and their bodies strewn around poisoned horse carcasses. Though separated by only decades, the ranchers of Hokkaido who worked under Dun lived in a world apart from their early modern grain-farming—and even horse-farming—countrymen: they lived in the modern world.

WOLVES TODAY

Scattered droplets of blood frozen in trampled snow marked the edges of a kill site along Antelope Creek, near Tower Junction, in Yellowstone National Park. My academic interest in the wolves of Japan had brought me from the archives of Hokkaido University, those poorly ventilated rooms filled with dust-covered echoes of Dun and the Hokkaido wolf, to this spot in the North American wilderness, where wolves had been reintroduced in 1995.[15] With the permission of the program's director, Douglas Smith, I participated in the Yellowstone Wolf Project Winter Study, working with field researchers to collect data primarily on three wolf packs: the Druid, Rose Creek, and Leopold. Walking clumsily through the crusted snow south of Antelope Creek, I recalled my recent trip to Japan and considered why a historian writing about extinct Japanese wolves needed to be at this site. I looked down to see scattered droplets of blood change to larger patches of trampled snow stained a deep red. The grass and sage poking through the snow had been snapped and ripped up by the weighted stotting of sharp ungulate hooves. On a crest overlooking Antelope Creek, the patches of blood grew an even darker red and the area reeked with a sweet gamey smell. Near these patches of fresh blood, piles of the soft belly fur from an elk littered the crest. At this spot, wolf feces and urine were everywhere: mixed with the bloodied snow and elk rumen, it created a grotesque collage of colors and smells.

Further on, tucked in a small stand of pines along the southern bank of Antelope Creek lay the mostly devoured carcass of a large bull elk (see fig. 22). When we approached, magpies and ravens took to nearby trees, while a single coyote fled up the opposite bank, only occasionally looking back over its shoulder. The hide of the elk was turned inside out, exposing its pink skin and a meatless rib cage. Its legs were broken and contorted. The so-called six Tower wolves, a splinter group of the (once powerful but now largely defunct) Rose Creek pack, had killed the bull, eviscerated it, and then dragged it to this spot along the creek. Its left eye had been poked out and eaten, perhaps by the ravens now squawking in the trees around the site, and the socket gaped, no eye left to pierce the forest and stare at the frozen Yellowstone sky. Around the carcass, the Tower wolves had sprayed trees with urine to mark the site, and it froze as a kind of brazen assertion of lupine ownership and territoriality. Wolf feces also lay scattered around the carcass, in the snow and draped over twigs. Later, I discovered the spot where the wolves had bedded down after gorging themselves. There the snow was matted down, brushed a faint red from the blood that no doubt covered their black and gray faces. Again, the site was marked with more feces and urine.

Simply put, I felt a sense of anxiety—a mixture of fascination, fear, and revulsion—about this kill site: the blood, feces, urine, devoured elk carcass, and nearby presence of wolves combined to spark something inside me, something both learned and instinctive, both cultural and beyond culture. It reminded me that, at least according to some cultural anthropologists, the researcher often carries the same "burden of anxieties, cultural values, and personal idiosyncrasies" as the object of research, and so the anxiety I felt at this kill site might have been akin to what ranchers experienced discovering a half-devoured horse in Ohio or northernmost Japan. In short, the researcher and object of research are made of the "same essence," and so the two "react to stimuli, at least occasionally, in the same way." Important for the historian of Japan, "irrespective of cultural differences

FIG 22. This is a wolf kill along Antelope Creek, near Tower Junction, in Yellowstone National Park. This large bull was brought down by the six Tower wolves, a splinter group of the once powerful Rose Creek pack, in November 2000.

or similarities, we are at least similar in nature in that we are culture creators, and symbol makers, though the content of our symbol systems may differ." This is true because the symbols created, whether by a historian at Yellowstone or ranchers in Japan, "may emerge from the wellspring of our unconscious and then, mediated through culture, be transformed by our conscious rational and cognitive faculties."[16]

It became obvious to me that those human societies with the strongest desire to kill wolves—those people who celebrate, quite literally, cultures of wolf killing—are agricultural societies with strong traditions of animal husbandry. Mainly, they are countries that practice industrial ranching, countries such as Australia (where the dingo was the victim of traps and poisons), South Africa (with its wild dogs and other canines), and the United States. Presumably, most ranchers from such countries have, at least once, stood over the half-devoured bloody carcass of a calf or foal, surrounded as it surely was by urine and feces; full of hatred and rage, they then created cultures and symbols by relaying their stories to others, stories that spread throughout society and shaped the opinion of people who had never even encountered wolves or other predators in the wild. Indeed, the culture of wolf killing might be a product of feelings of anxiety evoked at wolf kill sites around the world, which led me to believe that, staring at this dead elk along Antelope Creek, I had come to the birthplace of the Western obsession with wolf killing, the very one transported to Japan in the Meiji years by Dun and others.

WOLF KILLING IN EDWIN DUN'S DAY

Almost all late-nineteenth-century American ranchers (among whom, of course, we must include Dun) lived in what can only be described as a culture of wolf killing. Indeed, they sometimes took wolf killing to near pathological heights.[17] Take Ben Corbin, the self-proclaimed "Boss Wolf Killer." In his writings, Corbin's metaphors and narratives mixed nineteenth-century biblical language and modern ranching imagery, and they contrast starkly to early modern Japan's agrarian culture as represented by Kumazawa Banzan and others. Born in Virginia, Corbin's father had "hunted redskins with Daniel Boone," and in Corbin's mind, killing wolves similarly meant clearing the way for civilization by destroying another tribe on the American frontier. In 1883, just a decade after Dun first went to Japan, Corbin relocated to northern Dakota Territory, where high bounties for wolf "scalps" provided the incentive for him to perfect his wolf-killing skills. Corbin killed wolves for bounties; but, more importantly, he killed them for cultural rea-sons, to clear the way for the settlement of the American West by ranchers, whom he likened to the kindly shepherds of the New Testament.

In his book, *Corbin's Advice, or the Wolfer's Guide* (1900), Corbin yoked biblical figures to nineteenth-century notions of progress. Abraham, he argued, was "rich in cattle, silver and in gold—something like the ranch-men and stockmen of North Dakota." The pastoral life of the herdsman "preceded every other profession." Corbin, moreover, saw guarding live-stock as part of Divine Providence. "Largely my life has been spent in pro-tecting these flocks against the incursions of ravenous beasts of prey," he explained. "I know it is but a first step and the first step, which counts in the march of civilization." Corbin also drew on Social Darwinism to jus-tify ranching culture: "I can not believe that Providence intended that these rich lands . . . should be forever monopolized by wild beasts and savage men. I believe something in the survival of the fittest, and hence I have 'fit' for it all my life." He concluded: "Civilization is a fine thing, and it may spread itself like a green bay tree in the cities, and lordly mansions of the million-aires, with all their silks and broadcloths, but it has to have plenty of beef and pork and mutton—yes, yes, and wool too, and plenty of it."[18]

For Corbin, therefore, clearing the land of wolves became a raison d'être. In an arrogant self-revelation, he wrote, "The wolf is the enemy of civilization, and I want to exterminate him." In a deeply disturbing episode, he told of shooting a female wolf as she fled her den and then performing a "Caesarian operation" to remove the unborn pups from her cooling body. He said that there were "four of them alive and kicking," and two more dis-covered in the den. The female, it turns out, had been in the process of giv-ing birth to her young. In a twisted scene, Corbin "laid them down beside their dead mother for their first meals and this is according to Scripture, 'although you may be dead you yet shall live.'" He then killed them and took the carcasses to Bismarck, North Dakota, to try and retrieve bounties (which had been terminated).[19] With the Bible as his compass, notions of progress as his rudder, and the winds of civilization powering his sails, Corbin helped transform the wolf into a powerful nineteenth-century symbol of evil.

EDWIN DUN

This was Dun's world. A prominent nineteenth-century Ohio rancher, he oversaw land that his ancestors had farmed, and as his activities in Japan reveal, he saw wolves as a symbol much as Corbin did. As Corbin noted,

wolves were animals that only stood in the way of the "march of civilization." And Meiji modernizers would not tolerate any hindering of the march
of civilization. Wolf killing in Japan, after the Meiji Restoration of 1868, needs
to be seen in this light. Poisoning wolves was no longer a "strange way" to
kill them but an example of industrial efficiency.

Published in 1900, Corbin's book, and the many others like it, nicely illustrate the intense hostilities felt toward wolves in the United States in the nineteenth and early twentieth centuries. It was out of this milieu that Dun traveled
to Japan, where at the Niikappu ranch (in southeastern Hokkaido) he oversaw for the Kaitakushi a deadly program designed to eliminate wolves and
wild dogs with the use of strychnine, which had become a favorite tactic of
wolfers in the American West and other regions where industrial ranching
had become a prominent fixture of modern agronomy. It is worth exploring Dun's experiences in Japan in some detail because he is often celebrated
as the man who, as one official history trumpets, "saved the farm" at Niikappu from the ravages of wolves, which elevated him to the status of the
"father of Hokkaido agriculture."[20] Indeed, as an agricultural pioneer and
wolf killer, Dun was featured in Funayama Kaoru's *Zoku Otōsei* (Otōsei continued; 1975), the second installment to his popular 1969 novel, *Otōsei*. He
became an even more heroic figure, wielding a samurai sword on horse-
back, in the comic *Ōkami no hi: Ezo ōkami no zetsumetsuki* (Memorial to
the wolf: A record of the extinction of the Hokkaido wolf; 1994), with a narrative by Togawa Yukio and illustrations by Honjō Kei (see fig. 23).[21]

Dun was born in 1848, on a ranch near Springfield, Ohio, the grandson
of Scottish immigrants to the United States.[22] The family ranch employed
as many as twenty hired hands. The impression one gets looking through
his unpublished memoirs, *Reminiscences of Nearly a Half Century in Japan*,
is that Dun fancied himself as a hereditary rancher, the son of a landowner
in central Ohio, an area he called "Dun Plains," and a keeper of a family
charge to husband animals. His father and three uncles together owned about
15,000 acres of the "finest 'blue grass land' to be found outside Kentucky,"
where they raised short-horned cattle and racehorses.[23] Yet, despite the quaint
homegrown strokes with which Dun painted his early life in *Reminiscences*,
he actually came of age in the aftermath of America's first agricultural revolution, which lasted between roughly 1812 and 1830. As many historians
have argued, the United States, bolstered by scientific advancements, the
invention and perfection of a flood of new farming techniques and technologies, and the creation of new crops and livestock breeds, set a course
for agricultural industrialization that culminated in the 1920s.[24] Not sur-

FIG 23. A dramatic scene from Togawa Yukio and Honjō Kei's *Ōkami no hi: Ezo ōkami no zetsumetsuki* (Memorial to the wolf: A record of the extinction of the Hokkaido wolf; 1994). After wolves killed and ate horses at the Niikappu pastures, Edwin Dun, an employee of the Kaitakushi (Hokkaido Development Agency), says, "I'll use strychnine!"

prisingly, it was precisely this model of industrial agriculture that interested Meiji policymakers.

After high school, Dun enrolled in Miami College in Ohio to study law, but he abandoned his education after his eldest brother left the family ranch to seek a career as a civil engineer. Dun spent the next six years learning the ropes of ranch life. In 1871, after gaining considerable expertise in livestock, he and a cousin opened their own dairy farm, which, by his own accounts, prospered until the center of cattle breeding in the United States shifted westward, signaling tough times for Dun and other Ohio livestock owners.

All this changed in 1873, when the twenty-five-year-old Dun met Albert Capron, whom the Kaitakushi had entrusted to buy cattle and ship them to Japan. After their initial meeting, Capron later visited the Dun ranch, where he was shown the herds of shorthorns and other stock that Dun reckoned were best suited for the Kaitakushi enterprise. The two decided on about eighty head in all, including calves and heifers, which Dun promised to deliver to the Chicago stockyards. In Chicago the two men dined together at the Stock Yard Hotel, where Capron informed Dun that he had another commission to fill for the Kaitakushi. As Dun recalled, "It was to secure the services of some one well up in live stock breeding and handling as well as a practical farmer experienced in up to date methods in the United States." Eventually, Dun asked if he might be suitable for the commission, which, despite the low salary, seemed like a good opportunity "owing to the business depression in the United States." (Here Dun is probably referring to the panic of 1873.) Capron agreed and the young Dun proceeded to go to Japan to advise the Kaitakushi on "up to date methods" at its ranches.[25]

Dun left for Japan in June 1873 aboard the steamer *Great Republic* and was met in Tokyo by Kaitakushi officials. He talked with Kuroda Kiyotaka, governor of Hokkaido, and other high-ranking officials, who confirmed his provisional contract and reported on the status of the livestock held at the Tokyo experimental farms.[26] Dun spent two years in Tokyo at the Kaitakushi headquarters before leaving for Sapporo. In Tokyo, under the advice of Horace Capron, three experimental farms—the laboratories for scientific agriculture—had been set up, consisting of farmland, barns, corrals, and classrooms to train the future ranchers of Japan. In a country dominated for centuries by wetland rice cultivation, these experimental farms were an important part of Japan's embrace of modern agriculture, serving, quite literally, as a breeding ground for both livestock and Japanese ranchers. These experimental farms taught not only animal husbandry but often other kinds of agricultural science as well: new fruit trees and vegetables were planted, reproduced, and distributed throughout the country. Dun wrote in his memoirs that two million fruit trees from the third experimental farm in Tokyo had been distributed on the main island (of Honshu) alone.[27] It was these types of experimental farms that so dramatically altered the eating habits of nineteenth-century Japanese. Indeed, the year before Dun arrived in Tokyo, the Meiji emperor, both a repository of tradition and a symbol of modernity, ate beef reportedly for the first time.[28]

In a progress report of sorts, *Japan in the Beginning of the 20th Century*, the Department of Agriculture and Commerce boasted of its successes as of 1904. Although cattle had been in Japan from ancient times, it noted, and previous emperors even had established official pastures, these animals, as Katayama Sen remembered, were seen largely as beasts of burden and not widely consumed by the public. Starting in 1869, however, fifteen foreign cows had been purchased from the English at Yokohama, and this set in motion a program wherein the Meiji government would "spare neither money nor pains" to encourage the development of the livestock industry. In the case of horses, some regions, such as Morioka, had long bred them, and new foreign breeds, such as Arabians, had been introduced as gifts from foreign governments. In 1871, explained the ministry report, an American official was employed to oversee the horse-breeding program, while two Japanese officials traveled to the United States to learn the craft there. Swine and sheep also were imported in the early Meiji years. The department report explained that although "eating the flesh of animals [was] forbidden in former times [for] religious motives" (a notion, incidentally, that fails to hold up under close historical scrutiny), the Meiji Restoration had brought what department officials hailed as a "revolution in the Butcher business."[29] In 1900, butchers built 1,396 slaughterhouses throughout the country. At these slaughterhouses, between 1893 and 1902, employees dispatched over 1.7 million cattle to fuel the workers and soldiers of modern Japan.[30] This revolution came about through the efforts of Dun and others like him.[31]

Experimental farms also employed the technologies of scientific agriculture. Dun was impressed by the threshers that could yield one thousand bushels of grain per day and by the self-binding reapers, gang plows, corn planters, and "innumerable smaller machines and implements" that were all of the latest technology. In Tokyo, Dun quickly went to work organizing the ranching section of the experimental farms. The stables and barns, designed by Horace Capron, were "located in about the most unhealthy spot that could have been found," and so one of Dun's first accomplishments in Japan was to relocate the barns to more open areas where disease would be less likely to plague livestock. This was easy, recalled Dun, because the "detail of farm work and care of domestic animals had been drilled into me from childhood." In *Reminiscences,* Dun fancied himself not as a "college-bred, book-learned" expert, like some of his colleagues, but as a practical farm boy who could "swing a scythe or ax with the best." With no "graduated veterinary surgeon" in Japan at the time (although Japanese medical culture boasted a long tradition of Chinese-influenced veterinary training), Dun also found himself employing his homegrown skills and knowledge of animal anatomy to deal with the various health-related problems of the livestock.[32]

Along with steam trains and modern buildings, these experimental farms became evidence of Japan's modern transformation.[33] Not only did the Meiji emperor eat beef in 1872, but in September 1873 the emperor arrived at the Tokyo experimental farm in his "imported court carriage," part of his new image as a more visible European-style sovereign. Meiji luminaries such as Prime Minister Sanjō Sanetomi, Saigō Takamori, Ōkubo Toshimichi, Kuroda Kiyotaka, and Ōkuma Shigenobu attended him. The emperor inspected the livestock and machinery and was treated to a display of how the thresher worked. "His Majesty was well pleased with the exhibition," Dun remembered. That night he visited the emperor with Saigō, where he "made bows as instructed and backed out." One suspects that Dun's ease within such elite circles contributed to his later diplomatic career in Japan. In 1884, the Chester A. Arthur administration appointed Dun to the newly established post of second secretary in the American legation, and then in 1893, during the administration of Grover Cleveland, he was made chargé d'affaires. (Dun also married a Japanese woman in 1875, during his first visit to Hokkaido.)[34]

THE NIIKAPPU RANCH

From 1875 to 1883 Dun lived and worked in Sapporo, the capital of Hokkaido. There he shaped directly the agricultural development and landscape modification of the northern island. Through his version of equine zootechny—the science of horse breeding—that was practiced at the experimental farms of Hokkaido, Dun promoted the language, methods, and ideals of American animal science.[35] Refashioning Hokkaido in the image of the American West (or in Dun's case Midwest) meant altering horse-breeding practices in Japan. It also meant killing the wolves that threatened the survival of the enterprise.

Dun opposed raising sheep on the island because he concluded (erroneously) that the "arable lands of Hokkaido as well as Japan proper were more urgently needed for the production of food for the people than for the growing of grass for sheep." Moreover, Dun believed that the grasses of Hokkaido were poorly suited for ranching. Later, after an experimental batch of sugar beets failed, he also advised against growing that crop, arguing that Hokkaido "is too cold for the development of a high percentage of sugar."[36] Dun's most important task, however, was overseeing large experimental farms at Makomanai, Izari, Niikappu, and Shizunai. At Makomanai, south of Sapporo, "about 200 acres of wild land was cleared out and cultivated in corn, hay and various kinds of roots for food for the cattle." In Izari, south of Sapporo near Chitose, Dun and others "enclosed a fine bit of native pasture land about 2,000 acres in extent for the use of some of our horses." Niikappu, however, emerged as the centerpiece of the Hokkaido ranches run by the Kaitakushi. There Dun and others established a "great stud farm and ranch for the improvement of Hokkaido horses," a process achieved through the crossbreeding of Nanbu stallions, a famous Japanese breed from Morioka, with imported animals. Finally, just east of Niikappu, along the Shizunai River, an adjoining 300 acres were set aside to supply winter forage.[37]

The Niikappu ranch covered about 35,000 acres in all and was eventually separated into smaller pastures by post-and-rail fencing (see fig. 24). The ranch was situated between the Niikappu River, which served as the western border, and the Shizunai River on the east. It was about ten miles inland from the coast, and its total length was about fifteen miles, and the width somewhere between five and six miles. The southern section of the ranch was covered with grasslands about fifty or sixty feet above the banks of the rivers, while the northern section graded from hills into mountains, with ground cover consisting of scrub bamboo. Dun's first report on the Niikappu ranch, dated October 1875, highlights some of his early concerns about the location and breeding practices there. The first problem was that Niikappu spread onto the slopes of the Hidaka Mountains, and so the site was isolated and had a "great amount of useless land." Moreover, ranch hands allowed horses to roam within "the large extent of mountainous territory," where it became "impossible to find them." The horses, noticed Dun, ventured into the mountains in search of bamboo, and yet, by the end of winter, they still were "so poor that almost every bone can be seen." Dun laid out seven recommendations that he hoped would improve the location for horse breeding.[38] However, two years later, after another inspection undertaken in July 1877, Dun again voiced concern, writing that the size of the Niikappu ranch was "much greater than is necessary for the purpose for which it was, and is now intended, and the difficulty of managing and controlling a large number of horses roaming at will over so large an area of land is very great." He also commented once more on the poor winter cover and food provided at the site.[39] Even as late as 1881 he complained of Hokkaido stallions "running wild in the pasture, and mountains adjoining it," a problem that needed to be dealt with.[40]

Although Dun said nothing of wolves and wild dogs in these letters and reports, this type of situation—open pastures bordered by mountainous terrain and inhabited by weakened horses in the winter—was ripe for problems involving predators. In 1881, while inspecting the Shiriuchi Valley, on the Oshima Peninsula, as a possible site for raising sheep, Dun explained that open pastures bordered by mountains could invite predation from wolves and wild dogs.[41] Dun wrote that the Shiriuchi Valley, once drained, might make fine sheep country, but he cautioned that an elaborate fence would have to be erected because the valley had mountains on three sides: "If it were not for the wolves and dogs, it would only be necessary to turn them in their pasture, and leave them there, as soon as there was sufficient grass to keep them, but as there are dogs and wolves at Shiriuchi, it will be necessary to guard them at night: for this purpose pens with high, tight fences can be made in which the sheep can be kept at night."[42]

In time, after Dun recommended that horses from Izari be moved to Niikappu, the ranch developed into the center of Dun's horse-breeding activities.[43] In March 1878, he boasted of making the Niikappu ranch the "principal horse breeding establishment of Hokkaido."[44] He explained in his *Reminiscences* that ranch hands brought one thousand Hokkaido mares to Niikappu, along with about fifty of the best Nanbu stallions they could find. Moreover, he integrated four imported thoroughbred stallions into the breeding program. Dun discovered that about 90 percent of the Hokkaido mares bred to Nanbu stallions became pregnant, whereas only about 40 percent of those bred to imported stallions were with foal. Dun explained this as "owing to the difference of temperament of the native and thoroughbred" horses. In his reports to the Kaitakushi, he continued to push a regimen to weed out weak horses. This aspect of Western equine zootechny represented the "first principle in breeding any of our domestic animals," he explained.[45]

So Dun went to Japan in 1873 confident in the powers of industrial agriculture: the productive capabilities of modern farm technology, the benefits of certain crops and livestock, the transformative abilities of equine

FIG 24. The Niikappu ranch site, with its barns and post-and-rail fencing, looks as though it might be located somewhere in southwestern Montana. The United States provided the model for Japan's colonization of the northern island of Hokkaido. Courtesy of the Resource Collection for Northern Studies, Hokkaido University Library, Sapporo, Japan.

zootechny, and the fruits of scientific breeding. His counterparts, the Kaitakushi officialdom, eager to modernize their country, paid close attention to Dun's many lessons and adopted many of his suggestions, often with little question. As already noted, soon after Dun had arrived in Japan, the Meiji emperor even had visited one of the Tokyo experimental farms, witnessing the industrial magic of a mechanical thresher. Later, as an ambassador representing industrial agriculture, Dun attended an audience with the emperor and other government officials. This confidence in modern agriculture had deadly consequences for Japan's wolves. If Neo-Confucianism, Buddhism, and more native Shinto traditions shaped earlier Japanese attitudes toward wolves, then industrial agriculture born in the West shaped post-Meiji ones. Wolves hampered Dun's efforts at equine zootechny—a stone jammed in the wheels of industrial progress. They stood in the way of Japanese modernity. As wolf predation increased at Hokkaido's horse pastures, so too did the need to exterminate them, and Japanese cast wolves (or re-created them) in the image of "noxious animals" that needed to be exterminated.

WOLF KILLS

Despite the success of the Niikappu ranch (and Dun's lack of comment on the topic), wolves preyed heavily on the foals the first year.[46] Dun later wrote in *Reminiscences*:

> But to our horror we discovered that wolves with which that part of Hokkaido was at that time infested seemed competent to devour horse flesh rather faster than we could produce it. One lot of 90 mares with foals had been placed in an enclosure to themselves, [and] within a week to ten days they were rounded up but not a colt was with them. Every one of the 90 had been killed, their bones were scattered all over the place.[47]

In May 1878, the problems involving wolves, and feral Ainu dogs, still remained endemic at the Niikappu ranch. Scattered bones and decaying carcasses—potent symbols of the kill site—were the only evidence that the foals had existed at all. A report filed that month by Iwane Seiichi, an official on the scene at Niikappu, stressed that predation from wolves and dogs at Niikappu caused economic losses and threatened the basic viability of the ranch. Iwane wrote that the total number of horses at the ranch was 2,000, over 1,300 of which were Nanbu mares. Of these mares, reported Iwane, about 900 were actively producing foals every year. However, since the creation of the Niikappu ranch in 1872, only about 300 of these foals had survived to adulthood. Iwane explained, in no uncertain terms, that he believed wolves, bears, and Ainu dogs were killing and eating them. At the time of his report, every mare over three years old was pregnant, and of these about 400 had already dropped foals. "The wolves and dogs know this," wrote Iwane, and they enter the pasture looking for easy prey, and because the ranch is so vast, it is nearly impossible to stop the killing. "In the summer, the colts simply become wolf food," he wrote. In response, he made three recommendations designed to reduce the size of the ranch and fence it into plots for better management of breeding. Iwane admitted that it would be expensive, but he believed that in the end the returns from foal survival would pay for the fencing. He then went on to discuss his policy, already under way, that was designed to stem the losses from dogs and wolves.[48]

FIG 25. Horse corral and ranch hands at the Niikappu ranch site. Courtesy of the Resource Collection for Northern Studies, Hokkaido University Library, Sapporo, Japan.

Iwane's policy was as follows: First, all horses were rounded up at night and moved into smaller, fenced pastures (see fig. 25). Guards were placed at these pastures to ensure that no wolves or dogs broke in. Second, and most interestingly, Iwane proposed prohibiting the keeping of Ainu hunting dogs in the villages near the Niikappu ranch. Not only could dogs not be kept, but all wild dogs were to be hunted down and shot. However, in the nearby Ainu villages, many people refused to surrender their dogs and hid them when local officials came to their homes. This was also true of Ainu living in Shizunai. Considering the traditional place of dogs and wolves in Ainu culture (Ainu believed themselves to be born from the union of a goddess and a canine), some Ainu had even resorted to taking their dogs into the mountains and raising them there, away from the villages and watchful Kaitakushi officials. Frustrated, Iwane explained that even if Kaitakushi officials offered to buy dog pelts, in effect tempting Ainu to kill their own dogs, Ainu staunchly refused. The only solution, he argued, was to tell Ainu who wanted to keep their dogs to move and to hire special "dog killers" (*satsukensha*) to hunt down any remaining canines. Already incidents involving the killing of foals had been "in the many tens," he explained, and surely would continue unless something was done.[49]

The Kaitakushi was evidently slow to act, however, for this policy report was followed by an even more urgent request from Hosokawa Midori, another official at Shizunai. He too emphasized that the damage caused by bears, wolves, and dogs was "not a little." Hosokawa himself had seen a half-eaten horse near a barn at the Niikappu ranch, and he said that it was indeed a disturbing sight. The deeper pastures at Niikappu, he wrote, were teeming with wild animals, and so protecting the horses at the ranch proved extremely difficult. The ranch hands and horse caretakers tried to ward off the night attacks by wolves and wild dogs. In a scene that must have resembled a war zone more than a ranch, workers lit bonfires near the stables, while the deafening sound of rifle fire split the Hokkaido night. But wolves and dogs still managed to get into the stables and pastures and injure or kill horses. Hosokawa urged officials in Sapporo to get moving on Iwane's proposal to partition the land with fences and suggested that professional "wildlife hunters" (*yajūbōgyosha*)—armed with Western rifles and paid ¥5 per month—be brought in by no later than November or the ranch would be overrun by wolves. Even more than Iwane, Hosokawa painted a warlike picture of the conflict occurring at the Niikappu ranch, where local managers were at a loss about how to handle the striking onslaught of wolves and wild dogs.[50]

The war at Niikappu, at least as described by Iwane Seiichi and Hosokawa Midori, was completely out of control, a situation much worse than that seen in normal descriptions of wolf predation from the American West.[51] There probably was an unusually high concentration of wolves and wild dogs in the Hidaka region at this time. Inukai Tetsuo, an ecologist who studied Hokkaido's wildlife, postulated that many wolves and wild dogs had moved into the Hidaka region around 1878 as a result of a variety of environmental factors, some a product of Hokkaido's development, others related to climatologic conditions. Inukai suggested that the fate of the wolf and wild dog was tied to that of the deer. These canines came to the southeastern section of the island, to places such as Hidaka, where the Niikappu ranch was located, because deer habitually went there in the winter to escape the deep snowfall of the western section of Hokkaido. When they arrived, however, hunters had killed most of the deer, forcing the wild canines to turn to livestock. In the wake of the Kaitakushi's massive harvest of deer, in other words, habitual wolf and wild dog subsistence systems broke down. Naturally, they turned to the best alternative, the Niikappu ranch.[52]

TABLE 2. Export Numbers for Deer Skins, 1873–1881

	1873	1874	1875	1876	1877	1878	1879	1880	1881
Sapporo	47,784	52,895	70,024	58,034	33,708	24,529	35,036	7,659	6,080
Hakodate	—	—	—	109	51	96	68	408	1,364
Nemuro	7,262	5,603	6,398	6,450	10,173	5,844	11,281	4,591	2,762
Hokkaido Island	55,046	58,498	76,422	64,593	43,932	30,469	46,385	12,658	10,206

SOURCE: Inukai Tetsuo, "Hokkaidō no shika to sono kōbō" (The Hokkaido deer and its rise and fall), *Hoppō bunka kenkyū hōkoku* (Research reports on northern culture) 7 (March 1952): 26–27

NOTE: 398,209 skins exported between 1873 and 1881.

In the 1870s, the Kaitakushi had stepped up its harvest of deer as part of a broader policy of exploiting resources on Hokkaido (see table 2).[53] They established venison canneries such as the one in Bibi, near Chitose in central Hokkaido, where deer meat was canned and saltpeter extracted from the bones and blood, and where skins and antlers were exported as part of the burgeoning European and Asian trade in pelts and pharmaceuticals (see fig. 26).[54] Many Ainu, their previous hunting and fishing lifestyles lost in the wave of forced assimilation that swept Hokkaido after 1800, had little choice but to become professional hunters under the Kaitakushi.

As for the climate, the winters of 1878 and 1879 were extremely severe. Indeed, as table 2 suggests, the winter of 1879 was so harsh that in 1880 deer harvests dropped to less than a third of what they had been the previous year, putting even more pressure on Hokkaido's human and nonhuman predators. Dun explained that despite the heavy snowfall, deer might have pulled through were it not for the "great demand for hides and horns." During the winter, he wrote, "deer had collected in thousands in the most sheltered valleys and ravines where, owing to the deep snow, the Ainu on snow shoes overtook them easily and slaughtered many tens of thousands with clubs and dogs. In the Mukawa District alone—15 miles by 5—75,000 skeletons were counted in the spring [of 1879] by men sent by the government to ascertain the loss." The slaughter occurred the following winter as well, said Dun, and the "result was practical extermination."[55] After the exceptionally harsh winter of 1879, the Kaitakushi shut down the venison canneries for two years, but deer numbers were probably at historical lows by 1880–81.[56]

To a certain extent, deer killing in Hokkaido mirrored the North American experience of bison killing, although on a lesser scale. When the bison

FIG 26. Venison cannery at Bibi, near the present-day city of Chitose. Courtesy of the Resource Collection for Northern Studies, Hokkaido University Library, Sapporo, Japan.

slaughter rose to its height, wolves and other scavengers thrived on the availability of carrion, and wolf numbers probably spiked briefly. However, beginning in 1868–69, the year of the completion of the Omaha-Ogden stretch of the Union Pacific Railroad (and the Meiji Restoration in Japan), cattle gradually supplanted bison, and abnormally high populations of wolves turned to this new (and easier to catch) ungulate, causing high rates of depredation. Modeled after the development of the American West, the Meiji development of Hokkaido, under the Kaitakushi, resonated with similar environmental upheaval. Stanley Young, the wolf-killing czar of the U.S. Bureau of Biological Survey, made this point, as have environmental historians. Young wrote that "the American's taste for beef had to be satisfied," and so the bison and other wild game were eliminated in favor of cattle, which "brought the wolf into direct competition with the producer of prairie cattle, and it became a war to the bitter end."[57] Dun, just such a prairie cattleman, brought this bitter war to Japan, and the Niikappu ranch became his primary battlefield.[58]

For Dun, threshers, clean stables, fencing, winter forage, and controlled equine zootechny were part of the recipe for modern agriculture in Japan, but so too was wolf killing. In the nineteenth-century American West, wolf killing became a standard feature of modern agriculture: wolfers used rifles, traps, poisons, biological agents such as mange, and the technique of "denning" to rid the rangeland of *Canis lupus*. As a foreign expert, Dun brought these skills to Japan, and he employed them with ruthless efficiency.

DESTROYING THE HOKKAIDO WOLF

Even before Dun had arrived in Sapporo, other foreign experts, such as Benjamin Lyman, a geologist for the Kaitakushi, speculated that predators would need to be eradicated before successful ranching could be achieved on Hokkaido. While touring Hokkaido in 1874, a year before Dun arrived in Sapporo, Lyman wrote that the "presence of bears and wolves in the mountains (though far less numerous than the deer) will perhaps be of some hindrance to the introduction of sheep and even larger cattle; and perhaps it will be thought necessary to encourage still further their extermination by offering bounties, as is done in other countries."[59] Lyman submitted his comments to Horace Capron, who then forwarded the recommendations to high-ranking officials of the Kaitakushi. The Kaitakushi took Lyman's advice, and a complex bounty system, to be examined in the next chapter, ultimately was organized for the extermination of bears, wolves, feral dogs, and crows.

Dun was concerned mainly with horses, not sheep or cattle, but nonetheless he came to the same conclusion as Lyman. In *Reminiscences*, he lucidly described the Hokkaido wolf as an animal built to kill and so deadly to the ranching enterprise:

> The Hokkaido wolf is a formidable beast but not dangerous to man as long as other prey is to be had for the killing. During the winter months, at the time of which I am writing, they lived mostly upon deer which were very plentiful. During the summer their diet was principally horse meat. A full grown wolf weighs from 70 to 80 pounds, he has an enormous head and mouth armed with tremendous fangs and teeth. He is generally very lean but exceedingly muscular. Of a grey color in summer and greyish white in winter, when his fur is thick and long. His feet are remarkable for their size, three or four times larger than the feet of the largest dog which they resemble in shape, only the claws are much longer. Their large feet enable them to travel rapidly over deep snow that soon tires a fleeing deer that could easily run away

from his enemy when the ground is bare. They usually hunt singly or in couple but frequently the trail of a pack of four or five or even more is seen in the snow. They are widely scattered throughout the island as a rule but few in any one neighborhood.[60]

Under normal conditions the Hokkaido wolf remained "scattered throughout the island," a distribution shaped by wolf ecology. However, as Hokkaido's deer population plummeted in 1878–79, wolves redefined the boundaries of their subsistence activity. There is also a good chance that many of the canines that killed horses at Niikappu were in fact feral Ainu dogs (which are fairly wolfish in appearance). In Kaitakushi documents, officials used the Chinese characters for "wild dog" and "wolf" virtually interchangeably, sometimes together in a compound (rendered as *rōsai* or *rōyaken*), suggesting that not only wolves but also dogs threatened the Niikappu ranch. Like the Neo-Confucian taxonomists of chapter 1, even modern wolf killers, it appears, struggled to distinguish wolves from mountain dogs in Japan's ever evolving landscape. Whether he was writing of wolves or wild dogs, however, in the face of food shortages in 1878–79, Dun reported that "the large number of horses we had confined in a limited area attracted them from near and far." As Dun remembered, "After killing the colts in the outlying pastures it was not long before they began on the mothers. In fact the situation became so serious that it was up to us to exterminate the wolves or go out of the horse breeding business at Niicapu." Because hunting wolves was next to impossible, Dun continued, "we sent to Tokio [Tokyo] and Yokohama for all the strychnine to be had and fearing there was not enough for our purpose in those places, sent a supplementary order to San Francisco for more." In the end, Dun concluded, "We succeeded in getting enough to poison every living thing on the island."[61]

Though Dun relied on strychnine to "exterminate" Hokkaido's wolves, its use did not originate in the United States. It had long been used in Europe (and, apparently, early modern Japan) to kill predators and "noxious" birds such as crows and magpies. As wolfer Stanley Young wrote, however, as a "ready weapon in the hands of profit-seeking pelt hunters and enraged stockmen," strychnine did contribute to the near annihilation of wolves in the United States. Once in the American West, it became the "law of the prairie" that no carcass should be left without lacing it with the deadly poison, and as Young himself observed, that led to one of the "strangest and most lurid chapters in the history of killing mammals such as wolves."[62]

Although the use of strychnine did not originate in the American West, wolfers there certainly perfected the technique. And Kaitakushi officials, on the other side of the Pacific Ocean, were about to turn the page to this "lurid chapter" in Japanese history as well. Killing wolves with strychnine remained one of the favorite tactics of industrial ranchers and their wolfers around the globe. In the industrial mind-set, poison, at least when compared to building elaborate fences, paying trained herders, or placing steel traps, proved extremely cost effective. In the United States, the frontier journals and memoirs that celebrated the grisly deeds of wolfers are littered with vivid descriptions of the use of strychnine. To some, like William Edward Webb, it evidenced the "wonderful command that God gave [humans] over the other animals." In the 1870s, at about the same time Dun was in Japan, Webb recalled that in the American West, Mexican wolfers readily used the poison to kill wolves. They cut from a slain old bull "lumps of flesh about the size of one's fist, into which gashes were made, doses of the powder inserted, and the flesh then pressed together again." Once prepared, the lumps of meat "were scattered close around the carcass, and a few laid upon it." The next morning Webb discovered to his delight the effectiveness of the poison: "Twenty-three dead wolves were found," he wrote, "and the even two dozen was made up by a large specimen of the gray variety . . . who was exceedingly sick, and went rolling about in vain efforts to get out of the way." They later tied up this wolf and tortured it—behavior, one should add, common in the American West.[63] In the 1860s, Granville Stuart also documented strychnine use. He wrote that along with horses, flour, beans, sugar, coffee, salt, blankets, a rifle, plenty of ammunition, and a hunting knife, "a supply of strychnine" was all a wolfer needed for success in the West. Stuart, too, described slicing pieces of bison meat, inserting the poison, and then waiting for the next morning when a "poisoned carcass would often kill a hundred or more wolves."[64]

Joseph Batty, in *How to Hunt and Trap*, wrote in the early 1880s that after bison had been slain to be used for bait, their "skins are partially removed, bodies laid open and contents of the thorax taken out. The viscera and blood which settles is poisoned, the upper quarters are gashed with a knife, and strychnine is put in the incisions. The crystals soon dissolve and penetrate the flesh." The carcasses often freeze at night, so the wolves first feed on the "frozen blood from the thorax, and die in twenty minutes to an hour afterwards." With such techniques, boasted Batty, "Seventy-eight wolves have been taken in Montana in a single night with one buffalo."[65] Later, in 1897, with a flippant tone that betrays the nonchalance with which wolves were poisoned in the West, L. S. Kelly wrote that after shooting a bison cow, he noticed that wolves were watching him. "They were lined up in a row, as if they had been bidden to a feast," he wrote, "and were not particular as to the manner in which it was served. I proceeded to satisfy them." He cut open the bison and "loaded the carcass" in the "usual way" with strychnine, while the wolves "took an unusual interest in the work of preparing this bait." The wolves, Kelly speculated, probably had never seen a poisoned carcass and were accustomed to following Native Americans during their bison hunts. "Though uninvited guests," he wrote blithely, "I felt that their appreciation should not pass unrewarded." When he returned a couple days later, he discovered that "their beautiful carcasses covered the prairie."[66] This was the savage wolf-killing culture of the United States in the mid–nineteenth century, of which the "law of the prairie" was a part.

On June 20, 1878, Hosokawa Midori again wrote to Kaitakushi officials in Sapporo to inform them that predation of foals was continuous. Ranch hands and others tried to keep the wolves and dogs out of the pastures, but the "wild animals come in packs of hundreds" and were relentless. Officials hired two hunters, and they killed some wolves; the overall results were minimal, however. Hosokawa then noted that at Niikappu a new strategy would be adopted for the summer (from July 1 to September 30, 1878). He explained that ranch hands had decided to use strychnine to poison the wolves and dogs, and a request was dispatched to the Sapporo Hospital to obtain some. Four days later, Hosokawa again wrote regarding the plan to use strychnine, though his attitude toward the poison was cautious. He warned that the meat and skins from wolves and dogs poisoned should not be eaten or touched, and so local villagers should be told in advance to report any dead animals to officials at the Niikappu ranch. He requested 1.3 ounces of strychnine, but the Sapporo Hospital only had 1.06 ounces on hand, and so that would have to do.[67]

"We went to work systematically," wrote Dun in *Reminiscences.* Dun first organized a patrol of about twenty men on horses, and each individual was given a daily route. Each rider was given "chunks of poisoned meat to be dropped at likely places," and, adhering to the nineteenth-century law of the prairie, a "small bottle of strychnine to be used in case . . . the carcass of a murdered horse or colt was found." Dun wrote, in that same strangely nonchalant tone so familiar to the journals of wolfers in the American West, that in such a case the carcass "would be deeply slashed and a liberal allowance of our seasoning sprinkled within it. The success

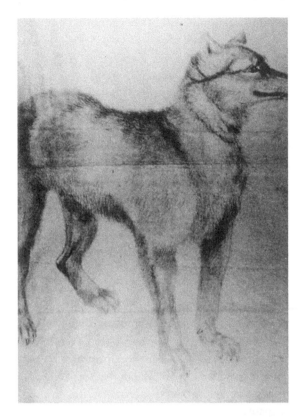

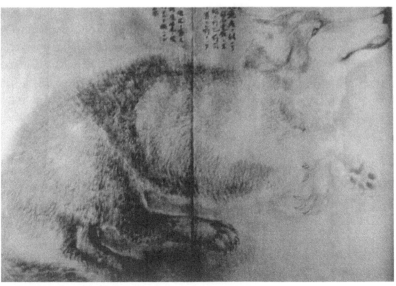

FIG 27 (*a–b*). Images of the Hokkaido wolf. Courtesy of the Hokkaido University Museum of Natural History, Sapporo, Japan.

of our systematic work was immediate and within a few months complete."
Dun continued,

> The first day's bag was five or six dead wolves found, probably others slunk
> away to die in places where they could not be found. Their bodies would
> usually be found near the poisoned carcass or bait, where if undisturbed they
> would remain gorging themselves until the deadly stuff began to work and it
> works very quickly. Often they would be found near water where they had
> gone to quench the terrible thirst the poison creates. Our first day's bag was
> our best. A few were bagged every day for a week or ten days, then only one or
> so occasionally. Then for weeks our bag would be nil until finally the beasts
> were wiped out. So within one summer and autumn we were freed from a
> pest that in the spring seemed very threatening to our enterprise. Hundreds
> of dead foxes, crows and an occasional Ainu stray dog were found near our
> plants which was of course unavoidable.[68]

The precipitous drop in deer numbers in 1878–79 had focused wolves in the
Hidaka region, where poisoned carcasses and bits of meat awaited the "murderers."[69] Kaitakushi officials could boast to Tokyo in March 1880 that the
strychnine campaign had been a success and should be continued. "Everybody knows the damage done by wolves on Hokkaido," wrote officials in
Sapporo; yet wolves, being fast and clever, were extremely difficult to kill.
Hunters could rarely anticipate from which direction they would come next.
So officials explained they had spread strychnine in chunks of meat around
the Niikappu ranch. Of four wolves that fed on the poisoned bait, two died
almost instantly, while the remaining two, having had a smaller dose, wandered away, but they too were soon reported dead (see fig. 27). Considering
the success of the program (and because the next horse-breeding season
was quickly approaching), officials wanted more strychnine from Tokyo.
They asked that it be shipped as quickly as possible.[70]

CONCLUSION

Despite the success of Dun's late-1870s chemical campaign against wolves
and wild dogs, he alone was not responsible for the extinction of the
Hokkaido wolf. The Kaitakushi's elaborate bounty system, the subject of
the next chapter, also played an important role. However, more interesting than the policies that led to wolf extinction, or the reasons why Dun
and other Westerners felt so compelled to "exterminate" (and not just cull)

wolves, is a deeper question. Why did some Japanese, having come from a culture that revered the wolf as the Large-Mouthed Pure God or reviled it as a "mad" demon only decades before, approve of the ruthless and "strange way" used to eradicate wild canines? To explain this about-face in Japanese attitudes toward wolves, we need to return one last time to the issue of Japan's vision of modernity in the early Meiji years. When foreign advisers such as Dun came to Japan, they brought much more than their expertise: they brought with them an individual aspect of the larger mosaic that would become Japan's modernity. Foreign advisers brought to Japan deeply rooted opinions about the promise of modernization that, when integrated into the Japanese education system, workplace, political values, and attitudes about agriculture and the natural world, laid the foundation on which the Japanese nation would be built.

It is understandable, given the logic of capitalism, that scientific and industrial agriculture proved such an important part of Meiji economic reforms and the colonization of Hokkaido. What is less understandable, however, is how the Hokkaido wolf, much like so many other human and nonhuman beings, could be so hastily sacrificed on the bloody altar of modernity. Historians have written at length about the social impact of Japan's early industrial age, when celebrated Japanese capitalists came to dominate world textile markets by forcing young women, called "factory girls," to labor under horrific conditions in damp, tubercular mills; when Ainu, or what remained of them in the late nineteenth century, labored at salmon and herring fisheries along the Hokkaido coast and were cheated, raped, and beaten daily by greedy overseers; or when brave young men boarded "factory ships" to catch and process crab in North Pacific waters, working in unbelievably dangerous conditions—all for the benefit of the nation or to make somebody else rich.[71] In the name of modernity and industry, we—and I say "we" because this is our world too— consistently prove willing to sacrifice members of our own species. Seen in this light, sacrificing another species, one without a voice with which to protest, seems unremarkable.

What is remarkable, however, is that when viewed in the context of wolf extermination on Hokkaido, after the Meiji Restoration Japan had come to resemble other modern industrialized nations—in this case the community of Western wolf-exterminating nations—almost more than it resembled its own premodern self. It is the historical discontinuity, not continuity, that stands out. This rather jarring lesson is one that historians need to consider when teaching their students about the real roots of modern Japan.

5

WOLF BOUNTIES AND THE
ECOLOGIES OF PROGRESS

By the late nineteenth century, the Ashio copper mine, located on the upper reaches of the Watarase River north of Tokyo, had stepped up production after a series of technological advances and became Japan's leading producer of copper. Shortly thereafter, the river turned a "bluish white" color and began killing crops located downstream; rotting fish floated belly up and strange sores festered on people's legs and feet. Sulfuric acid and other toxins in the water caused the effects. By the turn of the century, the Ashio mine (whose wealthy owners the Meiji government supported unconditionally) had created Japan's first major pollution disaster in the name of modernizing the country. But the mine also created one of Japan's first environmentalists. Tanaka Shōzō, who often walked along the silent, polluted river, fought tirelessly on behalf of the poor souls ruined by the Ashio pollution disaster. A Meiji philosopher as well as a politician, Tanaka was a wise man and once said of our relationship to nature and other living things:

> Humans should not be the supreme spirits of living things; they should rather be the slaves of living things [*banbutsu no dorei*], their servants and messengers. . . . They should dwell in the midst of nature and become its reflection, living at peace rather than in conflict with other creatures, correcting their own faults and nurturing their own energy, never cutting themselves off in solitude. . . . Thus they may grow close to the Spirit.[1]

One suspects that Tanaka, during one of his many strolls along the horribly polluted Watarase River and its surrounding poisoned rice paddies, came to believe that the Ashio disaster was evidence that the Meiji government's push to realize the two slogans of "civilization and enlightenment" (*bunmei kaika*) and "rich nation and strong military" (*fukoku kyōhei*) had forever transformed attitudes regarding nature in Japan. In modern Japan, nature evolved from a terrain of spirits whom the Japanese served as "slaves" to something to be subjugated through culture and industry, an order wherein the Japanese became "supreme spirits of living things." In the Meiji years, Japanese modernizers would no longer bow before such spirits as the Large-Mouthed Pure God or quake in fear before its rabid demon cousin; they decided to live "in conflict with other creatures."

In northernmost Japan, wolf killing became one manifestation of this repositioning of Japanese as "supreme spirits of living things." Wolf killing, perhaps more than other forms of environmental exploitation, elevated Japanese to spiritual and actual mastery over part of the natural world, in this case through the systematic erasure of one of the archipelago's largest carnivores. On Hokkaido, the Kaitakushi (Hokkaido Development Agency), through a combination of poisoning, hired hunters, and a bounty system, promoted the extermination of the Hokkaido wolf. And the plan came none too soon for village head Nadokoro Seikichi, who, as documents reveal, complained to Kaitakushi officials that wolves terrified Japanese settlers by howling at night dangerously close to villages in the Kamiiso District and by killing and eating livestock.[2] Both Japanese settlers and nearby wolves, it appears, were voicing in their own ways concerns regarding competition over the land and anxieties regarding their futures on the northern island.

A particularly intriguing episode exposes the lengths to which the Meiji government went, or at least considered going, to kill wolves on Hokkaido. In July 1879, Enomoto Takeaki, a high-ranking Hokkaido official, made a remarkable request to the Tokyo secretariat. It appears that in Enomoto's office hung a painting that depicted pigs being led behind a horse-drawn wagon somewhere in Russian Siberia.[3] As in an 1887 painting by Paul Powis and an engraving published in Frédéric Lacroix's 1845 *Les mysteres de la Russie*, the image of Siberian wolves pursuing horse-drawn wagons or sleds (and the accompanying stories of sacrificing wives and children to escape hungry wolves) was common in Russian folklore. This frightful imagery found its way into Robert Browning's poem "Ivan Ivanovitch" and Willa Cather's novel *My Antonia*.[4] In Enomoto's office painting, the pigs were depicted as squealing loudly, and the Siberian wolves, lured by the appetizing noise of the swine, were emerging from the woods. The painting also shows armed men in the wagon shooting the wolves. In a decision that evidences the sometimes absurd lengths that the Meiji government went to emulate the West, officials requested that the image, seen by Enomoto's office personnel as a blueprint of sorts, be copied so that the technique of dragging pigs and shooting wolves might be tried on Hokkaido. Officials later mused that the same method, if successful, might be tried with Hokkaido's brown bears as well.[5]

On Hokkaido, the organized wolf killing that ensued after the Meiji Restoration (although never to my knowledge in the form of squealing pigs being dragged behind horse-drawn wagons) did signal the emergence of new Japanese attitudes toward the natural world. Once inhabited by the native Ainu, Hokkaido had only recently been incorporated into the Japanese Islands. In this sense, seen from Tokyo's viewpoint, it was a primitive colonial space, a space without culture that needed the footprint of "civilized" people. Many Japanese had once worshiped the wolf as the Large-Mouthed Pure God and with talismans distributed by shrines, and so the wolf, for Kaitakushi officials, became associated with this primitive, pre-Meiji world; and its destruction not only facilitated the development of Hokkaido by cutting livestock depredation and thus yielding economic benefits but by dramatizing that Japan had become more like civilized nations such as the United States. As we have seen, poisoning Hokkaido wolves with strychnine similarly evidenced Japan's developing Western-style civility.

This chapter draws on a wealth of sources—some on Hokkaido's agronomy and bounty system, others on museum exhibitions and what is known of wolf ecologies—to argue that rather than viewing, on the one hand, Japanese ranchers as reacting to punishing economic hardships and, on the other, wolves as reacting to just as punishing ecological ones, in this case both the hominid and lupine species need to be seen as natural competitors struggling for survival on Hokkaido. On Hokkaido, forms of competition occurred that today's ecologists might recognize as "competitive exclusion," a process that often leads to the "displacement," and even the extinction, of certain species. Of course, this chapter does not propose some sort of all-inclusive theory regarding the role of ecology, or even biology, in shaping our historical hatred for wolves, as the economics of ranching stands out, at least in the historical sources, as the primary motivator for wolf extermination; but certainly, considering the circumstances on Hokkaido, and given that economics is part of our ecology, a more holistic, ecological perspective is worth keeping in the back of our minds as we work our way

through this chapter. If humans around the globe share cognitive dispositions to classify certain life-form categories, as argued in chapter 1, then certainly it is possible that they also share some sort of propensity to, under
the right circumstances, exterminate economic competitors such as wolves.

"In the struggle for existence one variety or species must often squeeze
another," writes Jonathan Weiner in his award-winning study of evolutionary research on the Galápagos Archipelago. "Nearest neighbors, closest cousins, will pinch each other very hard, generation after generation.
They collide because they are so much alike in equipment, instincts, and
needs."[6] In some respects, wolves can be seen as our "nearest neighbors"
and "closest cousins." Humans and wolves, though certainly blessed with
different equipment, actually do share many habits, instincts, and needs,
and when forms of "competitive exclusion" between the two species took
place on Hokkaido in the arena of economics (and then combined with
Japan's social and political goals of modernization), wolves had little
chance for survival as hunters pursued them in the cold forests of northernmost Japan. With every mating pair of wolves killed and every den raided by
Japanese and Ainu hunters, another potential generation of wolves was lost
forever; in this way, by the late 1880s, Japanese had replaced the natural legacy
of the Hokkaido wolf with an artificial one, represented by a handful of
patchy, lifeless specimens exhibited in a few museums.

THE ECOLOGIES OF PROGRESS

In 1875, the Meiji intellectual Fukuzawa Yukichi, in *Bunmeiron no gairyaku*
(An outline of a theory of civilization), wrote that "an essential feature of
civilized progress lies in endeavoring to intensify and multiply human enterprises and needs."[7] When first pronounced by modernizers like Fukuzawa,
such statements, at least within the context of the early Meiji era, represented
prescriptions for Japan's future progress, a cultural therapy that could help
transform a backward Japan into a civilized and enlightened nation and allow
it to shed the "semicivilized" customs of its Confucian past. His philosophy contrasts sharply with Tanaka's.

Illustrating the shift in political philosophies in early Meiji thought, in
Bunmeiron no gairyaku, Fukuzawa divided societies into three historical
stages: primitive (*yaban*), semicivilized (*hankai*), and civilized (*bunmei*).[8]
Primitive man, Fukuzawa wrote, "cowers before the forces of nature [*ten-
nen*] and is dependent upon arbitrary human favor or accidental blessings."
That is to say, primitive man remains "unable to be master of his own sit-

uation." Fukuzawa then explained that the triumph of culture over nature signals the emergence of the semicivilized stage. Nonetheless, these semi-civilized societies, cultured as they are, become "slaves to custom" (much as Confucian societies do) and lack the liberating forces of full-blown civ-ilized societies. Fukuzawa argued that complete mastery over nature and culture is part of the emergence of civilization, a stage wherein "men sub-sume the things of the universe within a general structure." In other words, when Fukuzawa traced the trajectory of human societies toward the real-ization of a civilized condition, mastery over nature was central to his think-ing.[9] Once more, as first mentioned in chapter 3, it is in the invention of a "general structure"—in a rational historical order; in a view of nature as an inanimate object; and in a recognized "universal" taxonomic system—that the modern nation discovers its power to rule people and to legitimately exploit the landscape.

On Hokkaido, the intensification of Fukuzawa's notion of "human enterprises and needs," the highly civilized act of generating "blessings" of one's own, at first took the form of ranching and other kinds of modern agriculture. However, as seen in chapter 4, predators could cause losses at horse ranches by killing and eating livestock, and the economic impact of wolves, not to mention bears, wild dogs, and even crows, should not be underestimated. Fukuzawa anticipated, however, such conflicts with ani-mals in Japan's struggle to become a more civilized nation. Progress begins, he wrote, when humans no longer "struggle for survival with other animals" but instead transcend to complete mastery over the natural world.[10] Of course, the hope of establishing a mastery over nature is precisely what Tanaka Shōzō had cautioned his countrymen against; but, as we know, Fukuzawa emerged as the hero of Meiji intellectual life, not Tanaka, and in the Meiji years and beyond, Japan pursued an agenda of harnessing or exploit-ing its natural resources for the exclusive purpose of strengthening the state. In terms of economic values, Kaitakushi officials understood wild animals on Hokkaido's ranches to be assaulting the highly civilized world of scientific agriculture, and so the Kaitakushi, not wanting to cower before such nat-ural forces, positioned itself as discipliner of nature and protector of indus-try, and to those ends brutally hunted these predators. In ecological terms, wild animals assaulted the horse pastures in the first place because they had come to depend on horse flesh for nutrition following the deer-hunting cam-paigns and harsh weather of the 1870s, which created a dearth of traditional sources of food, forcing them to turn to the postal stations and private stables scattered along the Pacific side of the island. When compared with reliable depredation numbers from the United States (predation by coyotes as well as wolves), those suffered by horse ranchers on Hokkaido appear to be quite high but also highly localized.

Archival records representing a period of a little over one year (from January 1877 to May 1878) reveal that 366 horses died as a result of predators, weather, and other, unspecified reasons in southeastern Hokkaido (see table 3). In an area that included Usu, Muroran, Horobetsu, Yūfutsu, Shiraoi, Chitose, Saru, Niikappu, and Shizunai Districts, officials discovered that wolves and bears had killed no fewer than 136 horses at both state-run and privately owned stables and postal stations, while another 230 horses had died from deep snow and "other" causes. Obviously, horse ranchers had bigger concerns than predation, such as winter weather; and one reason wolves preyed on horses at Yūfutsu and Chitose appears to have been because the latter were weakened by poor forage. But predation rates were nonetheless high, and the infrastructural costs were as important as the economic costs: when wolves sank their teeth into the hindquarters of postal-service horses they disrupted communication lines and hindered the colonization of the northern island at a time before the telegraph was fully up and running on Hokkaido.

By comparison, in 1892, predation by coyotes and wolves accounted for economic losses in the U.S. sheep industry that amounted to 3–7 percent in New Mexico, 5 percent in Nebraska, and 10–25 percent in Texas.[11] Of course, Japanese in Hokkaido husbanded much smaller herds than their American counterparts, but that should have made their horses easier to watch over. Nonetheless, depredation percentages in Hokkaido as represented in table 3 remained far higher, generally speaking, than in much of the United States, for environmental reasons that were discussed in the previous chapter.

Interestingly, despite Hokkaido's high depredation rates, in most districts weather and "other" factors still actually outpaced mortality caused by predators. Of the large number of dead horses, most died from hard weather or other reasons, not wolves and bears. This is an important point because, to again use the United States as a comparison, in the western states weather was also the primary cause of livestock mortality, although wolves often paid the price. As historian Bruce Hampton observes, American ranchers constantly faced natural enemies such as weather and disease that, despite the best efforts of local governments and stock growers associations, left the

TABLE 3. Horse Depredation and Mortality Usu District to Shizunai District
January 1877–May 1878

Date	Location (privately-owned or state-owned) (total number of horses)	Depredation	Weather	Other	Total dead (percentage killed by predators)
1877.10–1878.5	Usu District (69)	48	12	9	69 (69%)
1877.1–1878.5	Abuta District (59)	17	35	7	59 (29%)
1877.10–1878.5	Five wards of Muroran District (12)	3	9	—	12 (25%)
1877.8–1878.5	Muroran port (state) (37)	4	4	13	21 (19%)
1877.8–1878.5	Muroran port (private) (198)	30	32	30	92 (33%)
1877.11–1878.5	Horobetsu postal stations (164)	1	11	2	14 (<1%)
1877.11–1878.5	Horobetsu settler union (13)	—	1	—	1 (0%)
1877.11–1878.5	Horobetsu permanent residents, settlers and Ainu (private) (606)	2	39	—	41 (<1%)
1877.11–1878.5	Horobetsu temporary residents (70)	—	3	—	3 (0%)
1877.11–1878.5	Other Horobetsu (state) (13)	7	6	—	13 (54%)
1878.1–1878.5	Yūfutsu District, Tomakomai Station (state) (4)	—	—	4*	4 (0%)
1878.1–1878.5	Yūfutsu District, Tomakomai Station (private) (3)	3	—	—	3 (100%)
1878.1–1878.5	Yūfutsu District, Yūfutsu Station (state) (5)	—	—	5*	5 (0%)
1878.1–1878.5	Shiraoi District, Shiraoi Station (state) (11)	11	—	—	11 (100%)
1878.1–1878.5	Shiraoi District, Shiraoi Station (private) (4)	4	—	—	4 (100%)
1878.1–1878.5	Chitose District, Shimamatsu Station (state) (6)	6	—	—	6 (100%)
1878.1–1878.5	Chitose District, Chitose Station (state) (5)	—	—	5*	5 (0%)
1877.10–1878.5	Saru District (3)**	—	—	3	3 (0%)
1877.10–1878.5	Niikappu District	—	—	—	0 (0%)
1877.10–1878.5	Shizunai District	—	—	—	0 (0%)
1877.1–1878.5	Usu District to Shizunai District	136	152	78	366

SOURCE: "Usu Shizunai ryōgun nōji keikyō narabi ni imin gaikyō chōsho no ken" (Matter of memorandum regarding agricultural conditions and general outlook of immigrants in Usu and Shizunai districts) [1878.5.23], in Honka todokeroku (Records of home division reports) (A4-53-73). Hokkaido Prefectural Archives, Sapporo, Japan. I am grateful to Tanabe Yasuichi for sharing this and other archival documents.

*The document states that horses were killed by wolves and bears or by a lack of sufficient forage. I have read this to imply that a lack of proper forage weakened horses during the winter and made them easy prey for wolves and bears.

**Inconsistent record keeping between districts makes it unclear whether the "total number of horses" listed refers to the total number of living and dead horses or just the total number of dead horses. In many districts, there are more "total number of horses" than dead horses, and so I assumed for the sake of this table that "total number of horses" refers to both living and dead ones. Regardless, this discrepancy does not affect the percentages of horses killed by predators, which is the main point of this table.

ranching public feeling powerless. Although disease and weather might have been beyond the domination strategies of these nineteenth-century interests, they felt they could do something about wolf predation.[12]

We can conclude, then, that the formulation of bounty systems in the United States and Japan was motivated by a basic need to stem economic losses and, in Japan, communications disruptions that hindered the development of Hokkaido; but it was also motivated by a need to effect some change on the natural world, to show that civilized nations do not cower before the forces of nature, as Fukuzawa had insisted they did not. In the case of Japan, weather and equine epizootics remained beyond the reach of the civilizing hand of the Meiji government and its science, but wolves, bears, and crows could be extirpated, through the efforts of Edwin Dun and his strychnine campaign, a bounty system, and hunting. And so, starting in the late 1870s and early 1880s, the Kaitakushi organized a bounty system aimed at ridding Japan's natural landscape of large and small predators alike. Viewed from either vantage point—an economic or an ecological one—the Kaitakushi needed to destroy wolves.

THE AMERICAN MODEL AND
WOLF KILLING IN THE MEIJI PERIOD

For more than two centuries, the Japanese settlement of the lower Oshima Peninsula in southern Hokkaido, in the place known as Wajinchi (literally, "Japanese land"), had situated people near wolves and bears, and a limited amount of local state-sanctioned predator killing had taken place. During the period around 1804–22, for example, bears and wolves often wandered near the homes of Japanese settlers, and to deal with the potential threat settlers acquired certificates allowing them to borrow firearms and shoot these animals.[13] Then during the 1870s, under the Kangyōka, a division of the Kaitakushi charged with promoting the industrial development of Hokkaido, the killing of wolves, bears, and crows became entwined with the Meiji economic and cultural priorities of progress and development.

The process of creating an official bounty system in Hokkaido began in 1875. That February, officials proposed a licensing system wherein settlers were to report wolves and bears to village heads, who then assembled hunters to dispatch the animals. Kaitakushi officials required hunters to obtain licenses and they punished unauthorized hunting. In March 1877, because of the damage crows caused at fisheries and farms, officials started offering a four *sen* bounty per crow to settlers who killed crows with either poison or firearms. Later that same month, they extended the bounty to include crow eggs. Then, in September, they broadened this bounty system even further to include wolves and bears. Under this system, hunters brought in the ears of either bears or wolves and received a ¥2 bounty per pair.[14] The yield from this program is unclear, but in January 1878, the Kaitakushi official Nakamura Morishige wrote to the central office in Sapporo that, despite such bounties, bears and wolves continued to attack both settlers and domesticated animals and thereby complicated "progress" and "economic development" on Hokkaido. Evidently spurred by the news, less than five months later, in May 1878, the Kaitakushi's central office drew up a new bounty system, one largely modeled after that of the United States.

The United States provided a fine model for killing predators. When European settlers arrived in the New World, they wasted no time in trying to eliminate the wolf of the eastern seaboard (generally known as *C. l. lycaon*, although now thought by some to be the same species as the red wolf, or *Canis rufus*). Most of the earliest settlers came from the British Isles, where the English had exterminated their wolves by about 1500, while hunters killed the last wolf in Scotland in 1743 and in Ireland in the 1770s.[15] The colonists carried the pathology of wolf eradication across the Atlantic Ocean, and killing wolves became part of reshaping the New World into a Neo-European landscape.[16] Later, in the context of nineteenth-century imperialism, cultural borrowing, and wolf killing, Japan too became a kind of Neo-European landscape, one shaped, not by the hands of colonists, but by the hands of Meiji modernizers and their imported vision of the nation.

The American colonial pursuit of wolf eradication involved bounties. Throughout the 1630s and 1640s the colonies of Massachusetts Bay, New Plymouth, Virginia, and Rhode Island passed legislation that rewarded those who killed wolves with payment in either shillings or livestock. In May 1645, at the Massachusetts Bay Colony, a committee was formed to "consider of the best ways and means to destroy the wolves which are such ravenous and cruel creatures." Whether generated by this wolf-killing committee or not, ideas on how to erase the wolf from New England ranged from populating towns with specially bred dogs "for the destruction of wolves" to forcing local Native Americans to pay tribute in the form of wolf heads. In these early years, constables also often employed professional wolfers, such as a certain Roger Williams of Rhode Island. Paid thirty shillings per wolf head (money collected from local farmers, who paid "in proportion to their number of cattle"), Williams led a "grand hunt" to "extirpate" wolves from the colony of Rhode Island. He appears to have been only moderately successful, however, as wolves "continued to be a source of annoyance." Similarly, as early as 1648, New York colonists had initiated a bounty system so that "the breed of Wolves may be whooly rooted out and Extinguished."[17] As settlement spread westward, white settlers saw wolves as cattle-rustling outlaws who, as in a proposed 1856 amendment to Iowa's bounty law, "feloniously, maliciously and unlawfully, attack with intent to kill."[18] Civilized, law-abiding, and utterly disciplined societies required that these outlaw wolves be eradicated, as they openly threatened economic progress; but wolf cruelty to prey—the fact that they "attack with intent to kill"—also offended the sensitivities of law-abiding white people.

In the nineteenth century, with the emergence of complex new legal regimens, the killing of outlaw wolves became urgent, and it rose to new, and even more pathological, heights. By the 1870s and 1880s, cattle had largely replaced bison in the American West, and ranchers attributed some 5 percent of cattle losses to wolves (while they attributed around 8 percent to Native American raids and weather, somewhat similar to the situation on Hokkaido described above). As Hampton notes, although cattle ranchers could do little about weather, disease, poisonous plants, high taxes, shipping costs, and fluctuating livestock markets, they felt that they could do something about wolves. At first, individual ranchers paid wolfers to eliminate predators. Later, however, after the formation of powerful stock growers associations, membership dues covered wolf-killing costs. These nonstate bounty practices only became state ones when stock growers, pointing to the fact that they owned the most land and hence paid the most taxes, argued that county governments should pick up the tab for destroying wolves. Iowa initiated a bounty law in 1838, and other western states, including Colorado, Wyoming, the Dakotas, and Montana, followed suit. Wolf bounties ranged from a low of $5 in Wyoming in 1890 to as much as $400 in Wyoming and Colorado by 1910. In 1886, Montana wolfers severed the ears of wolves to collect between $9 and $18 a pair. With high prices put on their heads, these outlaw wolves were basically eradicated from the American West.[19]

Theodore Palmer, of the U.S. Bureau of Biological Survey, admitted that dealing with "injurious" animals such as wolves required "radical measures for their extermination," and bounties were one technique in the "warfare against predatory animals."[20] In Montana (and even in Yellowstone National Park) such radical measures included steel traps, rifles, strychnine, dynamite, fire, hunting dogs, biological agents such as mange, and the technique of "denning" (i.e., pulling pups from a den site with wire and hooks and then strangling them or using their whining cries to lure the parents within shooting range). Montana records reveal that between 1883 and 1918 alone, the severed body parts of over 80,000 wolves were turned in for bounties. Over all, in the twenty-five years before 1896, Palmer estimated that $3,000,000 had been spent in the United States by counties and states on the bounty system.[21] Not only at the county and state level, but in 1914–15 the federal government, under the Bureau of Biological Survey, became actively involved with wolf killing, placing more pressure on a canine that already was nearly gone from the western states. By a congressional act, the federal government made money available for "experiments and demonstrations in destroying wolves, prairie dogs, and other animals injurious to agriculture and animal husbandry."[22] During a year of operations, federal agents of the Survey killed 424 wolves and countless other predators. By 1941, when the U.S. government turned its attention to a new enemy, the Japanese, some 24,132 wolves had been killed by its agents.[23] The increase in the kill rate was partly due to the fact that in 1921 the Survey had begun the mass production of strychnine baits. A few isolated voices cried out to end the killing—typically voices from "university people," who, as Senator Thomas Kendrick observed, were "uninformed."[24] However, with the mass production of strychnine baits, and with public lands crawling with federal agents, Edward Goldman, the chief scientist of the Survey, could declare that wolves "no longer have a place in our advancing civilization."[25]

Neither did they in Japan's advancing civilization. In Hakodate, in February 1878, Japanese officials proposed an even more radical version of the bounty system, with wolf bounties as high as ¥15 and standardized exchange rates for bear, wolf, and otter pelts. As the length and number of documents reveal, officials spent a great deal of time and reams of paper formulating this bounty system, including discussions concerning the relative worth of wolves and bears. The decision ultimately made in Sapporo to set bounties for wolves higher than for bears was motivated by two lines of thinking. First, officials explained that bears were bigger and hence might be seen as being worth more, but Ainu hunters already had economic incentive to kill bears because bear rugs and pelts generated money (anywhere from ¥2 to ¥6, depending on their quality) while the meat was sold and eaten. By contrast, officials explained, wolves had no such market value and were not eaten, and therefore, without high bounties, Ainu hunters in particular would have no interest in killing them. Second, as discussed in chapter 2, Ainu hunters from the Hidaka region believed that they descended from wolves, and hence they may have regarded wolf killing as a kind of mythological patricide.[26] In a sense, one could argue, Kaitakushi officials used the bounty system to induce Ainu hunters to destroy their ancestral gods in favor of what would become in a matter of decades a new Japanese one under the ideology of the emperor system. Officials treated Ainu as imperial subjects in the context of the bounty system, and killing wolves, crows, and bears became, in the eyes of Kaitakushi officials, a sign of their commitment to the new Meiji order.

In May 1878, Nakamura Morishige, evoking the earlier nineteenth-century American experience as a model for wolf killing in Japan, pointed out that in Europe and the United States, authorities spent "big money" protecting settlers from the economic threats and physical dangers posed by wolves and bears. If the Kaitakushi too invested "big money" to protect settlers, argued Nakamura, it would contribute greatly to the prosperity of Hokkaido. So in May 1878, officials proposed a new uniform bounty plan for the entire island of Hokkaido similar to the early modern Japanese bounty plans mentioned in chapter 3 and the American ones mentioned above. The first measure raised the bounty for wolves to ¥7 and the bounty for bears to ¥5; officials later added wolf pups at ¥3 and bear cubs at ¥2. Another measure modified the method of collecting bounties. Professional hunters, of whom Ainu constituted a substantial portion, relied on the marketability of animal pelts to make their living, and so rather than cut off the ears as evidence of the kill (a process which, officials explained, made the animals less attractive to rug and pelt hawkers), hunters cut off the four paws instead. Hunters then brought the paws to the office of the village head, and he provided a stamp of authenticity. The hunter received payment from either a branch office of the Kaitakushi or a local ward office.[27] In January 1880, the Kaitakushi fine-tuned the bounty system, deciding that in the case of wolves, bears, and adult crows, the four paws or two legs should be brought to provincial, ward, or local offices, where officials would cut the limbs into pieces and then have the hunter bury them. In the case of crow chicks and crow eggs, officials asked hunters to bring in the entire carcass or egg for bounty collection.[28]

The 1878 bounty system, in which awards were offered as "imperial gifts" (*kashi*), continued through the 1880s, contributing to one of the more lurid, but less well known, chapters in the history of modern Japan (see table 4 and the appendix for bounty numbers).

KILLING CROWS

Given what biologists understand about crows and ravens, it should not be surprising that Kaitakushi officials included crows in the bounty system. It should also not be surprising that crows survived the assault, becoming, in the words of Ōta Shinya, a wild-bird expert, "kings of the city streets."[29] Two subspecies of crow live on the Japanese Archipelago, the common crow (Japanese, *hashibosogarasu; Corvus corone*) and the significantly larger, thick-billed, deep-voiced, more raven-like jungle crow (Japanese, *hashibuto-garasu; Corvus macrorhynchos*). Members of the family Corvidae, crows are intelligent birds that have a knack for adapting quickly to new situations. Highly curious, crows learn to eat nearly anything, including agricultural crops and herring, which qualified them in Japan as "noxious animals" (*yūgai dōbutsu*), ones that competed with human society in the Meiji ecological order. Like wolves and ravens, jungle crows, in particular, organize themselves into what University of Vermont biologist Bernd Heinrich has identified among ravens as a "dominance hierarchy." If we accept ecologist Imanishi Kinji's definition of culture as "socially transmitted adjustable behavior," then some corvids too possess culture of a sort, such as the ability to learn how to hunt.[30] Indeed, Eytan Avital and Eva Jablonka argue that intelligent animals such as crows "learn from others how to behave."[31] In the wintry forests of Maine, Heinrich actually taught his young ravens to hunt, and he noted that they readily modified this behavior. Following lengthy experiments, he concluded that ravens "evaluate and make choices."[32]

No doubt the crows of Hokkaido made similar evaluations and chose to live with wolves and other hunters, including bears and Ainu. Crows and ravens can be attracted to hunters, not simply prey, and so crows likely followed hunting wolves on Hokkaido. For crows, such strategies were central to their survival because even the largest crow cannot open a carcass, particularly if it is frozen. In Yellowstone, ravens are nearly always present at the time of the kill, flying over the spectacle of the chase like eager spec-

TABLE 4. Bounties for Noxious Animals *(yūgai chōjū)*
Killed on Hokkaido, 1878–1881

	Bear	Wolf	Wild dog	Crow	Crow eggs	Total
1878						
Number	217	38	496	11,342	2,038	14,131
Bounty	¥761.000	266.000	¥24.800	¥453.760	¥81.520	¥1,587.080
1879						
Number	419	59	257	15,099	8,823	24,657
Bounty	¥2,089.000	¥413.000	¥12.850	¥603.960	¥352.920	¥3,471.730
1880						
Number	607	76	457	14,152	7,060	22,352
Bounty	¥3,035.000	¥532.000	¥22.850	¥565.800	¥282.400	¥4,438.050
1881						
Number	717	121	461	15,176	3,699	20,174
Bounty	¥3,579.000	¥827.000	¥23.050	¥607.040	¥147.960	¥5,184.050
Total	1,960	294	1,671	55,769	21,620	81,314
Total bounty	¥9,464.000	¥2,038.000	¥83.550	¥2,230.560	¥864.800	¥14,680.910

SOURCE: Ōkurashō. ed., *Kaitakushi jigyō hōkoku* (Report on Kaitakushi industries), vol. 2 (Sapporo: Hokkaidō Shuppan Kikaku Sentā, 1885), 486.

tators. They follow wolves during the hunt because they anticipate being fed. Some biologists and hunters have even gone so far as to suggest that ravens identify prey for wolves and other hunters, squawking loudly to alert the predators of nearby herds of elk or deer. Heinrich, during his studies at Yellowstone, concluded that given a choice, ravens actually choose to be with wolves, leading him to characterize the corvids as "wolf-birds." "Maybe they had evolved with wolves in a mutualism that is millions of years old," he speculated, "so that they have innate behaviors that link them to wolves, making them [ravens] uncomfortable without their presence."[33] Whether or not the Hokkaido crow-wolf relationship was truly symbiotic, surely the loss of wolves (and later prohibitions aimed at Ainu hunting) severely reduced their access to food, encouraging them to turn to the fast-growing population of human providers, with their crops and livestock. Once crows began to form an ecological attachment with the nineteenth-century human inhabitants of Hokkaido, they came to compete with the Kaitakushi's plans for economic development, and bounty hunters killed tens of thousands of crows in the early Meiji years. Kaitakushi sources reveal that between 1878 and 1881 alone, settlers and hunters turned in no fewer than 55,769 dead crows and 21,620 crow eggs.

JAPAN'S WOLFERS

Hokkaido's crow and wolf hunters left only traces of their activities in the form of bounty records buried deep in archives. Records from between February 1880 and June 1881, for example, tell of Hayashi Kuninosuke collecting ¥28 for killing four wolves, Murakami Yōsuke collecting ¥21 for killing three wolves, and Higuchi Jūkichi collecting ¥12 for one bear and one wolf.[34] The use of firearms was one state-sanctioned method for their work. As early as January 1874, even before the organization of the formal bounty system, the Noboribetsu ranch, as if preparing for battle, made a request for several pounds of gunpowder and 200 percussion caps for the purpose of killing wolves and bears that ate livestock there.[35] Four years later, Ōshima Kunitarō, a local official from Mitsuishi, made a lengthy request for firearms to shoot wolves. Also, three Ainu from near the Kerimai River (Iraman, Igurushugi, and Yakichi) hoped to get their hands on the guns and licenses required to kill wolves.[36] Indeed, documents of this sort suggest that Ainu made up a substantial proportion of the wolfers on Hokkaido.

Killing crows, wolves, and bears in the name of the Meiji emperor could get confusing, however, as a report from August 1878 reveals. Early that year, Iida Ichigorō, of the postal service, and forty-two other hunters discovered a wolf den and killed the four pups inside. As per rules of the first bounty system, they received ¥5 for each of the pups killed. Kaitakushi officials expressed concern, however, because as of May, the wolf bounty had been raised to ¥7, and so it appeared that they had underpaid Iida and his hunters. Kaitakushi officials began questioning whether a wolf pup was worth as much as an adult wolf, eventually settling on a bounty of ¥3 per wolf pup.[37]

In dealing with such questions, the administration of the bounty system became even more bizarre by May 1880. When Ega Jūjirō, a Japanese wolfer from Oshamanbe, brought in four severed limbs from a wolf pup, local officials became suspicious, and so they asked Ega to bring in the entire carcass for inspection. When he did, officials discovered to their morbid amazement that the limbs Ega had cut came from a wolf fetus, which, officials explained, looked more like a mouse than a wolf.[38] Consequently, this incident raised serious questions about whether the wolf bounty system extended to fetuses. Some wolfers were not troubled by such matters, however. State-sponsored hunters like Mikami Hidetsuna and his party were funded by the Kaitakushi to hunt wolves. He sought out wolf dens, hid nearby, shot the adult wolves as they provided food for their young, took the pups from the den, brought them to a Kaitakushi office, and collected his bounty. The Kaitakushi paid Mikami eighty *sen* per day; his assistants received forty-two *sen*. Denning could be profitable for such men.

JAPAN'S ADVANCING CIVILIZATION

One set of Mikami's victims sheds light on the other ways that wolf killing was useful to Japan's advancing civilization. In May 1878, he and his hunters discovered a wolf den in the Komagatake Mountains near Hakodate. After hiding nearby, they eventually shot one of the adults and then took the seven pups from the den site alive.[39] The fate of these seven pups and one adult is worth pursuing briefly. Although the archival trail is extremely hard to follow, it appears that Kaitakushi officials decided to keep three pups in a cage for luring adult wolves during future hunts; as any seasoned wolfer could tell you, whining pups always manage to attract adults within shooting range. (This is because wolves raise pups communally in a practice called "allo-parenting," probably because of learned behavior and prompted by increased amounts of prolactin in their bodies.) Kaitakushi officials ordered that the remaining four be sent alive to Tokyo. Taxidermist Watanabe Shōzō pre-served the mother wolf and sent it to an unspecified museum. Earlier that year, in January, officials had expressed an interest in sending live wolf pups, captured at Sahara Village near Hakodate, to a Tokyo museum (*haku-butsujo*).[40] In May, four months later, officials argued that a museum served as an excellent location for the stuffed wolf specimen because it would "pro-mote knowledge among the citizenry" and thereby aid Japan's advancing civilization; but they also insisted that a museum was not a place for keep-ing live animals, and so the suggestion that a live wolf (perhaps one of the seven pups) be sent to the museum should be scrapped and the wolf imme-diately killed for reasons of public safety.[41]

It is not surprising that officials hired taxidermist Watanabe Shōzō to preserve the mother wolf and that it was placed in a museum to "promote knowledge among the citizenry." They selected other wolves for museums as well, such as one shot in the Nemuro area in 1881, although this wolf was lost at sea when the *Hakuhō-maru* overturned while transporting it to Hakodate.[42] Taxidermy is important in this context, as it is a technique of "effecting meanings," writes historian of science Donna Haraway. In her investigation of Carl Akeley's African Hall at the American Museum of Natural History, Haraway argues that in the context of the museum, taxidermy becomes "the servant of the 'real.'" Rather than being seen as art or the act of creation that it is, taxidermy presents animals within the "epistemological and aesthetic stance of realism." She continues, "what is so painfully constructed appears effortlessly, spontaneously found, discovered, simply there if one will only look." What the "stance of realism" reveals is "the single story" and "nature's unity," the natural hierarchy and order wherein humans stand comfortably on top while other creatures, such as wolves, find their "natural place" in the museum display.[43] Ensconced in museums, and sold to the public as natural knowledge, such wolves evidenced where Japanese stood in the greater natural dominance hierarchy of civilized nations.

Besides discussions concerning taxidermy, archival documents tell of proposals to send live wolves to Tokyo for exhibition at a "provisional museum." In June 1877, for example, officials planned to send two wolf pups captured near Esashi, at Otobe Village, to Tokyo aboard the *Hakodate-maru*.[44] As of June 16, 1877, Ichiki Masatane was caring for one of the pups in Hakodate (feeding it "raw fish"), while the Kaitakushi proposed buying the second to send to the museum.[45] Documents reveal an outbreak of official bickering two days later, however. Officials could not decide who should fund specialist Miyamura Ichitarō to care for the wolves en route to Tokyo, not to mention who should fund their shipment, and it remains unknown whether the wolves ever made it to Tokyo.[46]

The Tokyo provisional museum referred to in Kaitakushi documents was probably what would later become the Ueno Zoological Garden, the brainchild of Tanaka Yoshio, the zoologist mentioned in chapter 1. Born the son of a doctor in Nagano Prefecture, Tanaka had studied under the Western Learning scholar Itō Keisuke, who, as you may remember, was the father of the Linnaean classification system in Japan. In 1861, at the age of twenty-three, Tanaka traveled to Edo, where he served at the Bansho Shirabesho (later the Yōsho Shirabesho), a shogunal school devoted to deciphering Western texts and technology. There he helped develop a commerce (*bussangaku*) department of sorts. In 1867 he attended the World's Fair in Paris, where he visited the Musée d'histoire natural and the Jardin des plantes, developing an interest in Western culture and museums. He later held positions at the Tokyo Nanko (an early version of Tokyo University), where he was well placed to oversee the Koishigawa Medicinal Gardens and the Uchiyama Shimomachi Museum (a predecessor to the Ueno Zoological

FIG 28. Taxidermically preserved wolf at the Ueno Zoological Garden, Tokyo. Courtesy of the Tokyo Zoological Park Society, Tokyo, Japan. I am grateful to Ian Miller for locating and reproducing the image.

Garden). Like European menageries of the nineteenth century, the Uchiyama Shimomachi Museum, built across the street from the Tokyo Imperial Hotel, boasted many exotic sights, such as a bear room, a taxidermy room, and a greenhouse. As historian Harriet Ritvo suggests, captive animals held at such menageries served as "emblems of human mastery over the natural world."[47] Although a brown bear captured on Iturup Island (and sent to the museum by the Kaitakushi in 1878) was among the animals that citizens could gawk at as they wandered the museum, no records of Hokkaido wolves exist in documents related to the museum or the later Ueno Zoological Garden. Records do reveal, however, that a Japanese wolf (*C. l. hodophilax*) was exhibited at the zoo in June 1878, and one from Iwate Prefecture was exhibited between 1888 and 1892, a transfer from the Tokyo Educational Museum.[48] Moreover, photographs of a taxidermically preserved Japanese wolf survive in the Ueno Zoo archives, apparently the same specimen now housed at the National Science Museum (see fig. 28). Only an obscure reference to a dog (a gift from a Hakodate merchant) stands out in early documents related to the zoo.[49]

In light of these findings, one suspects that the pups captured by Mikami died while at Hakodate or during later wolf hunts. On June 8, 1878, officials in Hakodate argued that all the pups should have been set aside for shipment to Tokyo, considering how hard it was to capture wolves and that while awaiting transport some had already died.[50] Indeed, wild wolves were prone to sickness while in captivity. Previously, a female wolf kept at the Kaitakushi "provisional museum" had contracted a disease in October 1877, and officials called in foreign veterinarians to help care for the animal. In July 1879, moreover, despite the efforts of a veterinarian from the military horse division, a different wolf died of disease at the same museum.[51] While officials drained Kaitakushi coffers to fund wolf eradication in the wild with poison, guns, and traps, they spent other resources to keep wolves both stuffed and alive within the safe confines of museums and zoos. Victors in war do love their trophies, whatever their form.

THE NATURAL ECONOMIES OF WOLVES

Some scholars argue that among those animals we tend to hunt most relentlessly and kill most brutally are ones that, in one way or another, closely resemble ourselves. Obviously, such a hypothesis fails to explain the extinction of the passenger pigeon and the near extinction of the bison, but it is helpful for understanding, at least in part, the economy and ecology of wolf killing. For instance, Haraway explains that what made large primates targets in the eyes of nineteenth- and early-twentieth-century hunters was their "similarity to man," who was the "ultimate quarry, a worthy opponent." In other words, "the ideal quarry is the 'other,' the natural self."[52] Wolf hunters saw wolves as our "natural self" as well; they saw them as, socially and economically speaking, much like human beings. As just one example, after years of hunting, U.S. federal agent Bert Hegewa mused about the irony of wolf killing. "They're just the knowin'est brutes on four legs," he reportedly said. "Sometimes they're almost human."[53]

Charles Elton, an ecologist investigated in the next chapter, insisted that human societies and economies need to be seen as ecological in nature, as important parts of regional environments and not as distinct from nature. So, in the spirit of Elton, let us turn the table and say that wolves might be seen as social and economical creatures, as part of colonial financial systems and not distinct from our modern order. To this point in our discussion, we have spent two chapters talking about the Japanese economic activities on Hokkaido that brought them into conflict with wolves. But what was it about wolf society and economy that brought them into conflict with Japanese ranchers? It turns out that Meiji modernity not only forced changes on most Japanese but forced changes on wolves and other animals as well. What this means is that historical watersheds, though often political, social, and economic in nature, reverberate throughout the natural world as well, often in ways that modify the behavior of all living things; because wolves adapt to their surrounding ecologies, in a sense they too experience history and they too can modernize. For this reason, a wolf's-eye view should help us see the modern enterprises of horse farming, industrial deer hunting, and bounty systems—all inventions of human "artifice"—as "natural," much as with the forms of "artificial selection" discussed in chapter 1.

Competition at horse farms on Hokkaido was the product of modern wolf economies as well as Japanese ones.[54] Scholars know little about Hokkaido wolves because hunters exterminated them before naturalists took an active interest in the subspecies. Their numbers, however, appear to have remained limited for a variety of reasons. Kadosaki Masaaki and Seki Hideshi have surveyed the earliest Japanese manuscripts related to Hokkaido (called Ezo or Ezochi before the Meiji Restoration) for references to wildlife. Of forty-two documents examined, only six make reference to the Hokkaido wolf. This is probably because most early catalogs of the wildlife on Hokkaido focused on trade products, among which the wolf was not included. Being intelligent and shy animals, moreover, and certainly under no pressure to venture near Hokkaido's human populations, wolves tended to avoid people, and so Ainu and Japanese would have encountered them infrequently.[55] In 1781, Matsumae Hironaga, in his highly detailed *Matsumaeshi* (Matsumae record), provided one of these six references to wolves. He explained that wolves had crossed over from areas in the north, and that they lived in the mountain recesses of the northwestern part of the island, where they periodically caused injury.[56] Early records such as this one suggest that prior to the advent of modern agriculture on Hokkaido, real and visible economic competition between the hominid and lupine species was minimal, and so wolves, though no doubt killed on occasion, were not regarded as noxious enemies of progress. The Hokkaido wolf, because preindustrial Japanese barely knew it and Ainu hunted alongside it and worshiped it, remained a mountain deity that prowled the forests until the late 1870s and 1880s.

For subsistence, Hokkaido wolves ate a variety of land and marine animal life. In 1856, for example, Kubota Shizō wrote that a beached whale near Teshio, in far northwestern Hokkaido, had wolf and bear tracks surrounding it. At night, the predators came to feed on the decaying carcass.[57] In 1863, Captain H. C. St. John, while surveying the island's coastline, observed wolves feeding in the Nemuro area in eastern Hokkaido. Describing the massive number of herring that migrated to and from Hokkaido, he wrote that in "May and June this fish appears in incredible numbers. The straits, bays, and creeks appear alive with them. They are in such numbers, that those nearest the beach are pushed out of the water, and the shore for miles is thus kept constantly replenished with fresh fish; countless numbers of seagulls, eagles, crows, besides foxes, wolves, and bears, find an ever ready meal."[58] In other words, Hokkaido wolves could feed on a rich harvest of seasonal marine life as well as a fairly constant winter supply of deer. Until Japanese overexploitation of herring led to the fishery's near collapse by the end of the nineteenth century, wolves and humans had little reason to compete for that fish, either.[59]

Biologists believe that wolf relations with prey species such as deer (or carrion such as beached whales and herring) and social distance (the proximity of one wolf to another) shape pack organization and territoriality. Wolves are intelligent animals who boast what a chorus of scientists call "complicated purposive behavior" and "high order behavior," suggesting that they are capable of "abstractness, flexibility, complexity, foresight, mental representation, and insight into rudimentary means-ends relationships."[60] Much like humans, these canines construct rigid hierarchical group structures and a strong sense of group cohesion inside their territories. Indeed, wolf notions of territoriality are so powerful that they sometimes kill outsiders to defend it. They also implement strategies for maintaining group cohesion and territorial borders, mutually accepted forms of social ritual, and the basic social morality necessary to care for young and maintain mutual existence.

L. David Mech and Luigi Boitani, prominent wolf biologists, explain that the "basic social unit of a wolf population is the mated pair."[61] The mated pair forms the core of the pack, or a "group of individual wolves traveling, hunting, feeding, and resting together in a loose association, with bonds of attachment among all individuals." These social bonds can become so powerful that people see wolves more often together than individually, and these packs can consist of between two and forty-two animals (as will be discussed in chapter 6, in the winter of 2000, the Druid pack of Yellowstone numbered around thirty). Complex relationships among the smallest number of wolves required to locate and kill prey, the size of prey, the largest number of wolves that can feed on that prey, the number of individuals among whom social bonds can be formed, the amount of social competition that individual members can accept, and rates of pup and wolf mortality define pack size. Packs take a long time to form and prove resilient because, in essence, they are a kind of nursery: it takes several years to raise wolf pups. The psychological bonds among the members of a wolf pack, explains Mech, and the forms in which wolves openly express them are one reason it takes so long for packs to form and why the killing of pack members, such as the elimination of the alpha male or female, can be so disruptive. Wolves have "strong filial bonds," moreover, and for this reason packs often consist of "groups of related individuals."[62] Adolph Murie, in his early field research on the Denali wolves, concluded that "different wolf families seemed to have rather definite year-round home ranges which overlapped somewhat," meaning that oftentimes an "intolerance to strange wolves exists."[63]

Social bonds, continues Mech, take the form of a strict system of order; but perhaps more importantly, a "system of communication that promotes that order" provides the glue that bonds members together.[64] Scientists call this social hierarchy a "dominance order" (similar to the hierarchy among ravens mentioned earlier in this chapter), with an alpha male and female serving as the "basic element" and often as the core breeding pair of the pack. The alpha pair generally eat before the other members, for example, and often, but certainly not always, reproduction is confined to this couple.

As an example of wolves abiding by, but also rejecting, such "normal" lupine behavior as described by Mech and others, in the spring of 2000 Wolf 42, the former beta female of the Druid pack of Yellowstone, became alpha female after the pack ousted her sister, Wolf 40, in a coup d'état probably led by Wolf 42. What made the ouster so dramatic for Yellowstone Wolf Project Winter Study observers was that Wolf 40 had become notorious for abusing her beta sister, whose daily groveling got her nowhere. One morning, Wolf Project scientists discovered Wolf 40 dying in a ditch in the Lamar Valley from wounds she sustained during the coup. Illustrating the sustained cohesion of the pack, however, the other members raised the pups of the dead Wolf 40, and by winter, they too cruised effortlessly over the snow with the other Druids during elk hunts. At this time, the Druid wolves numbered nearly thirty animals, a fact no doubt made possible by the large number of elk in Yellowstone (hence providing a stable source of nutrition) and the close psychological bonds formed by the pack members (because most were related to each other).

Not just in Yellowstone, but wolves around the globe care for pups as a pack in the practice of alloparenting mentioned earlier. Murie was the first scientist to observe wolves at their dens in the wild. He noted of one pack in Denali that the members were "all concerned over the welfare of the pups."[65] This social care for young and the fact that wolves seldom use force to take food away from their pups have led some to speculate on "a primitive sense of morality in the wolf, one that might be expected to develop among social carnivores."[66] Though mutually caring, adults also negotiate dominance during what can be very violent fights and affirm it through a complex order of vocalizations, social behavior, and body language. These social messages might be likened to rituals. Prior to a hunt or at the conclusion of a hunt, for example, wolf packs commonly rally as a group and howl. Little is known about howling, admit biologists, but it might be an attempt to regroup following a prolonged hunt, an "advertisement of territory," an expression of anxiety, or, more likely, a mood-synchronizing activity, or what in the human world is called "ritual."[67]

In November 2000, I witnessed such mood-synchronizing activity when twenty-seven Druids howled near Jasper Creek in the Lamar Valley. I had first caught a glimpse of them around the headwaters of Crystal Creek, and they appeared to be heading for the Chalcedony and Opal area (near their rendezvous point), possibly to the south-facing slope of Norris Mountain to look for elk. What stands out in my mind was how howling and the accompanying "group rally" spread mutual good feelings among the wolves and served to strengthen social bonds, a fact made obvious by the tail wagging, muzzle licking, and playing that went on for several minutes. To say that such rituals spread "mutual good feelings" among wolves will, no doubt, lead some to caution against anthropomorphism. However, many of the basic hormones that blend together to contribute to human actions and feelings, such as oxytocin, epinephrine, serotonin, and testosterone, are also found in nonhuman animals. Similarly, the physiological pathways for emotions are among the phylogenetically oldest parts of the brain—the limbic system—which animals share with humans, leading Jeffrey Moussaieff Masson and Susan McCarthy to conclude, "From a purely physical standpoint, it would be a biological miracle if humans were the only animals to feel." If we rely exclusively on instinctual behavior, without taking into account at some level the emotional lives of wolves, we risk misunderstanding the motivation behind wolf behavior.[68] Indeed, animals such as wolves possess all the faculties required for complex emotional lives, and the Druid wolves, speaking with what scientists call a metalanguage—by playing and closely interacting with one another and howling—made this fact more than obvious as I stood in the Lamar Valley on that cold winter day. I could not help but wonder what Hokkaido's wolves had tried to say when they howled so close to Japanese villages, as reported by Nadokoro Seikichi. Among other reasons, their howling could have been an expression of anxiety or an advertisement of territory.

Wolves such as these howling Druids perceive their territories through an elaborate sensory array, and they appear to define territory in relation to the protection of hunting ranges and denning sites. Mech explains that wolves, like other animals, need "certain minimum distance between themselves," and because wolves always move, and have far-flung geographies of subsistence, wolf notions of territory exist both in space and in time, which biologists call spatiotemporal territory. Wolves map out the boundaries of these constantly fluctuating territories not only through eyesight but through auditory and olfactory senses as well. Such territories are important because it is within these realms, which are internalized through a complex web of visual, auditory, and olfactory input, that wolves hunt and interact with the world around them.[69] One might speculate that lupine perceptions of territory are at least as complex as the modern hominid's symbolic perceptions of it.

Because of their intelligence, the complexity of their pack life, and their predatory habits, wolves can easily be viewed as ecologically competing with humans. The result of this competition, however, at least as it is documented in historical sources, takes the form of economic losses rather than ecological ones, because that is how Meiji Japanese, as industrial people in a colonial context, perceived their interaction with and exploitation of nonhuman beings. The cattle, sheep, and horses that wolves no doubt viewed as living prey—with large eyes that carefully watched for the slightest movement in nearby stands of trees; with ears that twitched and listened; with nostrils that flared and snorted; and with hooves that kicked ever so sharply—were viewed by Japanese (and industrial humans in general) as living stock, or "livestock," economic investments on the hoof that they raised, rode, or, in the case of cattle and sheep, slaughtered and ate. Of course, the historical records from Japan and the American West only hint at forms of ecological competition, because what was ecological competition for wolves was economic for ranchers. Nonetheless, when wolves dragged down horses within the pastures of Niikappu, providing food for themselves and their young, they cut into the revenue of ranchers and the ranchers' basic ability to feed themselves and their families. Simply, in modern Japan, wolves really did threaten the ranchers' ability to survive.

When Edwin Dun and Kaitakushi officials claimed that wolves threatened the viability of horse ranching at Niikappu, they did not mean that wolves actually threatened to somehow displace people from Japan's northernmost island: to expel them in some apocalyptic Orwellian moment, an eruption outside the barn this time. Rather, wolves threatened the economic solvency of modern agriculture, which was, at least in part, the foundation of Japan's colonial enterprise on Hokkaido. Japanese survival on Hokkaido was dependent on making colonialism work economically: their ability to resist Russian expansion relied on it. When wolves threatened the economics of ranching—which, as we have seen again and again in the last two chapters, they absolutely did—they threatened Japan's burgeoning nation, as it existed on the newly acquired Hokkaido. Of course, the Japanese eradication of wolves on Hokkaido was motivated by financial concerns, but principally because we, as industrial beings, project our interaction with the natural environment, both past and present, through the lens of economics. Obviously, by creating categories such as "economy" and "ecology" we have tricked ourselves into thinking that the world can actually be contained in such quaint boxes, but the enduring lesson of environmental history is that it cannot be.

In other words, the struggles at the horse farms of Hokkaido, as depicted in this chapter, although expressed in economic terms, were also ecological and biological in nature, as both Japanese and wolves came to compete over the same large ungulates for survival. Once more, modernity reconfigured both human and lupine subsistence and behavior. But all this speculation raises intriguing questions regarding the specific nature of the historical interaction of wolf economies and ranch ecologies. Rather than try to answer these complex questions here, let us turn in the next chapter to a few Japanese ecologists who wrote on wolf extinction and the intermixing of human and wolf ecologies in the context of Japanese history.

CONCLUSION

The process of creating and killing the wolves of Japan was both cultural and ecological. Ecologically speaking, the wolves of Japan were created when they dispersed from Siberia to the landmass that, over geologic time, became the Japanese Archipelago, where they likely underwent a process of evolutionary dwarfing. But later Japanese taxonomists, born on that same archipelago, interpreted this evolutionary process through a variety of culturally ground lenses, ranging from Neo-Confucianism to Darwinism. In terms of their behavior, Japan's wolves really did chase, kill, and eat the deer and wild boars that roamed in grain fields throughout Japan. For such behavior, Japanese farmers anointed them deities—gods with "large mouths"—when in fact they killed and ate other beings, not because they were deities, but because they were hungry or had to feed their families.

Sometimes, as seen in chapter 3, these wolves contracted the rabies virus or other diseases and lost complete control of their neurological functions; they went crazy with the "mad dog disease," and over the centuries Japanese fashioned legends of demon wolves. Only an analysis of the epidemiology and symptomatology of disease and the creative experiences of making legends explain the myths and realities of Japan's wolves. Finally, Hokkaido wolves attacked, killed, and ate horses for ecological reasons—because the modern world they lived in presented declining deer populations, harsh winters, and the introduction of livestock to Hokkaido and because they were hungry, one of the most basic biological motivators of all. But just because we, as humans, see raising livestock and eating it as an economic activity does not mean that, in the eyes of every other living creature on this planet, it does not conform to the same biological logic that guides all other life on earth. Only in our ill-informed conceit do we think otherwise.

To be sure, as discussed in earlier chapters, cultural explanations remain the primary part of our story; but such explanations do not, on their own, identify the roots of historical agency: agency of both people and wolves as they interacted with one another in an increasingly limited amount of space. However, just as I have presented an economic and ecological perspective on wolf extinction, one that views competition between humans and wolves as fundamentally "natural," so too did Japanese ecologists as early as the 1930s and 1940s. The last chapter in this book explores this important body of scholarship as it relates to the development of Japan's discipline of ecology.

6

WOLF EXTINCTION THEORIES
AND THE BIRTH OF JAPAN'S
DISCIPLINE OF ECOLOGY

Today, the word "ecology" refers to the study of the interaction between a living thing (or things) and its environment, but such was not always the case. In 1902, illustrating this point, a debate broke out in the editorial pages of the journal *Science* on the origin and the meaning of the word "ecology." Those scientists who participated in the debate agreed that the word "ecology" originated from "œcology," but that much like other commonly used words, such as economy (once also "œconomy"), the *o* was eventually dropped. Participants also agreed that Ernst Haeckel, the German disciple of Charles Darwin, was the first to use the word "ecology" (actually *Oecologie*)—in his pioneering 1866 *Generelle Morphologie der Organismen*.[1] But more interesting than the humdrum quibbling over when and where the word "ecology" first appeared, or whether the *o* should be dropped from "œcology," were the attempts to fine-tune its actual meaning as an emerging scientific methodology and academic discipline.[2] Of specific interest to us are the attempts by zoologists to appropriate the word from botanists and apply it to the scientific study of animals.

In his letter to *Science*, botanist Charles Bessey pointed out that "ecology," although first used by Haeckel, actually came to the attention of botanists in the United States only in 1893, at the Madison Botanical Congress. "The word ecology has been in quite general use in the botanical world for the past eight years," wrote Bessey.[3] The botanist William Ganong, offering a preliminary definition, wrote that ecology "signifies the science of the adaptation of organisms to their surroundings, a field of study in which botanists have been more active than zoologists."[4] That zoologists would begin to explore the possibility of using ecology as a methodology for understanding the animal world at the turn of the century suggests that zoology had reached a sort of impasse in its ability to explain the relationship between animals, natural evolution, and the ties of an organism to its constantly changing environment. Perhaps entomologist William Morton Wheeler's thoughtful foray into the 1902 debate in *Science* best captures the sense of anxiety and excitement inherent in whether to apply the word "ecology" to the study of complex animals.

Wheeler began his thoughts with the observation that zoologists need a "satisfactory technical term for [the study of] animal behavior and related subjects," one like the "ecology" used by botanists to describe their scientific enterprise. Wheeler believed, however, that botanists should be left in "undisputed possession" of "ecology," because, quite simply, the term did not adequately fit the scientific goals of the study of animals. That is, when Haeckel first used the term "ecology," he saw it as describing an "economy of nature," suggesting, among other things, a "habitat" or the "dwelling or nest of an organism," and not what zoologists should be concerned with, which was the inner workings of the individual organism itself. As Haeckel wrote, "By ecology we mean the body of knowledge concerning the economy of nature—the investigation of the total relations of the animal both to its inorganic and to its organic environments."[5] However, Wheeler suggested that too many variables shaped an animal's behavior and its relationship with the environment, and hence animals could not be studied under the rubric of "ecology." Indeed, ecology did not work for zoologists because of "the great complexity of the zoological as compared with the botanical phenomena to be organized and methodized." Wheeler continued by explaining that botanists study the external influences on plants, "the effects of the living and inorganic environment on organisms which are relatively simple in their response." By contrast, zoologists investigate the internal nature or "character" of complex animals, "the expressions of a centralized principle represented by the activity of the nervous system or some more general and obscure 'archæus' which regulates growth, regeneration and adaptation, carrying the type onward to a harmonious development of its parts and functions, often in apparent opposition to or violation of the environmental conditions." Hence, as Wheeler suggested, "plant societies" are different from "social animals," implying that animals should not be studied (not to mention "organized and methodized" in the form of an academic discipline) within the framework of ecology. Instead,

Wheeler proposed "ethology" as a more adequate technical term, one that suggested an investigation of the "nature of instinct and intelligence" inherent in the animal world. Animals could not be understood unless their "physical and psychical" behavior was examined; only then could scientists uncover the nature of their relationship to "their living and inorganic environment."[6]

Basically, Wheeler (ironically, the first vice president of the Ecological Society of America) exposed his own reluctance to study animal species as "societies," to study them as one might plant communities. Rather, he viewed the animal world much as Charles Darwin had, as made up of competing individual "social animals" (a notion different from "animal societies"); animals, because of their internal "principle," modified their relationship to their species group and to the natural environment individually. Most, but certainly not all, zoologists in the West who studied animals at the turn of the century came to use Wheeler's term "ethology," and the study of animals developed into an investigation of individual animals and their interaction with one another, such as their reproductive activities and their ability to pass down individual hereditary traits.

Only three decades after the debate over the meaning of "ecology" in *Science*, Japanese scholars who defined themselves as ecologists began to fashion their own version of this new science, preferring, at least in their explanations of wolf extinction, the term "ecology" over "ethology" (as defined by Wheeler, anyway). That is to say, when crafting theories of wolf behavior and extinction, Japanese ecologists preferred to look at the influence that human activities and the environment had over animal societies as a whole and not, as Wheeler would have them do, look at social animals and the internal mechanisms that might be behind their individual struggle with the outside world. Even the renowned primatologist Imanishi Kinji fancied himself, at least in his early work, as an ecologist of wolf extinction. Theories regarding wolf extinction posed not just by Imanishi Kinji but also by Yanagita Kunio, Inukai Tetsuo, and Chiba Tokuji serve as useful vehicles for investigating the emergence of Japan's scientific discipline of ecology and continuing our discussion started in chapter 5. To a certain degree, Japanese ecologists were influenced by and remain reminiscent of those European and American ecologists Robert McIntosh has identified as part of the "holistic" or "organismic" tradition, a tradition he contrasts to that practiced by scientists concerned primarily with individual plants and animals.[7]

Similar to the organismic tradition in the West, Japanese ecologists focused on animal societies as organisms unto themselves, like aggregations of wolves or monkeys, and not on individual social animals. In Imanishi's case, a Confucian and Buddhist context that was linked directly to the Kyoto school of philosophy stood behind many of his scientific theories; but in his more scholarly writings on biological communities he frequently referenced the scientific observations of prominent European and American organismic ecologists. The influence of the Kyoto school on Imanishi meant that he and other Japanese ecologists tended to focus on the organic interrelatedness of the entire known world: all things, argued the famous Kyoto philosopher Nishida Kitarō, shared a common origin.[8] But the global organismic tradition stressed biological interrelatedness as well, which meant that Imanishi could seamlessly wed the distinctive Kyoto school philosophy to global trends in ecological science. Following Imanishi's example, Japanese ecologists came to position plant and animal societies as actors that shaped local ecosystems: they elevated the role of the ecosystem in causing, and the animal aggregation in experiencing, evolutionary change. In the case of Japan's wolves, such holistic natural forces caused their extinction.

PART 1: THE ORGANISMIC TRADITION AND JAPAN'S ECOLOGY

We begin by tracing the organismic tradition as it developed among animal ecologists in Europe and the United States, paying special attention to the scholarship of Karl Semper, Victor Shelford, Charles Elton, and Warder Clyde Allee. Then we will turn to the writings of two early Japanese ecological writers, Imanishi Kinji and Yanagita Kunio, examining how their theories on wolf extinction both fit into the tradition of Western organismic ecology and also differed in important ways. Though Japan's ecology developed a distinctive flavor to be sure, as with all things Japanese it is easy to overemphasize this distinctiveness; early Japanese ecologists resembled other ecologists around the world who came to view the ecosystem as a holistic living organism and humans as part of—not somehow above and beyond—that living organism.

One "big picture" question we should consider as we work our way through the more specific debate between Imanishi and Yanagita over wolf extinction is whether, considering the influence of European and American ecologists on Imanishi's vision of the biological community, there is really a distinctive "Japanese" form of ecology or whether Imanishi and his successors represent local variations on a global twentieth-century scientific trend. To answer this question, let us begin with a brief examination of the birth of organismic ecology in the West.

Animal Ecology and the Organismic Tradition

In the late nineteenth and early twentieth centuries, the "crystallization of ecology" occurred when botanists and zoologists blended aspects of physiology and natural history to form the new science; but many scientists still debated, in the pages of *Science* and other forums, the exact nature of the discipline of ecology.[9] From the *Science* debate forward, however, many ecologists came to focus their studies on plant and animal communities as a whole rather than study such communities as simply groups of distinct individuals. Vocabularies regarding the existence of organismic communities, such as references to the "biocœnosis" of oyster beds by the German zoologist Karl Möbius, distinguished "self-conscious" ecology from traditional zoology and botany.[10] Later ecologists, such as the botanist Frederic Clements, placed an even greater emphasis on organismic communities; but because this chapter explores the development of Japan's ecology as it relates to wolf extinction, let us confine our discussion here to the emergence of the organismic tradition as it applies to the study of terrestrial animals.

Interestingly, even though a zoologist, Ernst Haeckel, coined the term "ecology," his colleagues around the world only slowly (and in some instances reluctantly) adopted the new ecological perspective in their studies of animals. Eventually, however, some zoologists began to focus on the cohesive functioning of animal communities, but only in concert with their traditional investigations of physiology and morphology. As early as 1881, for example, zoologist Karl Semper, in *Animal Life as Affected by the Natural Conditions of Existence*, pushed zoology beyond the traditional analytical confines of physiology and morphology when he explored, if at times only by way of analogy, the organismic relationships between various animals and their environment. Early in *Animal Life*, Semper suggested the existence of body-like communities of living things, thereby recasting the traditional duties of physiology to serve as a literary metaphor for discussing the interdependency among individual plants and animals. He wrote:

> Although it is certainly true that the various animals inhabiting a country are not so intimately interdependent as the organs of the individual, the relations in the two cases may be very directly compared. The normal numerical proportion, mode of life, and distribution of animals would be altered or

destroyed by the extermination of one single animal, *just as the whole body suffers, with all its organs, if only one of them is destroyed or injured.* (Italics added)

Though he never used the term "ecology," Semper posited, in what might be read as a classic statement defining the organismic tradition, that the "fauna of a district thus takes the aspect of a vast organism whose separate members—the different species of animals—are living parts of the body."[11] Specific environments could be seen as living bodies of the natural world; plant and animal species comprised their interdependent internal organs.

Even more than Semper, Victor Shelford, an animal ecologist at the University of Chicago and, after 1914, the University of Illinois, utilized an organismic approach in the study of animals. As McIntosh explains, Shelford "took a different approach of concentrating on many species, their dependencies on one another, and their relation to the environment—in a word, communities."[12] Donald Worster agrees by pointing out that Shelford studied "the science of the *development* of communities."[13] In 1913, Shelford, in *Animal Communities in Temperate America*, harnessed the study of physiology and animal behavior to the lives of other local plant and animal communities, exposing how community equilibriums form and become disrupted. He defined the new science of ecology as the "study of all organisms of an area, from the point of view of their relations to each other and to their environment." Like Semper, he likened animal communities to organic machines of sorts when he wrote: "Communities are systems of correlated working parts. Changes are going on all the time as a sort of rhythm much like the rhythm of activity in our own bodies related to day and night."[14]

In the 1920s, Charles Elton, who was influenced, along with many later animal ecologists, by Shelford's studies, focused the attention of animal ecologists on the role of global environments in the distribution of certain animal species. The first animal ecologist to serve as president of the British Ecological Society (1932) and, one year later, founder of the *Journal of Animal Ecology*, Elton summed up the discipline in the second line of his classic *Animal Ecology* by explaining that ecology "means scientific natural history."[15] Elton wrote that, traditionally anyway, zoologists had sequestered themselves in laboratories, where they "occupied their time with detailed work upon the morphology and physiology of animals." These laboratory zoologists found it a "disconcerting and disturbing experience to go out of doors and study animals in their natural conditions." By contrast, ecologists study animals in their natural conditions, drawing connections between animal physiology and behavior and their interaction with the environment. Basically, Elton explored the forces that determined animal populations, including fluctuations and density stabilities and instabilities caused by ecological invaders.

Similar to Imanishi, moreover, who described his own approach as "animal sociology," Elton too might be characterized as an animal sociologist. Elton submitted that animal ecology should be studied in much the same manner as human ecology and vice versa. "Human ecology," he wrote, "has been concerned almost entirely with biotic factors, with the effects of man upon man, disregarding often enough the other animals amongst which we live." Turning the tables and labeling the ecologist—as a human being—as just one more "biotic factor" to consider, Elton suggested that too much attention has been given to strictly human-generated forces in both human history and economics (which should remind us of the discussion in chapter 5). By contrast, with animals, "Attention has been concentrated on the physical and chemical factors affecting" them. He concluded that "we are now in a position to see that animals live lives which are socially in many ways comparable with the community-life of mankind."[16] Eventually, Elton focused on animal populations, community ecology, and the development of the ecological niche: energy—both ingesting it and expending it—became Elton's measure of interaction among animals within a given environment. Indeed, he once described organisms as "connected energy transformers" that came together to create "loop channels in the ecosystem."[17]

It turns out that, in the early twentieth century, evoking analogous and real connections between human and nonhuman animal societies was far from strictly an ecological issue. As historian Gregg Mitman argues, when animal ecologists focused on the relationships between human and nonhuman societies in the early twentieth century, they often sought to elucidate troubling questions regarding human conduct in an age of fascism, communism, and global war. "By studying animal and plant societies," writes Mitman, "many ecologists hoped to bring biological understanding to problems confronting human society in what seemed to be an acutely troubled time."[18] At the University of Chicago, in particular, a distinct brand of ecology emerged, one that developed basically independent of Darwin's evolutionary doctrine and that contributed to the organismic tradition; its emphasis on geography and the holistic evolution of animal societies also resembles the thinking of some of Japan's ecologists. Specifically, Warder Clyde Allee, who became an assistant professor at the University of Chicago in 1921, advanced theories regarding animal aggregation that emphasized a "cooperative world," where "social evolution mirrored that of ecological succession" and where "metaphors of progress" were discussed with little attention to Darwin's theories of hereditary mechanisms in evolution. By the 1930s, for many Chicago ecologists, an animal population came to be studied as a "distinct physiological and evolutionary unit."[19]

Importantly, less than one decade later, Imanishi developed similar theories regarding the role of animal societies in evolutionary change and even speciation. In *Seibutsu shakai no ronri* (The logic of biological societies; 1948), written to be "a little more scientific" than his earlier and more famous *Seibutsu no sekai* (The world of living things; 1941), Imanishi repeatedly referenced the scholarship of Shelford, Elton, and Allee to bolster his own points regarding the holistic structure of biological societies. Imanishi, as his scholarship attests, was plugged into discussions of ecology as they developed at Chicago and elsewhere. Imanishi was as familiar with the work of these organismic ecologists as he was with the writings of the Japanese philosopher Nishida Kitarō: he was part of a global community of like-minded scholars.[20]

In Chicago, the legacy of Victor Shelford lived on well after his departure. His belief that organisms, even the smallest living creatures, interacted with and transformed their environments (despite the leviathan powers of geological change), shaped community ecology at Chicago for decades. Allee, a liberal pacifist who sought to denounce war on biological grounds, carried Shelford's banner, and like Elton, he believed that the same natural laws that governed human communities also governed nonhuman ones. For Allee, writes Mitman, "Human society was but an extension of the associations that existed throughout the animal kingdom—associations that arose as a consequence of the mutual benefit group life provided in the individual's struggle with its environment."[21] In short, nonhuman societies offered real models for social life (other than, of course, family life): the reason animals gathered in aggregations was to respond to specific types of environmental conditions or threats. Allee argued that such aggregations proved so cohesive that "natural selection can act on them as units" rather than as groups of individuals; evolution, he wrote in a letter to a colleague, "is a process that takes place on the group level only."[22] In *Principles of Animal Ecology*, a text coauthored by Allee, the introduction explains: "We view the population system, whether intraspecies or interspecies, as a biological entity of fundamental importance."[23] This "biosociological" approach inherent in *Principles of Animal Ecology* reminds one of both Elton and Japanese ecologists in their approach to animal ecology.[24]

Even though Allee and others at the University of Chicago tried to "biologize human sociality," as Mitman argues, they were certainly not fascists or racists (although some accusations were made); and neither were Imanishi and other early Japanese ecologists, even though, in the 1930s and 1940s, Japan's military leadership had moved in that direction. In Japan, emphasis on animal societies and the integration of human societies into discussions about a harmonious nature—viewing communities and hence nations as syncretic organisms unto themselves—should not be confused with the cultural chauvinist invention of the first half of the twentieth century of a Japanese "oneness with nature" or a "love of nature," though surely ecology and such political ideologies interacted and played off one another. Just at the historical moment when most Japanese political ideologues were abandoning Darwinian notions of nature and evolutionary change in favor of a unique Japanese explanation, bold ecologists such as Inukai Tetsuo continued to treat wolf extinction in the universal context of ideas about progress and development. But Inukai was one of only a handful of writers to do so.

As Julia Adeney Thomas argues, in the 1930s a radical transformation occurred in how many Japanese understood nature and their relationship to it. Thomas points out that during the late nineteenth century, scholars such as Katō Hiroyuki had subscribed to the presumed universalism of Social Darwinism and regarded Japan's evolution as following the same natural trajectory as that of other nations. But political events, in particular the Triple Intervention following the Sino-Japanese War (1895) and the Treaty of Portsmouth following the Russo-Japanese War (1905)—both of which partially robbed Japan of the spoils of battle—convinced Katō and the political Right that the West had commandeered all the seats at the table of progressive nations. This, writes Thomas, required Japanese pundits to create a new form of nature in their political discourse, one "configured in reaction against social Darwinism and in conformity with the requirements of national pride."[25] In the 1930s, with right-wing publications such as the *Kokutai no hongi* (Principles of national essence) preaching a Japanese "love of nature" (*shizen no aisuru*) and Japan's "exquisite harmony" (*bimyō no chōwa*) with the natural world, ultranationalist narratives undercut Darwinian ones of natural competition in explaining Japan's relationship to other nations. Many Japanese ecologists writing in the 1930s and 1940s viewed the human relationship to the natural world in similarly holistic ways; nonetheless, the lessons they often taught were not a "love of nature" or "exquisite harmony" but the unharmonious implications of wolf extinction.[26]

Importantly, when I characterize Japan's ecologists as more concerned with animal societies as holistic actors and as more comfortable with integrating human societies in their models of natural change, this phenomenon is partly distinct—even if born out of the same Confucian milieu that emphasized people's social roles over their individual ones—from discussions of a Japanese "love of nature" so prominent in the 1930s and thereafter. As Imanishi's engagement with the scholarship of Shelford, Elton, and Allee demonstrates, Japan's ecology also needs to be seen as part of a global scientific trend, one that thrived in the early twentieth century and sought to understand animal (and in turn human) societies as evolving and, to a certain degree, harmonious organisms.

Sacred Mountains and Japan's Arcadia

Our story begins in 1946. With famine, urban destruction, and mass dislocation, it was a dreadful year in Japan. Nonetheless, the forty-four-year-old naturalist Imanishi Kinji (1902–92), after returning from military duty, managed to put the final touches on his *Dōbutsuki* (Animal chronicle), in which he explored the extinction of the Japanese wolf from the perspective of an ecologist.[27] While he was writing the essay, Imanishi, as if to heed Elton's charge to get out of the overly sterile laboratory and breathe the fresh air of the natural environment, subjected himself to Japan's mountainous terrain and to Mongolia's wolves when developing his theories on wolf extinction. Imanishi believed that only by entering the mountains and forests, and thereby interacting with nature on a more personal level, could he, as an ecologist, discover the sensibilities within himself required for interpreting nature. Certainly, this reminds one not only of Elton but of even earlier naturalists, such as Henry David Thoreau, who wrote, "The true man of science will know nature better by his finer organization; he will smell, taste, see, hear, feel, better than other men. His will be a deeper and finer experience."[28]

Imanishi had started his study in the 1930s and early 1940s by investigating reports of wolf sightings near mountain villages and by climbing peaks where such sightings had occurred, all in an attempt to authenticate that the Japanese wolf was really extinct. He also linked his own life experiences and childhood memories with the story of the Japanese wolf, remembering that while pursuing his passion for mountaineering in the remote mountains of Yoshino, home to some of Japan's most sacred peaks in the Shugendō tradition, the nagging question of whether wolves still lived in the area tugged at him. Such thoughts, he recalled, ignited his lifelong passion for studying animals. More specifically, as Imanishi explored Japan's most sacred mountains as a young mountaineer and naturalist, he came to believe that understanding the extinction of the Japanese wolf was a problem for the discipline of *seitaigaku*—ecology.[29]

Imanishi recalled that when he was a child, many people around him still talked about wolves. They told lurid stories of wolves digging up and eating corpses from cemeteries; some people even carried daggers when they left to travel mountain roads, fearing sudden encounters with the dangerous Okuriōkami, the "sending wolf" (see chapter 2). One particular event from Imanishi's childhood found its place in *Dōbutsuki*, an event replete with the symbols of industry and war that had, by the 1940s, replaced wolves in Japan's communities. Imanishi recalled that one morning all the adults from his neighborhood had gathered to listen to the distant cries of what they believed to be wolves. "I'll never forget the tension on their faces as they spoke with one another," he wrote. However, the distant cries that the adults thought belonged to wolves really belonged to a nearby factory. In their imaginations, these people pictured the shadowy figures of wolves; they heard their lonely cries split the crisp mountain dawn when in fact there was only a factory siren beckoning people to Japan's industrial age.[30]

Imanishi linked his wolf studies with his passion for mountaineering (in 1931 Imanishi had founded the Academic Alpine Club of Kyoto): "Right after I graduated from college, like everybody else I wanted to explore all kinds of things, and so I climbed mountains like I do to this day, not simply as sport, but I went to the mountains with the thought of seeing and listening to the various things around me."[31] While climbing, listening, and observing, all critical activities in creating what Worster calls the ecologist's need for an "organic unity" with nature, Imanishi often selected for climbing those mountains where wolves had once lived: he hoped to better understand wolf extinction by establishing an "organic unity" with the mountains themselves.[32] Imanishi also set out for local archives and offices in towns and villages in the Suzuka Mountains of Shiga and Mie Prefectures, asking local elders questions about wolves. In the process, Imanishi discovered a sad national trend, one consistent throughout the regions he visited: although a large number of wolf sightings had occurred in the late 1860s and early 1870s, they appeared to have tapered off between the 1880s and early 1900s. It was, he concluded, evidence of wolf extinction, though isolated sightings and reports as late as 1934 led Imanishi to believe that wolves had survived longer than the "official" extinction date. He encountered mountain people, for example, who told stories of footprints, much larger than those of a dog, left in the snow at Ōdaigahara (Nara Prefecture) and the Seryō Mountains on Shikoku Island. On hearing such reports, Imanishi became convinced that investigating such sites, not laboratory work, was the key to understanding the dynamics of wolf extinction.[33]

Before Imanishi could begin crafting his theories of wolf extinction, however, he needed to meet the challenge posed by the ethnology of Yanagita Kunio, who, before Imanishi published *Dōbutsuki*, had formulated his own theories regarding wolf extinction. Imanishi did make the case that ecology was a better lens than ethnology for viewing dramatic natural events like wolf extinction; but traces of ethnological methodologies, in particular the reliance on folkloric and historical sources, became defining characteristics of Japan's practice of ecology. So before diving into the nuts and bolts of Imanishi's theories on wolf extinction, let us turn briefly to the scholar whom many hail as the father of Japanese folklore studies, a scholar who might also be included among the pioneers of ecology.[34]

Yanagita Kunio (1875–1962) and Japan's Ecology

In *Dōbutsuki*, Imanishi declared that investigating the extinction of the Japanese wolf was a job for ecologists. In making such a declaration, however, he was not simply sounding a rhetorical device. Rather, his oblique claim to academic territoriality was intended as a response to Yanagita and his writings on wolf extinction. Imanishi lamented that trained biologists had yet to take up the issue of wolf extinction in Japan. He noted that if any scholars had taken an interest in the issue it had been the ethnologists. In particular, Yanagita, in *Koen zuihitsu* (Miscellaneous writings on the lone monkey; 1939), had offered the first compelling explanation for the disappearance of the Japanese wolf.

In chapter 2 we learned that Yanagita colored his early writings with stories of animals and the sacred mountains they inhabited, and so it should not be surprising that in *Koen zuihitsu* he used such stories to explain environmental change and animal extinction in rural Japan. Interestingly, Yanagita exposed an anxiety born from the rapid importation of European ideas into Japan, particularly that of Western-style individualism. Much as Yanagita had argued in other writings that Western ideas and Meiji economic reforms had threatened to weaken Japan's traditional rural values and social cohesion, so he, drawing on the same Confucian and Buddhist milieu that Japan's ecologists did, argued of wolves that habitat loss and an increase in the human population had led to the breakdown of their all-important "pack life" (*mure seikatsu*). In other words, the loosening of the threads in the fabric of wolf social cohesion—the birth of the lone (i.e., "individual") wolf—was the primary reason for the disappearance of the wolves known by preindustrial Japanese.

By 1939, at the age of sixty-four, after abandoning careers as a bureaucrat, a delegate to the League of Nations, and a journalist, Yanagita had gone into semiseclusion and refocused his attention on collecting and documenting rural folklore. In *Koen zuihitsu,* Yanagita mustered this extensive body of folklore to recast vivid upland stories to document the slow breakdown of wolf society. Indirectly, he cautioned his countrymen of the threat presented by European ideas to their national cohesion. Somewhat reminiscent of Elton, Yanagita established an implicit link between the fate of wolf society and the fate of Japanese society, an intellectual expression of an organic parallelism between human aggregations and nonhuman ones. He believed that the story of wolf extinction was a rural one: the Japanese settlement of mountainous areas had caused environmental changes in wolf habitat and that these changes forced wolf packs to dissolve and lone wolves to strike out on their own. Once wolves were on their own, he argued, this instinctively social creature struggled to survive and eventually disappeared from the Japanese Archipelago. One implied lesson: Japanese too must maintain their natural and national cohesion in the face of external pressure or they too might become extinct like the wolf.

Yanagita conducted the research for "Ōkami no yukue" (Traces of the wolf), the chapter in *Koen zuihitsu* that explored wolf extinction, largely in the Yoshino area of central Japan. Much like *shugenja* (mountain ascetics) and Imanishi, he immersed himself in Japan's most sacred mountain landscape and site of Japan's ancient imperium, which was also the last known home of the Japanese wolf. Yanagita discovered that in Yoshino the last reports of wolves increasingly told of lone wolves, forcing him to conclude that after centuries of human rural development, any semblance of normal pack life among wolves had all but vanished by the dawn of the early modern age. As evidence, he explained (ignoring the history of rabies altogether, incidentally) that most injuries caused by wolves in the past hundred years or so had been the product of lone wolves. Yanagita created a model wherein injuries caused by wolves could be linked to the collapse of their social order, which in turn could be linked to centuries of rural development and the subsequent destruction of traditional wolf habitat.[35]

More precisely, Yanagita arrived at the conclusion that wolves had never really disappeared from the Japanese Archipelago at all and that somewhere in the mountains of Japan wolves still lived. Moreover, he argued that the hereditary legacy of the Japanese wolf could be found in Japanese dogs. He wrote that "even if only a small trace, the blood of [the Japanese wolf] circulates in domesticated dogs [*kai inu*], because if presented with an opportunity, village dogs [*sato no inu*] return to the mountains, where I think it is likely that they mixed with [wolves]."[36] Yanagita's speculation regarding wolf-dog hybridization had profound and lasting implications for theories related to wolf extinction. Ecologists ranging from Inukai Tetsuo and Naora Nobuo to Chiba Tokuji, Hiraiwa Yonekichi, and even historians such as myself have never really managed to craft theories of wolf extinction without returning to the issue of wolf-dog hybridization. Hybridization with dogs, argued Yanagita, or at the very least the dissolving of wolf packs and the dispersal of lone wolves that came to live more like dogs, ranked as the most important factor in the eventual disappearance of the pure "wolves" of the preindustrial era.[37]

"With the cultivation of new lands," argued Yanagita, "there was a reduction of mountainous areas" available for wolves. He continued,

> This was accompanied by a limited availability of food among other things; with a multiplying number of mouths to feed [relative to food supply], some wolves had to strike out on their own. As is probably the same with other animals, the sole reason that wolves instinctively form packs is for catching food and mate selection. But mate selection occurs for a short time once a year, and working against the impulse to form packs was the scarcity of food, which acted as a barrier to forming one large pack. The necessity for many companions thus disappeared. This appears to be one reason for the emergence of so many lone wolves.[38]

Yanagita asserted that the destruction of rural habitat resonated in other ways as well. The cultivation of new crops and the destruction of mountainous areas were accompanied by an increase in meat eating among Japanese. To provide meat for household tables, villagers hunted more frequently (and hence intensified competition over sustenance, similar to the situation described in chapter 5), which resulted in fewer prey such as wild boars and deer for wolves. Simultaneously, the growing number of human inhabitants in these newly cultivated areas provided wolves with a new, and profoundly dangerous, source of food in place of disappearing game. Ranch animals as well as discarded horse carcasses at *uma sutedokoro*, or horse-dumping areas, provided a ready source of food for lone wolves. Human settlement also provided human gravesites, such as those near Kyoto in the classical and medieval periods, which wolves easily dug up. For Yanagita, if the first stage of wolf extinction, during the centuries around 1600, occurred with habitat loss and hybridization with dogs, then the second stage, after 1800 or so, occurred when a sharp reduction in naturally occurring foods, such as deer and wild boars, and the availability of more easily obtained human-generated food further eroded the cohesion of wolf society.[39]

The near elimination of key prey species and the emergence of new human-generated foods meant that wolves, with less territory to hunt on and often on the brink of starvation, directly competed with human societies for available meat protein. On occasion, these hungry wolves turned their attention to women and children, at least when their usual prey species and carrion proved unavailable. Such shifts in wolf hunting habits represented the deathblow for all wild wolves in Japan. Wolves and mountain dogs thus were transformed from "beneficial animals" (*ekijū*) into extremely dangerous ones in the human imagination: Japanese attitudes toward the wolf changed as a result of ecological shifts. In the popular imagination, the wolf came to be seen no longer as a divine animal, as represented by the Large-Mouthed Pure God, but rather as an object of fear and hatred. Once more, ecological changes affected cultural ones and vice versa. This transformation of attitudes toward wolves, which Yanagita could document with folklore, led to an increase in wolf hunting throughout Japan, which contributed to their demise.

Yanagita's theories are indeed compelling, but the ideas that the formation of large packs is "normal" wolf social behavior and that lone wolves were somehow abnormal served as only a shaky pillar on which to construct an entire model. It was at this pillar that Imanishi, the trained ecologist, took careful aim.

Imanishi Kinji and Animal Sociology

That Yanagita situated pack dissolution at the core of his wolf extinction theory raised important questions in Imanishi's mind, not the least of which was whether wolves really do, under normal conditions, live in large packs. For Imanishi, such questions regarding canine social behavior fell within the purview of what he called *dōbutsu shakaigaku*, or "animal sociology," an approach similar to Elton's and, in Imanishi's case, one designed for the study of *shushakai*, the notion of "species society." He later called this the "specia," a term he first coined in his 1941 philosophy of biology, *Seibutsu no sekai.*[40] That is to say, Imanishi believed that questions regarding normal animal behavior should be answered by studies of comparative ecology and animal sociology and not solely by a folkloric investigation of changing rural Japanese values.

In the course of Imanishi's long career, the notion of the organismic species society, or "specia," evolved into the core assumption with which he studied animal aggregations. He placed himself squarely in the organismic tradition discussed earlier when he wrote in *Seibutsu no sekai*, "The gathering of members of a species is not simply an aggregation, but a shared life." He argued for a view of nature wherein "we ultimately have to recognize the dominance of the species over the individual."[41] Echoing the "cooperative world" theory of Allee, he continued, "People often associate the term society with a group phenomenon, but in a broad sense it is a place of shared living for members of the society. Biologically speaking, it is where the individual reproduces and sustains itself."[42] Exploring Imanishi's early writings reveals that he had begun to elaborate on the implications of the species society and "habitat segregation" quite early in his career and that his ideas on habitat segregation influenced his research on wolf extinction and his more scientific scholarship on biological communities.

Imanishi needs to be seen as within the global organismic tradition of ecology; but, at least in part, his ideas were influenced by his contact with Nishida Kitarō of the Kyoto school of philosophy, who also emphasized the wholeness and interconnectedness of the universe. All things, Nishida wrote, are expressions of the same reality, and this philosophy provided Imanishi with an epistemological underpinning for the specific brand of ecology he imported from the West.[43] In Japan, scholars, pundits, and politicians had kicked around ideas regarding the wholeness and interconnectedness of members of national communities for decades. It should not be surprising, for example, that Itō Hirobumi, the father of the Meiji Constitution of 1889, had defined the Japanese polity in organismic terms. In his eyes, Japanese society too functioned as an organismic whole—"like the human body," he wrote—with the Meiji emperor serving as a kind of spirit-mind.[44] None of the scholars discussed in this chapter were lackeys of Japan's imperial state ideologies; but they were certainly not outside the intellectual traditions from which the state drew to create these ideologies, either. Imanishi's ecology can only be understood as a thoughtful melding of indigenous Japanese and global ideological trends.

Imanishi began his discussion of Yanagita's theory by pointing out that many animals, even if living in social aggregations, do not necessarily form packs, questioning what he saw as Yanagita's blanket approach to the biology of mammal behavior. Evoking the social habits of tropical primates and hoofed ungulates of the arid plains, he explained that wolves, unlike these creatures, live differently than herd animals—not all animal societies are the same (something that Yanagita had more or less claimed). Imanishi then turned to a discussion of wolf whelping practices and how the nature of raising pups meant that female wolves, under normal conditions, often become lone wolves when caring for their young. Unlike the young of monkeys, which cling to their mothers, or kangaroos, who keep their offspring in a handy pouch, or even ungulates, whose young become active almost immediately after birth, in the case of wolves, large pack movement with pups can be a problem. For this reason, female wolves build dens, and so, at least on occasion, they need to live more solitary lives. Imanishi concluded that wolves formed packs while females raised their pups, say from spring through autumn, and males (and nonbreeding females) made up the majority of such groups. At the very least, wrote Imanishi, it is erroneous to argue that pack life is the only normal condition of wolves, because raising pups must be considered normal too. Where Yanagita had turned to folkloric accounts from Yoshino to argue that wolves formed packs under normal conditions, Imanishi looked to "animal sociology" to argue that wolf biology was more complex.[45]

Ecology and Organic Unity

North American wolves can still give us insight into the story of Japan's extinct wolves. On November 11, 2000, Yuka (my wife and partner) and I watched as twenty-seven Druids walked along the southwestern edge of the Lamar Valley in Yellowstone. The Druids had formed a single-file line, following Wolf 21, formerly of the Rose Creek pack but now the alpha male of the Druids after the death of Wolf 38 outside the park, and Wolf 42, formerly the beta female but now the alpha after her sister, Wolf 40, was ousted. (Sadly, Wolf 21 died during production of this book, and Wolf 42 died in skirmishes with Hollie's pack in the winter of 2004, also while this book was under way.) Most of the twenty-seven Druids were pups, meaning that biologically related individuals made up the overwhelming majority of the pack. In other words, the Druids basically formed an extended family, the kind of natal associations stressed by Imanishi and later wolf scientists. In this case, however, the extended family was around thirty animals. The pups represented litters from three females, including the late Wolf 40, whose insubordinate sister, Wolf 42, raised her own dead sister's pups. Interestingly, the wolves that orchestrated the coup that killed Wolf 40 felt compelled, perhaps for social reasons or because of elevated levels of prolactin, to raise the pups as their own. This was the recent history of the Druids, a history certainly on my mind as we watched them gather and howl at Jasper Creek.

Only five days later, on November 16, 2000 (my first day working with the Yellowstone Wolf Project Winter Study), I arrived in the Lamar Valley around 6:00 A.M., having driven a Montana State motor-pool car from Bozeman in temperatures hovering around twenty degrees below zero. From Dave's Knoll, a point off the road looking westward toward Buffalo Plateau and the Slough Creek drainage, author and wolf biologist Rick McIntyre and I watched the Druids wake up, gather and howl, and then proceed northward up the far slope toward Mom's Ridge. They then stopped and lay down, sleeping for the better part of the day, as they are prone to do. Eventually, I struck out on my own and caught up with the Druid Peak research crew. Using telemetry (Wolf Project biologists have fitted many of the wolves with radio collars), they also had located the Druids on the eastern slope of Mom's Ridge, up several hundred yards from Slough Creek. I spent the entire day with this crew on a point on the south side of Northeast Entrance Road, where it parallels the Lamar River. Through our spotting scopes we identified a kill up the slope from the wolves. Three Rose Creek wolves, members of the neighboring pack, actually discovered the kill and fed on it for nearly an hour (see fig. 29). But the Druids, notorious killers of Rose Creek wolves, never woke up from the sleep that consumed the better part of their day. Later that afternoon the wolves woke up, howled, and, led by Wolf 21 and Wolf 42, headed out of the Slough Creek area toward Junction Butte. We caught a glimpse of them at a point near the Lamar River and again as they ascended the eastern slope of Junction Butte, but we lost sight of them while they were on top of the butte. Then, just as the sun set and it was becoming too dark to observe them, and too cold to hold binoculars anyway, they appeared on the frozen ponds on the butte's eastern slope. All twenty-seven wolves were still together. It seems evident in retrospect that the Druids could keep a pack of so many together not only because they were basically an extended family with strong kinship attachments to one another but also because, at the time, they were killing an elk a night. So there was plenty of food to go around.

FIG 29. Photograph of Druid elk kill along the frozen banks of the Lamar River. In November 2000, the Druids killed about an elk a night to feed their members.

Imanishi too confronted wolves in the wild and chose to write about those experiences. A year before the end of the Pacific War (between May 1944 and June 1946) he departed Japan and left for Kalgan, Mongolia, for military research and survey purposes. Once in Mongolia, he soon noticed that the inhabitants of Kalgan had painted large white circles on the earthen walls of their homes. On closer inspection he realized that virtually every home had such markings. When locals told him that the circles served as charms to keep away wolves, Imanishi wrote, "once again, in my memory, wolves had been resurrected." For a time, Imanishi lived among Mongolian wolves (*C. l. laniger*), hearing stories of wolves eating the corpses of people killed in floods, of wolves attacking and eating small children, and of people who carried clubs when they walked around at night because of their fear of wolves. One night, he remembered, while some children played in the hills behind his house, they said they saw the figure of a wolf; but Imanishi, who investigated the sighting, never actually saw wolves in Mongolia until he began traveling in the backcountry with survey teams.[46]

Imanishi sought real contact with wolves, hoping to integrate his observations with his theories regarding wolf behavior and extinction. So, in Mongolia, when he did see wolves (nine times in all), he carefully documented those encounters. Imanishi then supplemented his sightings, which occurred between autumn and spring, with the observations made by Roy Chapman Andrews, which were made during the seasons when Imanishi was unable to observe Mongolian wolves.[47] Two wolves seen together constituted the largest "pack" observed by Andrews during his travels. Even when Andrews claimed to have seen more than two wolves in the same vicinity (he saw four on one occasion), Imanishi discredited such accounts as evidence of packs because Andrews had not actually seen the wolves together. Imanishi concluded that, when supplemented with the observations of Western naturalists, Mongolian accounts strongly suggested that wolves do not need to form large packs in the summer months.[48]

As for Imanishi's own observations, between November 4 and April 18 of 1944–45, he spotted seventeen wolves, the largest pack being four canines that he observed together in December and January, the depths of the Mongolian winter. Imanishi wrote that related individuals probably made up most of these packs. Wolves are different from dogs, he explained, and it takes pups as long as eighteen months to mature (scientists now think it actually takes much longer), whereas most dog pups mature rather quickly. Thus, Imanishi's Mongolian packs probably represented natal families. Imanishi insisted, drawing on a combination of research and personal observation, that "normality" for wolf social behavior was not necessarily the large pack, even in the winter months. As he explained, a lone wolf, under the right circumstances, might take down larger game such as deer on its own, meaning that large packs were not necessarily required for hunting. Indeed, my own observations confirm this, as Wolf 40, the former alpha female of the Druids, was notorious during her reign for single-handedly dragging elk down by their throats; as anyone who witnessed her skill and speed can attest, the experienced Wolf 40 was a deadly hunter even on her own (see fig. 30).

Imanishi then took direct aim at the accounts in Yanagita's *Tōno mono-gatari* (Tales of Tōno), explaining that packs of about thirty wolves observed in autumn could be explained in terms of winter pack formation. But reports of 200–300 wolves trying to kill and eat packhorses between Sakaigi and Wayama Pass, even if remotely reliable, tell a completely different story.[49] Such large packs were not the "normal" conditions of wolf society but rather a response to profound environmental upheaval. For Imanishi, "like the cornered mouse who bites the cat," packs of hundreds of wolves, if they ever occurred, represented an irregular response, early signs of coming wolf extinction.[50] In other words, large packs evidenced in Imanishi's mind a dangerous abnormality: the Japanese wolf could no longer adjust, at least within familiar aggregation patterns, to changing environmental conditions in Japan. Japanese wolves had reached the limits of their adaptability to the changing ecological circumstances of the Japanese Archipelago: in essence, these wolves had come to live within the changing ecology of our human history. Such large packs, reasoned Imanishi, represented a last-ditch effort to survive in the modern world.[51]

What the debate between Yanagita and Imanishi demonstrates is that creating the Japanese wolf did not end after the canine had already died off. Rather, creating Japan's wolves became embroiled in early-twentieth-century debates regarding ecology and, in the period immediately before and after the Pacific War, in lessons relevant to the social and ecological structure of human societies, including nations. Importantly, the creation of the Japanese wolf continued apace decades after its extinction. Moreover, tracing the development of Imanishi's ecology has shown us that debates regarding wolf extinction in Japan did not take place in intellectual isolation; they were not characterized by a Japanese distinctiveness. Rather, Imanishi drew on a blend of native philosophies and global scientific theories to create the wolves of Japan and explain the conditions of their early-twentieth-century extinction; and the legacy of Imanishi's wedding of

FIG 30. On November 23, 2002, this black female from the Druids chased a spike bull elk for more than two miles before finally giving up. As can be seen from the photograph, the first snow of the year had whitened the Lamar Valley, and the young bull elk had yet to be weakened by Yellowstone's long winter. Even though this wolf failed to catch the elk, she certainly tried to catch it, evidencing that wolves do not necessarily form packs in the winter solely for the sake of bringing down larger prey. Strength in numbers helps, though.

native and global ideas, as well as his basic receptiveness to (or at least engagement with) ethnological explanations, became a hallmark of much of Japan's later wolf extinction theories.

PART 2: HISTORY AND ETHNOGRAPHY IN JAPAN'S ECOLOGY

In this section let us turn to the writings of two other Japanese ecologists who wrote on wolf extinction, Inukai Tetsuo and Chiba Tokuji. Although one can hardly deny that Inukai was an ecologist, since he taught ecology and zoology at Hokkaido University, Chiba Tokuji is a self-styled geoethnographer. But one point that should be made is that, because of the legacy of the animal writings of Yanagita, ethnologists in Japan have continued to produce thoughtful scholarship on ecological matters, and Chiba is certainly no exception. When writing on wolves, Inukai and Chiba often donned their historian's and ethnologist's caps, because if humans really were part of an organismic natural whole, and if these same humans lived in societies akin to the ones other animals lived in, then certainly historical sources and folklore offered important insights and lessons regarding ecological change, in particular regarding wolf extinction. In retrospect, because of their penchant for integrating human-generated sources and natural science, these scholars wrote what today might be identified as environmental history.

Inukai Tetsuo (1897–1989) and the Historical Ecology of Extinction

Imanishi struggled with the demons of Yanagita when theorizing about wolf behavior and extinction. He tried to separate, with biological methodologies and comparative zoologies, the new science of ecology from the anthropocentric worldview of ethnology. But, despite such efforts, Japanese ecology, at least with regard to wolf extinction theories, retained a strong ethnological flavor: indeed, it had to because wolves were extinct and this forced scholars to turn to oral and written sources for information as opposed to that gained by observing them in the wild. The ecologist Inukai Tetsuo had focused on the extinction of the Hokkaido wolf before either Yanagita or Imanishi had written their essays, and he wove threads of folklore and history into the larger tapestry of his theories on the natural world in a way that came to characterize much of Japan's early ecology.

In a 1933 article published in the journal *Shokubutsu oyobi dōbutsu* (Botany and zoology), Inukai, a thirty-six-year-old professor at Hokkaido University, placed the disappearance of wolves within the two historical contexts of "progress" (*shinpo* or *hattatsu*) and "economic development" (*kaitaku*).[52] By placing Hokkaido in these temporal contexts, Inukai could explain wolf extinction as the natural outcome of Hokkaido's transition from the "uncivilized land" (*mikaichi*) of "about forty years ago," prior to the 1890s and the changes brought by Meiji modernizers, to its present condition as an island "under development." Inukai, unlike Imanishi, rejected space for linear notions of time; moreover, he universalized the situation of the Hokkaido wolf by explaining that its fate looked much like that of other animals around the planet. By the early twentieth century, indeed, prominent ecologists had started to recognize that wildlife extinctions often occur in the wake of unbridled development.[53] Normally, wrote Inukai, tracing wildlife extinction is complicated, if not impossible, because of the sheer historical and ecological breadth of the process. But the case of the Hokkaido wolf, he argued, is different.[54]

Inukai explained, reminiscent of Fukuzawa Yukichi's theories on civilization discussed in the previous chapter, that in the case of Hokkaido, after the Meiji Restoration and the subsequent spread of "modern culture" (*kindai bunka*), Japanese transported the island, in one decisive action, out of what he called a "primitive era" (*genshi jidai*) into a "cultural era" (*bunka jidai*). Hokkaido became a temporal site for exploring ecological change. Prior to the Kaitakushi's colonization of Hokkaido, the island had existed in a natural and primitive condition, one where wolves presumably thrived; but after 1868, the advance of the Japanese transformed the island into a cultured and developed space in a process that could be traced through ecological, ethnological, and historical records. Inukai explained that because Hokkaido was relatively small and animals were isolated and unable to migrate—and because of the coherence of the historical documents kept by the Kaitakushi—scientists could piece together more easily the puzzle of wolf extinction. Seen as something like a controlled ecological experiment designed to test the dangers of modernization when unleashed on a natural site, Hokkaido provided a case study of the rapid, and quite ruthless, impact that global modernity could inflict on the natural world.[55]

Inukai believed that nature and culture formed an organic unity and that their parts acted in concert with one another in any given ecosystem, and so he began his story of wolf extinction as an ethnologist or historian might, by exploring Ainu oral traditions and Kaitakushi historical documents for traces of wolves. Inukai turned to the archives, the traditional office of the historian, to answer ecological questions. As hunting-and-gathering people, he explained, the Ainu had developed a keen understanding of the life rhythms of Hokkaido's large and small animals; but Inukai also investigated Japanese documents, such as Kaitakushi surveys, using them to tie together wolf killing, prey-species movement, seasonal anomalies in animal behavior, and wolf decline. Inukai's first step in understanding the broader ecological processes that had led to wolf extinction was understanding the wolf's part in the unfolding story of human progress.[56] The Hokkaido wolf, too, experienced history.

Working under the assumption that culture and nature are not independent categories, Inukai listed three primary reasons for the extinction of the Hokkaido wolf. First, he asserted that the bounty systems and extermination campaigns of the Kaitakushi led to wolf killing throughout Hokkaido, particularly during seasons when wolves were most vulnerable. Second, Edwin Dun and his strychnine campaign in the Hidaka region also decimated wolves during a vulnerable time.[57] Finally, drastic declines in deer numbers, largely the result of unfavorable weather and intensified Japanese deer hunting, had further jeopardized the Hokkaido wolf. With these three factors as a framework, Inukai drew on the extensive documents left by the Kaitakushi to reconstruct a historical geography of wolf killing for the two years of 1881 and 1886 (see map 3).[58]

Inukai argued that the first lesson taught by this historical geography was that in 1881 hunters principally killed wolves in the southeastern section of Hokkaido, and disproportionately during the winter. In the winter, deer migrated to the Pacific coast of Hokkaido, where there was less snowfall and better access to forage; carnivores such as wolves knew this and they followed the deer on their easterly migration. Meanwhile, the Kaitakushi had stepped up its harvest of deer. With the establishment of venison canneries such as the one in Bibi, near Chitose, workers began canning meat, extracting saltpeter from the blood and bones, and exporting deer antlers to Asian countries as pharmaceuticals and deer pelts to European markets for clothing.[59] Combined with the exceptionally harsh winter of 1879, moreover, which led to such drastic declines in deer numbers that the Kaitakushi shut down the venison canneries for two years, deer numbers probably stood at historical lows around 1880. Weaving together human history, animal behavior, and climatologic change—and, importantly, casting the environment as an organismic whole, much as Semper, Shelford, Elton, and Allee had—Inukai explained that the fate of the wolf was tied to the fate of the deer, human progress, and other factors related to the ecosystem they inhabited. He argued that wolves came to the Pacific coast of Hokkaido, to places such as Oshima, Hidaka, and Nemuro, looking for deer, where instead they found modern ranches and postal stations like the ones described in the last two chapters; there they turned to the horses and sheep raised at these sites. For this reason, wolf killing, at least in 1881, was largely confined to the Pacific coast side of Hokkaido.[60]

In contrast, documents from 1886 suggest that hunters killed most wolves in the Oshima, Tokachi, and Ishikari regions of southwestern Hokkaido, to this day the island's most settled areas. As deer numbers plummeted with overhunting (and as livestock ranches in the southeast became better protected), wolves began scavenging near human settlements for pets, garbage, and other human-generated sources of food. In the wake of the

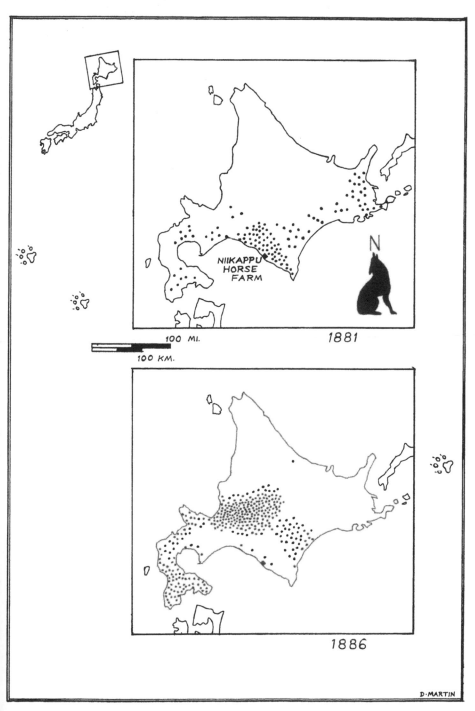

MAP 3. Hokkaido wolf kills in 1881 and 1886.

Kaitakushi's massive harvest of deer and the settlement of southeastern Hokkaido, wolves struggled to feed themselves, and in the process of looking for food they put themselves in harm's way or starved to death, which, along with relentless hunting and poisoning, contributed to their destruction.[61]

In his essays on the Hokkaido wolf, in other words, Inukai, much as the Chicago school did, focused on the geography of animal societies as well as their histories. He speculated on how the onslaught of human progress erased an animal that once thrived on Hokkaido, one that was once worshiped by the native Ainu inhabitants. Seen from this perspective, Inukai, like Imanishi, was an ecologist who made few distinctions between the human and non-human worlds when discussing certain kinds of ecological upheaval. To be sure, Inukai viewed modern culture as disrupting age-old ecological equilibriums on the "primitive" northern island, but neither Imanishi nor Inukai alienated humans from the ecosystem entirely, as their strong ethnological and historical methodologies attest. Inukai began his writings like any ethnologist or historian might, by discussing the place of the wolf in traditional Ainu culture; he then turned to its relationship with Japanese progress and development in the Meiji years. In this respect, Inukai's writings represent an important step in the further development of the discipline of ecology in Japan.

Among wolf extinction theorists, however, perhaps no other scholar utilized history and geography to the extent that Chiba Tokuji did. Rather than evade the ghost of Yanagita, Chiba celebrated it, drawing on ethnology, history, and geography to construct new versions of ecological change, ones that bring us closest to understanding the complex reality of the disappearance of Japan's wolves on a regional level.

Chiba Tokuji (1916–) and the Geography of Wolf Extinction

In 1995, Chiba Tokuji published *Ōkami wa naze kietaka* (Why did the wolf disappear?), a book that focuses on the history, geography, ecology, and present condition of much of Japan's most prominent wildlife.[62] The book drew on Chiba's earlier geographic and ethnographic studies and, most germane for our discussion, included the chapter "Nihon ōkami doko e" (Where did the Japanese wolf go?). Chiba, following graduation from the Department of Literature at Tokyo Higher Normal School, was largely self-taught: his early work focused on cultural and ethnic geography, while his later work, the hallmark of which was a series on hunting, explored the folklore of Japan's disappearing hunting cultures such as the Matagi people of Akita. In many respects, as an ethnographer of the mountain people, Chiba gravitated toward the same culture that had attracted Yanagita decades earlier during the writing of *Tōno monogatari*.

Chiba included scattered discussions of wolves and wolf extinction in his series on hunting cultures; but his most comprehensive treatment of the topic is the chapter in *Ōkami wa naze kietaka*. The first section of the chapter focuses on the disappearance of the Hokkaido wolf and largely reiterates points made by Inukai.[63] Chiba expressed some mild concern with Inukai's geographic model of wolf killing (specifically, Inukai's comparative geography of wolf killing between 1881 and 1886), speculating that the canines killed in the Ishikari region in 1886 were probably feral dogs and not really wolves. But Chiba, like Inukai, relied on history: in particular, he did so when he shifted his focus southward to the Japanese wolf and investigated records from the two early modern domains of Suwa and Kaga. So let us begin our exploration of Chiba's theories of wolf extinction with his treatment of Suwa domain.

Chiba began by explaining that in ancient times, Japanese south of Tsugaru Strait revered the wolf, but that rabies epizootics in the eighteenth century had demoted the wolf from its once divine status to a dangerous creature. While researching Suwa domain, Chiba discovered public notices and letters from the early 1780s that cautioned people about rabid wolves in the area. Despite such warnings, records tell of wolves injuring people near the village of Otsukoto. Wolf attacks in Suwa, specifically an attack that occurred in the seventh month of 1799, caught Chiba's attention. On investigating primary documents related to the attack, Chiba discovered that in the incident one man, Shimono Yoshiemon, received mortal wounds from a wolf near the Ushirozawa Bridge. Two other men, Jirōbei and Kinji, also received wounds, but they managed to kill the rabid wolf in the ensuing fight. Documents explain that one evening in the eighth month villagers lit 108 lamps for the two *ujigami* (tutelary deities) enshrined at Suwa domain and recommended that additional lamps be lit elsewhere. What caught Chiba's attention, though, was that villagers did not perform this ritual for the poor Yoshiemon. Rather, they performed the ritual for the dead wolf because, explained Chiba, the ancient notion that the wolf was a sacred animal continued to linger, and so they offered prayers in recognition of its divine spirit. Chiba wrote that the 108 lamps likely symbolized an attempt to erase the 108 Buddhist carnal desires (*bonnō*) from the local village cooperative.[64]

Much like Inukai, then, Chiba began his exploration of wolf extinction in historical documentation; but unlike Inukai, Chiba moved even deeper into the ecological and cultural nuances of the documentation, particularly in his analysis of the *Suwa-han goyō heya nikki* (Official chamber diary of Suwa domain). This document says that in the fourth month of 1702, at the time when Shogun Tsunayoshi's "clemency for living creatures" orders were in full swing, several wolf attacks occurred in Suwa domain. At the risk of offending Tsunayoshi, domain officials dispatched hunters to kill the problem wolves after local officials reported them running wild near the Yamaura-Kinshu area. Domain authorities provided hunters with a rice stipend and, in addition, for each wolf killed, one *ryō* in gold. In this particular instance, hunters came up empty-handed; but on the twelfth day of the sixth month, two hunters, Gorōsaku of Haniharada Village and Hikobei of Nishi Yamada Village, managed to kill one wolf each. On the same day, moreover, hunter Nakamura Yoshiemon killed two wolves in the forests of Kannon-dō, near Minami Ōshio Village; while on the twenty-second day of that same month, hunter Zenemon, from Otsukoto Village, killed one wolf. Despite Tsunayoshi's call for clemency, and the fact that the shogun often meted out severe punishments for those who disobeyed his orders, wolf killing continued in Suwa.[65]

People killed wolves in Suwa; but wolves also killed people there. Chiba provided documentation from 1702 for several wolf attacks that spanned the fifth through the seventh months. These lunar months, he argued, were critical in wolf ecology, as they were the months when wolves whelped and started raising their young. Here are the cases Chiba mentioned, listed by lunar month and place (when known), but with all years combined (1702):

1. On the eighteenth day of the fifth month, wolves killed the fifteen-year-old son of Jinbei from Shiozawa Village.
2. On the ninth day of the fifth month, the twelve-year-old daughter of Ichizaemon, of Kita Ōshio Village, was bitten by wolves but recovered when a physician treated the wounds.
3. On that same day, wolves killed a servant of Morozumi Roku-rōbei, a fourteen-year-old girl.
4. On the eleventh day of the fifth month, wolves killed a fifteen-year-old girl employed by Tsuchihashi Sagoemon.
5. On that same day, wolves bit the thirteen-year-old daughter of Shichirōbei, of Kita Ōshio Village, leaving her "half dead, half alive; whether she will live or not is unknown."
6. On the fourth day of the sixth month, wolves bit the sixteen-year-old maidservant of Denhachirō in Kami Kuwahara Village.

7. On the seventh day of the sixth month, wolves killed the twelve-year-old son of Itokaya Niidashichizaemon (or Niida Shichi-zaemon from Itokaya Village, the reference is unclear).

8. At noon that same day, wolves killed the thirteen-year-old daughter of Hachiemon from Serigazawa Village.

9. On the fifth day of the sixth month, wolves bit in three places the twenty-four-year-old servant of Magoemon from Minami Ōshio Village.[66]

Chiba noted that the 1702 wolf attacks resulted in five people dead, one person left on the brink of death, and three other people with minor injuries. More important than the sheer number of wolf attacks was that young daughters and female servants constituted the majority of the victims, and that the attacks occurred in the fifth through the seventh months of the lunar calendar. It is at this point that Chiba stopped asking historical questions of the documentation and began asking more ecological ones.[67]

The villages where these wolf attacks occurred sit at the western foot of Mount Tengudake, a peak in the Yatsugatake Mountains (Nagano Prefecture). Japanese created these villages (which are situated 3,333 feet above sea level) out of wilderness areas sometime in the seventeenth century. As the villages expanded, villagers cut more and more wild grasses and felled ever more trees on nearby hillsides for fertilizer and fuel and to make room for more dry-field crops, a process that encroached on the habitat of wildlife. Commonly, it was young servants who ventured into the mountains to cut wild grasses for fertilizer or feed and who fell victim to wolf attacks. Pointing to the expansion of the consumer economy in the late seventeenth and early eighteenth centuries (the Genroku era, 1688–1703), Chiba wrote that this period of expansion in Suwa corresponded precisely with the rise of Genroku commercialism. He continued, "You might say that wolves looked like they might lose when they became entangled in the competition for existence inherent within [an expansive] human society; and when the existence of wolves was threatened, they sought to protect their young at this time." In a slightly problematic conclusion (Genroku commercialism did not everywhere lead to population growth or demographic change in Japan's hinterlands), Chiba pitted wolves and humans in a struggle for survival in the Yatsugatake Mountains, connecting this struggle to Japan's expanding commercial economy and mountain development in the vibrant Genroku years. This is why so many people had been attacked in 1702: wolves were protecting themselves and their young from expanding human settlement.[68]

To strengthen the argument for linkage between wolf attacks, the ecology of wolf reproduction, and rural growth, Chiba next turned to Kaga domain. As with his exploration of Suwa domain, he looked to historical documentation, in particular a collection of source materials entitled *Kaga-han shiryō* (Historical sources on Kaga domain), as a way to outline his ecological argument. This collection of documents was generated largely by the Maeda family, who oversaw a large domain that sprawled across the Kaga, Noto, and Etchū regions, and contains records spanning nearly two hundred years. The documentation was detailed enough for Chiba to create a regional comparative model, much as Inukai had done in the case of Hokkaido. Referring to a table on the chronology and geography of wolf attacks that he drew up (see table 5), Chiba first divided contact between humans and wildlife into four main phases. The first phase, from 1658 to 1709, witnessed repeated human contact with wolves, but also with wild boars and deer. However, few incidents of wild-animal-related injuries occurred. Several winters of heavy snow marked the second phase, from 1709 to 1750, and no incidents of wildlife attacks occurred. Chiba speculated that with the heavy snow, wild boars and deer probably fled Kaga domain for the warmer region around Wakasa Bay and wolves followed them.

In the third phase, between 1758 and 1801, Chiba traced a growing disjunction between where people spotted wolves or where wolves attacked people and where people spotted or had their crops damaged by wild boars and deer. To this point, a logical geographic and seasonal correspondence between sightings of wolves and wild boar sounders and deer herds suggested that wolves lived off these prey. But after 1758, a troubling disjunction occurred in the record as wolves began a shift toward new prey: servant girls and domesticated animals. For Chiba, this trend represented the beginning of the end of the Japanese wolf. In the fourth phase, between 1802 and 1850, this trend became more visible, with wild boar and deer being found in more reports from the southern and northern sections of Kaga domain while wolves remained concentrated in the more densely settled central section (see map 4). Chiba argued that this regional separation of wolves and their earlier prey species coincided with a change in the behavioral disposition of the wolf. People recognized this change and left traces of it in the historical documentation. It was at this time, Chiba pointed out, that the Chinese characters for "wolf" (*ōkami*) stopped being used by Kaga chroniclers, and "big dog," or *ōinu* (not to be confused with *oinu*, the "honorable dog," mentioned in chapter 2), became standard for referring to wild canines in the region.[69]

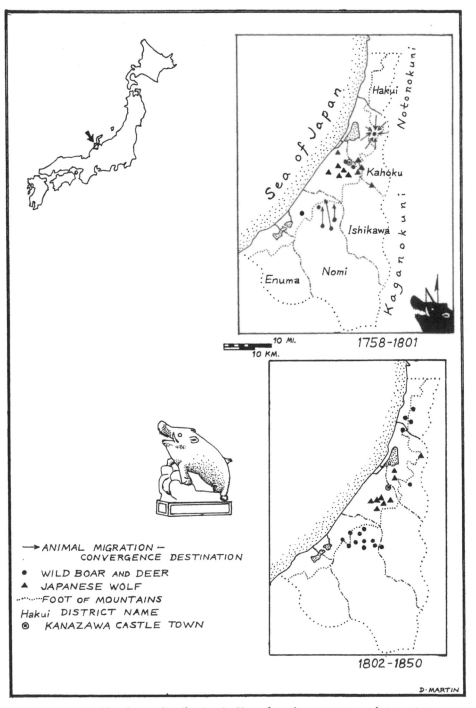

Sea of Japan

Hakui

Notonokuni

Kahoku

Ishikawa

Kaganokuni

Nomi

Enuma

10 MI.
10 KM.

1758-1801

→ ANIMAL MIGRATION –
   CONVERGENCE DESTINATION
• WILD BOAR AND DEER
▲ JAPANESE WOLF
······FOOT OF MOUNTAINS
Hakui DISTRICT NAME
⊚ KANAZAWA CASTLE TOWN

1802-1850

D·MARTIN

MAP 4. Wolf and prey distribution in Kaga domain, 1758–1801 and 1802–1850.

TABLE 5. An Abbreviated Chronology of Injuries Caused by Wild Animals
in the Region at the Foot of Mount Shira

Year	Event	Remarks
1658	Orders given to drive away wild boars and deer	Heavy snow
1668	A wolf injured eleven children in southern Noto	
1674.6	Indication that a wolf was killed at Nonoichi	
1682.7	Wolves chased from both Nomi and Ishikawa Districts	Heavy snow in 1681
1692.2	At Kanazawa, samurai banished for killing wolf	
1693.7	Young Atakaura girl killed by wolf when cutting wild grasses	
1697.3	Pig loose in Kanagawa	
1701.5	Wolf killed at Gosho Village, Kahoku District	
1709.4	Cancellation of Tsunayoshi's clemency orders	Heavy snow in 1710
1734–36	Heavy snow three years in a row	
1740–42	Heavy snow	
1750.12	Wild boar killed at Kanazawa castle town	
1758.12	Wolf killed at Tsuchishimizu Village, Ishikawa District	
1762.1	Wild boar killed at Kanazawa castle town	Heavy snow in 1761
1763	Wolves at villages at the headwaters of the Sai River	Heavy snow in 1756
1766.7	In Ishikawa District, wolves caught in pit traps; in Nomi District alone, sixty-seven wolves caught	Heavy snow in 1763–64
1772	Firearms lent to kill wild boars at villages in Noto	
1774	Wild boars and deer running wild from previous year	
1776.10–12	Many wild boars and deer reported in Nomi District	Heavy snow in 1777
1789	Large herds of deer killed as they fled heavy snow in the mountains	Heavy snow
1794	A man who killed a wild boar given rice as reward	Heavy snow in 1796
1797.9	In order to kill wild boars in Noto, orders given to destroy habitat	Heavy snow in 1797
1801.1	Orders given to prepare for wild boar and deer hunt	
1815.5	Wolf killed at Yachi Village, Kahoku District	Heavy snow in 1812
1818.6	"Big dogs" chased away from Kahoku and Ishikawa Districts	Heavy snow in 1818
1819	Gunpowder and shot requested to drive away wild boars and deer from Kamashimizu Village, Nomi District	
1828	Wild boars and deer damaged crops in Kaga and Noto	Heavy snow in 1829
1830.8	"Big dogs" chased away from Kahoku and Ishikawa Districts	Heavy snow in 1832

Year	Event	Remarks
1834.5	"Big dog" came to feed on corpses at Kanazawa famine relief hut	Heavy snow in 1836 and 1841
1848.7	"Big dog" killed on Tsurugikaidō circuit	
1849	Report given on animals killed by hunter	

SOURCE: Chiba Tokuji, *Ōkami wa naze kietaka* (Why did the wolf disappear?) (Tokyo: Shin Jinbutsu Ōraisha, 1995), 180.

The change in the name for the wolf in Kaga domain—from "wolf" to "big dog"—corresponded with more deep-seated transformations in Japanese attitudes toward the wild canine. Chiba explained that the damage most often caused by herds of deer and wild boars occurred in grain fields and other cash crops because, quite simply, they ate and trampled them. For this reason farmers had long seen the wolf, which preyed on these ungulates, as a "beneficial animal" (*ekijū*); but as the wolf began to attack people, farmers began to see the canine as a "harmful animal" (*gaijū*) instead. Chiba argued that seen alongside the changing ecology of the wolf, such transformations in Japanese attitudes toward the canine move us one step closer to understanding the looming extinction of Japan's wolves.

The first wolf attack mentioned in the *Kaga-han shiryō* was in 1668, with eleven people injured by wolves in the Noto region. However, local officials reported the incident about a month late, which angered the Kaga lord, who later meted out punishments to local officials. In 1674, after a child was attacked by wolves in western Kanazawa, local officials immediately dispatched hunters; and as village placards reveal, hunters scattered beef and horsemeat in the forests and fields to lure wolves within shooting or trapping range. Hunters employed such baiting techniques once more, in 1682, when they drove off wolves from the Nomi and Ishikawa Districts. Even with techniques such as these, however, hunters killed wolves infrequently, but one suspects that the will to kill wolves grew steadily with every discovery of a half-eaten servant girl. As Chiba reported, when officials discovered the dead body of an Atakaura girl killed in the seventh month of 1693, they found that a wolf had completely severed her carotid artery. In what must have been a grisly scene, reports of the incident say that even as the father of the girl came to gather her remains, the wolf lurked on the opposite bank of the nearby river.[70]

Even in the face of such provocation, hunters proved incapable of ridding the region of its wolf problem. Chiba was struck by the fact that Kaga possessed a notoriously high excess of human labor (and hence fewer draft animals) but that hunters still used beef and horsemeat to lure wolves. He asked the obvious question about the origins of such hunting techniques in a region with so little livestock. In perhaps one of the most interesting sections of his analysis, he argued that hunters probably borrowed the technique of baiting from Europe, specifically from the Nanbanjin, or Southern Barbarians: the Iberian missionaries who had stayed in Japan between the 1540s and 1630s, only to be evicted by increasingly paranoid Tokugawa shoguns. Such speculation led Chiba on a comparative survey of Japanese and European (specifically French) wolf-hunting practices.

In France, because of the reliance on livestock in European food culture, wolves and humans confronted each other in an all-out conflict over this important source of protein. To kill wolves, the French nobility organized massive hunting parties, which the monarchy gave special access to lodging, food, and labor, often at the cost of imposing cruel hardships on commoners. Such hunting parties lured wolves with live and dead farm animals. When wolves approached, hunters either shot them or unleashed trained hunting dogs to kill them. In Japan, by contrast, livestock did not become a central feature of the food culture until the twentieth century, and so there was a fundamental "ecological difference" (*kankyōteki na sōi*) between seventeenth- and eighteenth-century France and Japan. Chiba's point shows nicely how Japan's ecologists incorporated human-generated landscapes and cultures—livestock ranches or other farms and food culture—into their understanding of the holistic ecosystem; Chiba viewed food cultures, gastronomic taxonomies, agrarian landscapes, and attitudes toward livestock as all part of the natural ecology and an organic part of his story of wolf extinction. Chiba noted further that unlike in France, wolf hunting in Japan was not a monopoly of the elite and that common people often participated in such hunts. He also pointed to records suggesting that hunting parties respected commoners. One receipt, from Gosho Village in Kahoku District, signed by Okamoto Shichirōzaemon and Takakuwa Jinsuke, documents how wolf hunters properly paid lodging and other expenses on the fourteenth day of the fifth month in 1701.[71]

By the mid–eighteenth century, Kaga hunters began to notice differences in the wolves being killed. To begin with, like the canine killed on the fourteenth day of the twelfth month of 1758 at Tsuchishimizu Village, wolves were not appearing, or being shot, in the fifth through seventh months, as they had in the past. Rather, they appeared at ecologically odd times, such as in the winter, when Japanese encounters with wolves were traditionally rare. Reports also describe these canines as "unusual." For example, on the sixth day of the first month of 1763, a lone wolf attacked three people, somewhat out of season, near the headwaters of the Sai River. In 1766, only three years later, wolf attacks began to take place in nearly epidemic proportions, though most of these attacks did occur during the traditional whelping and pup-raising months. The first of these wolf attacks resulted in the death of a child at Gokuden Village in the sixth month of that year. A month later, at Rengeji Village, wolves killed a fourteen-year-old boy and a seventeen-year-old girl. In response to these attacks, a hunting party of 102 members set out to shoot wolves and dig pit traps. On investigating historical documents related to these hunts, Chiba discovered that these hunters found a variety of canines—some large but some smaller than local domesticated dogs—which, in Chiba's mind, pointed to the beginning of "ecological ruin" (*seitaiteki ihen*) for the Japanese wolf. That simultaneous to these bizarre hunting results chronicles began calling these canines *ōinu*, *yamainu*, or *noinu*, "big dogs," "mountain dogs," or "wild dogs," supports the theory that the Japanese wolf was in the midst of changes sparked by human development. In short, Chiba argued that the Japanese wolf underwent a process of hybridization (*zasshuka*) with dogs in mid-eighteenth-century Kaga.[72]

According to Chiba, real behavioral differences can be identified between Japanese "wolves" and "big dogs" or "mountain dogs." Take the example of a female canine spotted in 1766. Chroniclers reported that this particular animal lived in a bamboo grove in the southern section of Kanazawa, where she cared for two pups. She refused to reveal herself during the day, but at night she returned to feed her two pups. She never injured anybody, and over time, she became a kind of local attraction, with people traveling long distances to feed her. She apparently became acclimated to the presence of people. Chiba considered it unlikely that an actual wolf (as opposed to a hybrid mountain dog) would raise her young so close to a major human settlement. At about the time this mother wolf appeared, rumors began spreading among villagers about how mountain dogs had arrived in Kaga. Some people ventured that they were driven over from the Echizen region; others guessed that they crossed over mountains from the south. Regardless, in the late 1760s, these mountain dogs formed a pack of some fifty to sixty animals at the headwaters of the Sai River; and with the appearance of this pack, more traditional observations, ones that described wolves chasing wild boars and deer and occasionally injuring and killing villagers as they collected wild grasses, disappeared from the chronicles of Kaga domain.[73]

If one believes these historical accounts, concluded Chiba, then a hybrid canine replaced the Japanese wolf, meaning that hybridization stands as one plausible cause of the extinction of the Japanese wolf. (This would also mean that Yanagita was not entirely off base with his suggestion that, metaphorically, the blood of the Japanese wolf still pumps through the veins of Japanese dogs.) Chiba asserted that the upheaval of the Great Tenpō Famine of the 1830s made this fact most visible. Throughout the early 1830s, as tens of thousands of people in Kaga died in the wake of poor harvests and harsh weather, "big dogs" began raiding mass graves, feeding on the victims of the famine. Running alongside these "big dogs" were other dogs (described in documents as *sato inu*, "village dogs," and *kai inu*, "domesticated dogs"). Chiba speculated that no doubt the smell of the large number of decaying bodies attracted all sorts of dogs from miles around. Indeed, the problem became so intense that authorities placed foot soldiers (*ashigaru*) around mass gravesites. There, much like the ranchers of Shizunai discussed in chapter 4, they fired blanks into the air to frighten away both "big dogs" and "village dogs." Not even the poor and famished souls who sought shelter in relief huts (*hiningoya*) were spared the onslaught of hungry dogs. Less than a century before the complete disappearance of the Japanese wolf, the entire family *Canidae*, at least in Kaga domain, had entered a state of disorder and confusion, leading to, as Chiba concluded, the eventual extinction of the Japanese wolf.[74]

CONCLUSION

In some respects, this final chapter has brought us full circle to our opening discussion in chapter 1 of taxonomic debates regarding Japan's "wolves" and "mountain dogs." As Yanagita and Chiba have demonstrated, however, the emergence of the mountain dog is not only a taxonomic issue; rather, it is an ecological issue that may offer important clues as to why the Japanese wolf became extinct. Naora Nobuo, featured in chapter 1, had concluded the same thing regarding wolf-dog hybridization, even though he remained more concerned with questions of classification. For all these scholars, the emergence of the mountain dog was, nonetheless, just that—an emergence—the creation of a new type of wild canine, a product of human-generated changes that swept Japan during the early modern period. Whereas wolves were prone to avoid people, "big dogs" or "mountain dogs" reportedly attacked them, leading to changes in how people viewed wild canines throughout the Japanese Archipelago; and because "mountain dogs" became

Japan's wolves after the introduction of the Linnaean system in the nineteenth century, real wolves were blamed for the behavior of their more aggressive cousins. Creating wolves, such as the creation that occurred with the taxonomic debates outlined in chapter 1, contributed directly to their killing.

The enduring lesson that Chiba and the other scholars that have been discussed in this chapter teach us is that human-generated changes, particularly those that occur during important historical watersheds such as the Genroku period, should be seen as anything but "artificial" or distinct from "nature," because they can lead directly to the creation of new environments and, on rare occasions, even new kinds of beings. Historical shifts lead to, in this sense, ecological events and vice versa. In turn, this holistic lesson reminds us that many of Japan's ecologists were of the same ilk as the ecologists discussed at the outset of this chapter. Their writings on wolf extinction reveal that they believed that all things remain deeply intertwined in the broadly spun web that is our natural world. One could argue, moreover, that this belief in ecological holism, when combined with their reliance on folklore and historical documentation, qualifies Japanese ecologists as some of the earliest environmental historians, even though they never fancied themselves as such.

This chapter, by examining a sample of writings on wolf extinction, has also traced the intellectual background of at least part of Japan's discipline of ecology, demonstrating that Japanese ecologists, at least those who wrote on wolves, tended to view regional ecologies as holistic and, as did such philosophers as Nishida, to view all things—whether "artificial" or "natural"—as sharing the same origins and hence being interconnected with one another. The Japanese scholars discussed in this chapter saw hybridization of domesticated dogs and wolves, human development of rural areas, agrarian cultures and attitudes toward ranching, historical progress, not to mention other human-generated factors, as shaping the environment: as ecological and hence critical to understanding the dynamics of wolf extinction. Even if Japanese did not hunt or poison all of Japan's wolves to extinction, other historical forces—Genroku "commercialism" and Heiji "modernization," for example—altered Japan's ecology for the worse, at least from the vantage point of wolves, and ultimately destroyed them.

EPILOGUE

During the return flight from my research trip to Japan, as I settled in my seat and buckled up, I spied a *SkyMall* shopping catalog tucked in the pocket in front of me. Inside, among the electric cat-litter boxes and other senseless merchandise this airline shamelessly peddles to its clients, I found an advertisement for Sony's new robotic dog, named Aibo. Reading the short blurb under Aibo's cute, puppy-like picture, I learned that, with tactile sensors in its head, feet, and tail (not to mention three articulation points in its head, twelve in its legs, and one in its antenna), this is the most advanced robotic dog ever. Like a real dog, Aibo gives audio cues to its owner to signal when it's sad or happy or even when it wants to play, and it obeys seventy-five spoken commands. I once read that "Aibo clinics" have sprouted up throughout Japan where owners can take their robotic dogs when they break down. I also read that some people have become extremely attached to these robotic dogs, becoming clinically depressed when something happens to them. One Aibo clinic in Nagano Prefecture, for example, reportedly received a box of sweets with a note that said, "Thank you for curing our baby."[1]

With such minion contraptions, I thought cynically, pretty soon we won't need real creatures around us at all. With my trip to wolf-related sites in Japan still fresh in my head, I began to recall once more the sad fate of Japan's flesh-and-blood wolves. As the airplane ascended into the sky I descended into an imaginary world where a Japanese landscape crawled with Aibo-like creatures that resembled wolves, only with jerking, unnatural movements; but it was a landscape where all such creatures obeyed their human masters absolutely. I see now that Aibo represents the anti-wolf, a nightmarish product of our collective technological imagination: one giant step beyond the human-generated Large-Mouthed Pure God of Mitsumine and the "big dogs" and "mountain dogs" of Kaga domain. At last, I thought, growing more cynical by the minute, some Japanese have realized the dream of a large portion of the human species, and with it they can assert supreme control over the natural world by destroying it and then mimicking it in mechanized, and infinitely more docile, forms with emotional lives (brought to life by our own technological wherewithal) that parallel our own. Indeed, human history, as seen by most environmental historians, is more or less the tale of our attempts to understand and control the natural world, to recraft it and refit it so as to provide for our ever-expanding nutritional and energy needs. As a species, we have simply reached the point where we can no longer tolerate a nature that does not bend to our every demand—the survival of all six billion plus of us increasingly depends on it.[2]

Perhaps the history of Japan's extinct wolves can be tied to Aibo. Maybe Japan's war against wolves, and a major component of the extinction of these animals, was about understanding and controlling nature: forcing wolves to bend to, rather than letting them rebel against, our every expectation. Whether Japanese placed wolves in the physical and metaphysical categories of "this world" and the "otherworld," enshrined them in sacred mountain sites such as Mitsumine, hunted and killed them in ceremonial events designed to cleanse the natural landscape during New Year's ceremonies, or outlawed their existence in bounty programs geared toward total annihilation, Japanese controlled wolves within the confines of cultural expectations, ceremonies, and legal regimens. Unlike Aibo, a machine that is a human construct, wolves are not (even if we do "construct" them by perceiving them in different culturally specific ways), and so they march to their own behavioral drummers, often rejecting, sometimes with outright brazen self-confidence, our attempts at controlling them. One lesson that the fate of these wolves teaches us is that sometimes the human species can control certain creatures only by destroying them and then ensconcing them in museums or re-creating them like Aibo. (In Montana, we controlled wolves by violently decimating them in the nineteenth and early twentieth centuries and then reintroduced them with radio collars into the controlled setting of Yellowstone National Park. Park officials track their radio signals and then shoot them from helicopters when they get "out of control" and behave badly.) In Japan, rivers can be cemented over and dammed, forests can be cut down and replanted with commercial cedar, and dogs can be transformed into cold, sterile robots, but wolves—those most wild and excellent creatures—refuse to bend to our commands. Wolves sprint after game with exhausting speed, negotiate dangerous obstacles with remarkable agility, drag down huge elk and other game with frightening strength, and then bring meat to their young and care for their families with enviable compassion. Yet, despite possessing such admirable characteristics, they continually prove so fragile in the face of the demands of our species. They simply will not bark and sit when we tell them to, let alone evacuate their territories so that we can create safe pastures to raise our highly successful "minion biota" such as sheep, horses, and cattle.[3]

My Aibo-inspired theory of human attempts to understand and control wolves is backed by some slightly outdated research. Stephen Kellert, a scholar interested in human-animal relations, has compared attitudes toward wildlife in the United States, Japan, and Germany. With respect to the eight categories of attitudes that he examined, Japanese were found to be the most "negativistic" (i.e., "Primary orientation on avoidance of animals due to indifference, dislike, or fear") and "dominionistic" (i.e., "Primary interest in the mastery and control of animals") of these three major industrialized nations. Moreover, in an example that highlights the place of culture and aesthetics in Japanese attitudes toward animals, Kellert found that "Japanese appreciation for animals was generally restricted to species possessing unusual aesthetic and cultural appeal in certain highly controlled circumstances." One respondent to Kellert's questions answered that Japanese prefer "an artificial, highly abstract, and symbolic rather than realistic experience of animals and nature." Moreover, Japanese scored lowest in the series of questions devoted to "knowledge about animals," though they did tend to be knowledgeable about animals that possess practical value for people: Japanese know our "minion biota" well.[4]

Neither Aibo's existence nor Kellert's statistics are easily transported back in time, but obviously, as Japanese came in closer contact with wolves and as wolves killed servants and livestock in early modern Kaga and post-Meiji Hokkaido, they departed the "artificial, highly abstract, and symbolic" settings of popular images and ceremonies and competed directly with humans in a more "realistic experience," and this situation could not be tolerated and so wolves were hunted and poisoned to extinction. The way that an understanding of—and control over—wolves and other animals was imagined and wielded in Japan changed over time as Japanese society did, and so perhaps Aibo represents the newest incarnation of an age-old attempt to understand and control animals. Japan's experience with the wolf represents an attempt to conquer nature: the need to understand, control, and, in some instances, even mimic it. Still, as persuasive as this Aibo-inspired understand-and-control argument may sound, an exploration of changes in Japanese culture and society only tells half the story of why Japan's wolves disappeared. Really, it is only our arrogance that has tricked us into believing that we can completely understand animals such as wolves. As we have seen, the extinction of Japan's wolves was the result of ecological processes as well, ones often initiated by people but which then spun out of control. Take rabies epizootics, for example. Rabies infected Japan's canine population, altered the behavior of wolves, and thrust them into violent confrontation with one another and human society. In turn, people reinterpreted wolf behavior and labeled it "mad"—not understandable or controllable— and chroniclers included stories of wolf attacks in some of Japan's most prominent travel lore. Kaga domain's "big dogs" behaved this way as well. In this sense, wolf extinction in Japan was the product of agencies facilitated, but never fully managed, by the human population.

From the moment the ancient Siberian wolves set paw on what became the Japanese Archipelago, they constantly evolved to meet the changing demands of their island home, and some of these evolutionary changes, such as their reduction in size and their ability to live in closer proximity to people and pets, eventually left wolves vulnerable to extinction. Increasing human settlement and subsequent land development encroached on wolf habitat, introduced dogs to areas inhabited by wolves, and likely led to sustained hybridization among Japan's canines. This hybridization then weakened the bloodline of wolves, changing them genetically and physiologically, making their behavior not really wolflike at all in some circumstances. These "big dogs" were more comfortable around human settlements than real wild wolves, as with the female who whelped her pups in a bamboo grove near Kanazawa. This proximity to the human population in places such as Kaga domain further altered wolf evolution and behavior. As these hybrid canines fed on the bony carcasses of dead farm folk and animals during upheavals such as the Great Tenpō Famine, they exhibited behavior that made people revile rather than revere them, illustrating where the pathology and biology of animal behavior and the culture of human perceptions intersected to contribute to wolf killing and later wolf extinction.

I have also argued that competing human and wolf economies and ecologies contributed to the elimination of Japan's wolves. One element that stands out in the nineteenth-century eradication of the Hokkaido wolf is the competition between Japanese and wolves over available space and sources of nutrition. Wolves traditionally subsisted off the deer herds, beached whales, and fish that either lived or washed ashore on Hokkaido, and so when Japanese intensified their extraction of these resources in the context of the colonization of Hokkaido and turned wilderness into horse pastures, the hominid and lupine species started to compete for the same sources of nutrition. For this reason, wolf-human conflicts at the ranches of Hokkaido were never casual affairs—that is, they never took the form of the odd foal being killed by wolves, or hunters setting out poison to kill a wolf or two. Rather, they were more warlike in nature, with wolves killing hundreds of horses and threatening the basic viability of the equine enterprise itself, while pasture guards, armed with rifles, discharged them and lit bonfires to protect their livestock. This is why in the late 1870s Edwin Dun ordered enough strychnine to "poison every living thing on the island," because a war of exclusion, one with cultural and ecological roots, was being waged on Hokkaido.

But, other than to appease a nagging sense of guilt, why should we care about wolf extinction in Japan? To answer this last question, let us briefly shift to the example of North American wolves one final time.

CONSEQUENCES OF WOLF EXTINCTION

In July 1992, researchers working for the U.S. National Park Service submitted a highly detailed and multifaceted report to the U.S. Congress on whether wolves should be reintroduced to the Yellowstone region under the guidelines mandated by the Endangered Species Act. The report, entitled *Wolves for Yellowstone?* covered topics ranging from the historical evidence for the existence of wolves in Yellowstone and the sociology and economics of wolf reintroduction to the future management of wolves and wolf interaction with the broader Yellowstone ecosystem. The report made the case that wolves had traditionally lived in the park until federal hunters eliminated or exiled them, and that ungulate populations could more than sustain the return of this "top predator." Although many local ranchers and hunters opposed wolf reintroduction to the Yellowstone region (many still do), others did not, and so the report paved the way for the early stages of reintroducing wolves to Yellowstone.[5]

In July 1993, the U.S. Fish and Wildlife Service in Helena, Montana, drew up a draft environmental impact statement (EIS) regarding the controversial proposal to reintroduce wolves to the Yellowstone region, as well as central Idaho. The Yellowstone region, which includes Bozeman, covers some 25,000 square miles (76 percent of which is federal land) and is home to roughly 95,000 ungulates and is grazed by another 412,000 head of livestock. Once federal officials had reintroduced wolves to Yellowstone and the wolf populations recovered, the EIS explained, elk numbers might be reduced by 5–30 percent, deer by 3–19 percent, moose by 7–13 percent, and bison by 15 percent. The EIS speculated that the presence of wolves in Yellowstone would not affect the 14.5 million recreational human visitors who come to the region each year (the EIS actually projected an 8 percent increase in the number of visitors), though big-game hunters, who kill about 14,300 ungulates each year, might witness a decrease in the number of female elk, moose, and deer available. More importantly, the EIS claimed that wolves would cause minimal damage to local livestock (a projected nineteen cattle and sixty-eight sheep per year) because of the large number of natural prey species.[6] By the summer of 1993, after accepting public comment throughout Idaho, Montana, and Wyoming, the federal government undertook one of the most controversial environmental initiatives ever (indeed, the wolf reintroduction initiative was almost as dramatic as the federal government's decision to annihilate wolves in the first place): it relocated wolves from Canada to the Yellowstone region and central Idaho.

In January 1995, Yellowstone Wolf Project biologists placed fourteen wolves (three family groups) from Alberta in three chain-link pens in Yellowstone, and they were released into the wild in late March. The next year, park biologists released another seventeen wolves from British Columbia (four family groups) in Yellowstone. After biologists relocated another thirteen wolves (ten of which were pups) from northwestern Montana, captured during state wolf-control programs, the wolf population had reached 116 wolves in seven packs by the autumn of 1998.[7] Today, there are hundreds of wolves in and around Yellowstone. Although nowhere near their historic numbers, these figures might be read as one of the most successful conservationist efforts ever. But it was not only conservationist ideals that fueled wolf reintroduction in Yellowstone. Rather, biologists and park officials had come to realize that early-twentieth-century wolf extermination in Yellowstone had upset ancient ecological balances, and booming ungulate populations were overgrazing and changing the Yellowstone landscape. Overgrazing by elk, one visible consequence of wolf extermination, is turning Yellowstone into a desert. Fenced enclosures of thriving aspen groves in the Lamar Valley, for example, locations where biologists exclude elk to determine just how bad overgrazing has become, illustrate this point to all Yellowstone visitors.

In Japan, wolf eradication probably had very similar consequences. As early as the 1740s, for example, in the far northeastern domain of Hachinohe, wolf killing may have severely upset ecological balances between wild boars and farmers of upland cash crops such as soybeans, who had already begun to reel under the combined influence of a burgeoning commercial economy and harsh climatologic changes. Throughout the 1730s and 1740s, peasants in Hachinohe intensified slash-and-burn farming in the hills surrounding the domain's castle town, clearing new swaths of land by felling and burning local forests to make room for a mono-crop regimen of soybeans. They shipped the soybeans to urban areas like Edo (present-day Tokyo) and Osaka, bringing desperately needed funds to the struggling domain treasury. Once the soil became exhausted in these plots, peasants abandoned them for freshly developed ones. In the areas left behind, tuberous plants such as wild yams and arrowroot thrived, and so did wild boars, and they ate the tubers and found shelter in the thick underbrush that thrived near the forest edge.

When gathering tubers during times of dearth, peasants competed with wild boars over sources of sustenance, and thousands (about 10 percent of the population of Hachinohe) died in what Hachinohe chroniclers called the "wild boar famine" of 1749.[8] Historian Kikuchi Isao argues that, along with the Little Ice Age, slash-and-burn farming techniques, and mono-crop agriculture, wolf hunting in the far northeast also might have facilitated the explosion in the wild boar population by eliminating the region's top predator.[9] It is a hard case to prove considering the virtual absence of historical sources; but the fact that chroniclers reported few wolf sightings in the eighteenth century, whereas they had been reasonably common in the previous century, suggests that wolf populations, at least in Hachinohe, were declining.

Much more recently, in some of Japan's most pristine national parks, such as Nikkō in central Honshu, biologists have reported the effects of overgrazing by deer. To address this problem, in November 1998, the Japan Wolf Association and a Nikkō conservation group jointly hosted a symposium, published in the journal *Forest Call* less than a year later, on deer overpopulation and the possible effect of wolf reintroduction in Nikkō. One of the participants, Koganezawa Masaaki, of Utsunomiya University's School of Agriculture, summarized the detrimental impact of booming deer numbers in the riparian forests of Nikkō. He explained that within the prefectural boundaries of Gunma and Tochigi (where Nikkō is located), a space of about 1,553 square miles, in 1998 there lived an estimated 12,000 deer (nearly eight per square mile, a higher ungulate ratio than in Yellowstone). More specifically, Koganezawa reported that in the Nikkō Bird and Animal Conservation Ward (Nikkō Chōjū Hogoku), deer density increases to 12.5 deer per square mile. To put these numbers in perspective, if Yellowstone had the same ungulate density as the Nikkō Bird and Animal Conservation Ward, it would push Yellowstone's estimated 95,000 ungulates to an unsustainable 312,500. Koganezawa also noted that booming deer numbers have led to an increase in damages to agricultural and forestry lands. Indeed, the ¥250,000 in damages in 1988 had increased twenty times one year later. Koganezawa stressed that mild winters, not wolf extinction, had caused the increases in deer numbers, but obviously, the fact that there are no predators like wolves in Nikkō has made the situation worse.[10]

Earlier, in 1992, a team of ecologists had set up a study site in Nikkō at Senjūgahara, west of Lake Chūzenji at the foot of Mount Shirane, to determine more precisely what kind of impact deer were having on Nikkō's herbaceous communities. In 1998, they offered a preliminary online version of their study's results. Senjūgahara, the study site, is a fluvial floodplain, where old-growth riparian forests largely consist of Japanese elm and oak. Much like the situation in the Lamar Valley in Yellowstone, researchers in Nikkō erected fenced enclosures in Senjūgahara to determine what impact deer overgrazing was having on the flora by excluding deer from certain areas. What they discovered was that grazing had virtually eliminated scrub bamboo from the study area by 1994, and only two ragwort-like species, *Senecio nemorensis* and *Senecio nikoensis*, served as ground cover. Importantly, both these species belong to a genus that, when eaten, can have a cumulative poisonous effect on the liver, and so deer do not eat these plants. In terms of trees, debarking during the two years between 1992 and 1994 had damaged 13 percent of the trees in a single plot. Deer had debarked some 50 percent of the Japanese elm and Amur cork trees; and 17 percent of the elm and 4 percent of the Amur cork trees had died from the damage.[11] As in Yellowstone, wolf extinction (when combined with other forces such as climate) has also disrupted ecological balances in Japan's only "wild" places.

But perhaps the most haunting consequence of wolf extinction in Japan has been the arresting silence that deafens these wild places. The anthropologist John Knight, who has written on efforts by the Japan Wolf Association to reintroduce wolves to Japan, has told the story of an experiment conducted by members of the Nara Prefecture Wildlife Protection Committee to determine whether or not Japanese wolves are really extinct. As Knight explains, reported sightings of wolves, hearing wolf howls, and the discovery of wolflike tracks and scat have led some naturalists, including Ue Toshikatsu, to venture that wolves did not become extinct until recently.[12] In fact, several recent books have been published that catalog reports of wolves or that dispute the 1905 extinction date.[13] Specifically, Ue based his claims on stories told by woodcutters, arguing that wolf extinction actually occurred in the aftermath of the Pacific War, when Japan's industrial recovery led to massive deforestation and the destruction of wolf habitat. As late as 1935, Ue's father and other woodcutters reported seeing or hearing wolves in the mountains where Ue and his family lived. For Ue, the industrial lumber industry, not hunters or rabies, killed off the last wolves of Japan sometime after 1945.

To confirm whether wolves might still live in the deep mountain recesses of Japan, members of the Nara Prefecture Wildlife Protection Committee traveled to the sacred mountains of Yoshino (near where I began this book, at the wolf memorial in Higashi Yoshino) with tape recordings of howling Canadian wolves. They planned to play the recordings of wolf howls all night in the hope of luring out "the most territorial of animals." Although those conducting the experiment located themselves at strategic locations in the mountains, no wolves responded to the tapes. The sacred mountains of Yoshino, once the site of a vibrant wolf population, proved silent. Knight actually participated in a repetition of this experiment, only this time in the mountains of Saitama Prefecture near Mitsumine Shrine. Carefully placed tape recorders picked up the noise made by deer, birds, and even some macaques, but no wolves were heard.[14]

To me, these are terrifyingly sad stories about concerned Japanese listening carefully for wolves in the sacred mountains of their country but hearing none. They tried to locate an important beat in the pulse of their natural world and national heritage but found it missing. What these people heard was not only the silence of wolf extinction but a sample of the greater global silence that awaits our species as a whole if our relentless rape of the environment and its creatures continues at its current pace.

APPENDIX

Wolves and Bears Killed and Bounties Paid
by Administrative Region, 1877–1881

TABLE A1. Sapporo Branch Office

Location	1878 Bear	1878 Wolf	1879 Bear	1879 Wolf	1880 Bear	1880 Wolf	1881 Bear	1881 Wolf
Ishikari District								
Sapporo	35	3	74	14	75	9	67	3
Ishikari	0	0	1	2	0	0	10	0
Atsuda	0	0	0	0	0	0	0	0
Rumoi	0	0	3	0	2	0	8	0
Kamikawa	12	0	0	0	0	0	0	0
Uryū	4	0	0	0	1	0	0	0
Sorachi	1	0	0	0	1	0	0	0
No. killed	52	3	78	16	79	9	85	3
Bounty paid	¥260	¥21	¥390	¥112	¥395	¥63	¥425	¥21
Shiribeshi District								
Otaru	0	0	1	0	0	0	1	0
Takashima	0	0	4	0	4	0	0	0
Oshoro	0	0	0	0	6	0	0	0
Yoichi	0	0	0	0	0	0	0	0
Furubira	0	0	1	0	4	0	0	0
Biku	0	0	11	0	6	0	0	0
Shakotan	1	0	4	0	5	0	1	0
Furuu	0	0	2	0	5	0	3	0
Wakanai	4	0	4	0	11	0	43	0
No. killed	5	0	27	0	41	0	48	0
Bounty paid	¥25	0	¥135	0	¥205	0	¥240	0

Location	1878 Bear	1878 Wolf	1879 Bear	1879 Wolf	1880 Bear	1880 Wolf	1881 Bear	1881 Wolf
Iburi District								
Abuta	3	2	21	0	22	0	26	0
Usu	1	4	9	2	5	1	3	0
Muroran	2	0	0	0	0	0	1	0
Horobetsu	0	0	2	0	6	0	0	0
Shiraoi	1	0	15	1	9	1	17	0
Yūfutsu	5	5	52	19	74	20	112	9
Chitose	14	1	14	1	40	7	60	4
No. killed	26	12	113	23	156	29	219	13
Bounty paid	¥130	¥84	¥565	¥161	¥780	¥203	¥1,095	¥91
Hidaka District								
Saru	49	6	73	15	92	19	139	29
Niikappu	0	0	2	0	1	0	1	1
Shizunai	4	1	24	2	2	0	22	3
Mitsuishi	0	0	0	0	2	0	5	11
Urakawa	0	0	0	0	5	0	18	1
Samani	0	0	0	0	1	1	20	4
Horoizumi	0	0	0	0	1	1	5	7
No. killed	53	7	99	17	104	21	210	56
Bounty paid	¥265	¥49	¥495	¥119	¥520	¥147	¥1,050	¥392
Tokachi District								
Hirō	0	0	0	0	4	1	10	0
Tōroku (?)	0	0	0	0	0	0	3	1
Tokachi	0	0	0	0	32	1	23	6
Nakagawa	0	0	0	0	0	0	12	2
Kawanishi	0	0	0	0	4	1	20	1
Kawahigashi	0	0	0	0	0	0	2	0
Kamikawa	0	0	0	0	0	0	1	0
No. killed	0	0	0	0	40	3	71	10
Bounty paid	0	0	0	0	¥200	¥21	¥355	¥70

	1878		1879		1880		1881	
Location	Bear	Wolf	Bear	Wolf	Bear	Wolf	Bear	Wolf
Kitami District								
Esashi	0	0	4	0	1	0	0	0
Sōya	0	0	9	0	0	0	0	0
No. killed	0	0	13	0	1	0	0	0
Bounty paid	0	0	¥65	0	¥5	0	0	0
Teshio District								
Tomakomai	0	0	0	0	0	0	0	0
Rumoi	1	0	1	0	54	0	0	0
Mashike	3	0	0	1	0	0	0	0
No. killed	4	0	1	1	54	0	0	0
Bounty paid	¥20	0	¥5	¥7	¥270	0	0	0
Total wolf kill	223				Total wolf bounty		¥1,561	
Total bear kill	1,579				Total bear bounty		¥7,895	

TABLE A2. Nemuro Branch Office

	1878		1879			1880		1881			
Location	Bear	Bear cub	Bear	Bear cub	Wolf	Bear	Wolf	Bear	Bear cub	Wolf	Wolf pup
Nemuro	0	0	0	0	1	0	4	4	2	21	1
Kashō (?)	0	0	2	0	0	0	0	0	0	0	0
Akkeshi	12	4	39	2	0	10	0	1	0	6	4
No. killed	12	4	41	2	1	10	4	5	2	27	5
Bounty paid	¥60	¥8	¥205	¥4	¥7	¥50	¥28	¥25	¥4	¥189	¥15
Total wolf kill	37					Total wolf bounty				¥239	
Total bear kill	76					Total bear bounty				¥356	

TABLE A3. Hakodate Branch Office

Location	1877 Bear	Wolf	Wild dog	1878 Bear	Wolf	Wild dog	1879 Bear	Wolf	Wild dog	1880 Bear	Wolf	Wild dog	1881 Bear	Wolf	Wild dog
Kameda	1	0	0	9	8	3	9	0	23	14	1	0	13	0	0
Kamiiso	0	—	—	6	—	—	0	—	—	4	—	—	6	—	—
Kayabe	12	7	—	24	1	—	8	1	—	20	0	—	5	0	—
Matsu-mae	3	1	0	5	0	303	2	0	227	20	7	452	4	1	432
Hiyama	0	0	0	7	0	95	5	0	7	30	1	5	8	6	29
Nishi	0	4	—	0	7	—	4	0	—	4	0	—	2	0	—
Yamako-shinai	13	0	—	3	0	—	11	0	—	27	1	—	21	0	—
Kudō	—	—	—	—	—	—	—	—	—	—	—	—	—	—	—
Setana	3	—	—	0	—	—	0	—	—	0	—	—	4	—	—
Futoro	2	—	—	4	—	—	1	—	—	1	—	—	4	—	—
Shimamaki	0	—	—	5	—	—	5	—	—	1	—	—	5	—	—
Suttsu	0	—	0	1	—	95	0	—	0	1	—	0	0	—	0
Utasutsu	0	—	—	1	—	—	0	—	—	0	—	—	4	—	—
Isoya	0	—	—	0	—	—	0	—	—	0	—	—	1	—	—
No. killed	34	12	0	65	16	496	45	1	257	122	10	457	77	7	461
Bounty paid	¥68	¥24	0	¥289	¥112	¥2,480	¥225	¥7	¥1,285	¥610	¥70	¥2,285	¥385	¥49	¥2,305

Total bear kill	343	Bear-related bounty payment ¥1,577
Total wolf kill	46	Wolf-related bounty payment ¥262
Total wild dog kill	1,671	Wild-dog-related bounty payment ¥83,550

NOTES

INTRODUCTION

1. For more on the Washikaguchi (now called Higashi Yoshino) killing and wolves in Nara Prefecture, see Nara Kenshi Henshū Iinkai, ed., *Nara Kenshi: Dōbutsu-shokubutsu* (The history of Nara Prefecture: Animals and plants), vol. 2 (Tokyo: Meicho Shuppan, 1990), 31–36. I am aware that recently Yoshiyuki Mizuko and Imaizumi Yoshinori have argued that the last Japanese wolf was actually killed in Fukui Prefecture on August 3, 1910, at the Matsudaira Agricultural Experimental Station. However, considering the complexity of taxonomic debates regarding the Japanese wolf, it is hard to make any definitive statements regarding whether the canine killed in Fukui was really a Japanese wolf, particularly when the determination is based on physiological characteristics gleaned from a photograph that is nearly a century old. See the *Daily Yomiuri,* February 29, 2004, online edition: www.yomiuri.co.jp/newse/20040229wo71.htm. I would like to thank Luke Roberts for bringing this story to my attention.

2. The Japanese version of the Japanese Wolf Memorial reads: "About the Japanese Wolf Memorial. Until the early Meiji years, a large number of Japanese wolves lived on Honshu, Shikoku, and elsewhere. They then became rare and, according to records, the last wolf captured in Japan, a young male, was killed in Higashi Yoshino. At that time, at the Hōgetsurō Inn in Washikaguchi, an American zoologist dispatched by England to collect East Asian animal specimens bought the last wolf from hunters for just over ¥8, and the last wolf became a specimen at the British Museum of Natural History. This specimen has become an important source in the zoological records of Japan, the main islands, and Washikaguchi. Now, from the hands of Professor Kubota Tadakazu, of Nara University of Education [Nara Kyōiku Daigaku], comes this gallant life-size bronze figure of the last Japanese wolf, howling from the highest point of the mountains. This statue symbolizes the desire

of Higashi Yoshino—a hometown tucked amid green and water—to protect nature." The shorter, English version reads: "This is the statue of a Japanese wolf. A young male Japanese wolf was captured in Higashi-yoshino Mura in the 38th year of Meiji (1905). This specimen, having been recorded and collected in Washikaguchi, Mainland, Japan, has been preserved forever as a precious zoological record. We built this as a precious inheritance which adorns the cultural history village."

3. Wolves will not have federal protection under the Endangered Species Act for long. When I was finishing this book, state planners in Idaho, Montana, and Wyoming were actively putting the final touches on management plans for wolves, despite the fact that the canines are still nowhere near their historic numbers in any of these states. It appears that only in Yellowstone will wolves in the lower forty-eight states be safe from ranchers, trappers, and perhaps even trophy hunters.

4. For a discussion of the disturbing depopulation of Japan's upland communities, see John Knight, *Waiting for Wolves in Japan: An Anthropological Study of People-Wildlife Relations* (Oxford: Oxford University Press, 2003), 20–47.

5. Hiraiwa Yonekichi, *Ōkami: Sono seitai to rekishi* (The wolf: Its ecology and history) (Tokyo: Tsukiji Shokan, 1992). Another interesting and often personal wolf history is Fujiwara Hitoshi's *Maboroshi no Nihon ōkami: Fukushima-ken no seisoku kiroku* (The elusive Japanese wolf: A record of life in Fukushima Prefecture) (Wakamatsu: Rekishi Shunjū Shuppan Kabushikigaisha, 1994). For more histories and ecologies by Hiraiwa Yonekichi, see *Inu no seitai* (The ecology of dogs) (Tokyo: Tsukiji Shokan, 1989); *Inu to ōkami* (Dogs and wolves) (Tokyo: Tsukiji Shokan, 1990); *Watashi no inu* (My dog) (Tokyo: Tsukiji Shokan, 1991); *Inu no kōdō to shinri* (The actions and mentality of dogs) (Tokyo: Tsukiji Shokan, 1991); *Neko no rekishi to kiwa* (History and strange stories regarding cats) (Tokyo: Tsukiji Shokan, 1992); *Inu o kau chie* (Wisdom for raising dogs) (Tokyo: Tsukiji Shokan, 1999).

6. See also Imagawa Isao, *Inu no gendaishi* (A modern history of dogs) (Tokyo: Gendai Shokan, 1996); Taniguchi Kengo, *Inu no Nihonshi: Ningen to tomo ni ayunda ichiman nen no monogatari* (The dog in Japanese history: A tale of walking as a friend of humans for ten thousand years) (Tokyo: PHP Kenkyūjo, 2000).

7. Here, I am thinking of well-known animal journals, such as Inukai Tetsuo, *Waga dōbutsuki* (My animal chronicle) (Tokyo: Kurashi no Techōsha, 1970); and Ue Toshikatsu, *Yamabito no dōbutsushi: Kishū Hatenashi sanmyaku haruaki* (The animal journal of a mountain person: Spring and fall in the Kishū and the Hatenashi Mountain Range) (Tokyo: Shinjuku Shobō, 1998). See also studies of animal symbolism, such as Nakamura Teiri, *Dōbutsutachi no reiryoku* (The spiritual powers of animals) (Tokyo: Chikuma Shobō, 1989); Nakamura Teiri, *Tanuki to sono sekai* (The raccoon-dog and its world) (Tokyo: Asahi Shinbunsha, 1990); Nakamura Teiri, *Kitsune no Nihonshi* (The fox in Japanese history) (Tokyo: Nihon Edita Sukūru Shuppanbu, 2001). For animal histories, see Tsukamoto Manabu, *Edo jidaijin to*

dōbutsu (The people of the Edo period and animals) (Tokyo: Nihon Editā Sukūru Shuppanbu, 1995).

8. Charles Darwin, *The Expression of the Emotions in Man and Animals* (New York: D. Appleton and Co., 1872; reprint, Chicago: University of Chicago Press, 1965), xii–xiii, 12 (all page numbers cited are to the reprint edition).

9. Ibid., 60, 84, 96, 144.

10. Ibid., 166, 200, 213, 243–44.

11. Konrad Z. Lorenz, *King Solomon's Ring: New Light on Animal Ways* (New York: Harper and Row Publishers, 1952; reprint, New York: Time-Life Books, 1962), xxviii, 28, 86 (all page numbers cited are to the reprint edition).

12. Ibid., 88–91.

13. Jeffrey Moussaieff Masson and Susan McCarthy, *When Elephants Weep: The Emotional Lives of Animals* (New York: Delacorte Press, 1995), xx, 3, 22–23. See also Jeffrey Moussaieff Masson, *Dogs Never Lie about Love: Reflections on the Emotional World of Dogs* (New York: Crown Publishers, 1997; reprint, New York: Three Rivers Press, 1997).

14. Masson and McCarthy, *When Elephants Weep*, 27.

15. Peter Singer, *Animal Liberation* (New York: Avon Books, 1991); Steven M. Wise, *Rattling the Cage: Toward Legal Rights for Animals* (Cambridge, MA: Perseus Books, 2001); Gordon Brittan, "The Mental Life of Other Animals," *Evolution and Cognition* 5, no. 2 (1999): 105–13; Gordon Brittan, "The Secrets of Antelope," *Erkenntnis* 51 (1999): 59–77; Steven M. Wise, *Drawing the Line: Science and the Case for Animal Rights* (Cambridge, MA: Perseus Books, 2002).

16. Stanley Paul Young, *The Last of the Loners* (New York: Macmillan Co., 1970), 12, 87, 124. While not attempting to re-create their mental lives, other authors have also told their stories: Roger A. Caras, *The Custer Wolf: Biography of an American Renegade* (Boston: Little, Brown, and Co., 1966); R. D. Lawrence, *In Praise of Wolves* (New York: Ballantine Books, 1986); James C. Burbank, *Vanishing Lobo: The Mexican Wolf and the Southwest* (Boulder: Johnson Books, 1990); Rick Bass, *The Ninemile Wolves* (New York: Ballantine Books, 1992); John A. Murray, ed., *Out among the Wolves: Contemporary Writings on the Wolf* (Vancouver: Whitecap Books, 1993); Robert H. Busch, ed., *Wolf Songs* (San Francisco: Sierra Club Books, 1994); Rick Bass, *The New Wolves: The Return of the Mexican Wolf to the American Southwest* (New York: Lyons Press, 1998). For wolf histories, see Barry Holstun Lopez, *Of Wolves and Men* (New York: Simon and Schuster, 1978) and Bruce Hampton, *The Great American Wolf* (New York: Henry Holt and Co., 1997). Karen R. Jones takes a wolf's-eye perspective in *Wolf Mountains: A History of Wolves along the Great Divide* (Calgary: University of Calgary Press, 2002). In his Ph.D. dissertation, Jon Coleman argues that history, culture, and biology must be considered in concert when trying to explain the degree of violence inflicted on wolves by human beings; both species, argues Coleman, have through the ages sought to pass down genetic, cul-

tural, and material legacies to their descendants. See Jon Thomas Coleman, "Wolves in American History" (Ph.D. diss., Yale University, 2003).

17. Edward Abbey, *Desert Solitaire: A Season in the Wilderness* (New York: Touchstone Book, 1968), xiv.

18. Wallace Stegner, *Wolf Willow: A History, a Story, and a Memory of the Last Plains Frontier* (New York: Penguin Books, 1990), 4, 12, 14–15, 18, 23.

19. On the Sixth Extinction of modern times, see Les Kaufman and Kenneth Mallory, eds., *The Last Extinction* (Cambridge: MIT Press, 1986); Richard Leakey and Roger Lewin, *The Sixth Extinction: Patterns of Life and the Future of Humankind* (New York: Doubleday, 1995); John H. Lawton and Robert M. May, eds., *Extinction Rates* (Oxford: Oxford University Press, 1995).

1 / SCIENCE AND THE CREATION
OF THE JAPANESE WOLF

1. The Japan Wolf Association (Nihon Ōkami Kyōkai) publishes a newsletter entitled *Forest Call*, which often features articles on the role of wolves in creating ecological balance. "Thinking about a better connection between forests, wolves, and people" pleads the heading of each issue. For the first issue (*sōkangō*), see *Foresuto kōru* (Forest call) 1 (August 1994): 1–14. For excellent coverage of the wolf reintroduction effort in Japan, see John Knight, *Waiting for Wolves in Japan: An Anthropological Study of People-Wildlife Relations* (Oxford: Oxford University Press, 2003), 216–34.

2. For a recent treatment of the relationship between wolves and dogs, see Raymond Coppinger and Lorna Coppinger, *Dogs: A Startling New Understanding of Canine Origin, Behavior, and Evolution* (New York: Scribner, 2001).

3. *Kōchi shinbun* (Kōchi newspaper), September 19, 2002, evening edition. See www.kochinews.co.jp/0209/020919evening04.htm.

4. "Nihon ōkami ka? Yaken ka? Kyūshū no sanchū deno mokugeki sōdō" (Japanese wolf? Feral dog? Debate over an observation on Kyushu), *Asahi shinbun* (Asahi newspaper), November 22, 2000, evening edition.

5. For thoughts on the rise of ethnic nationalism in twentieth-century Japan, see Kevin M. Doak, "What Is a Nation and Who Belongs? National Narratives and the Ethnic Imagination in Twentieth-Century Japan," *American Historical Review* 102, no. 2 (1997): 283–309. On the rise of the "emperor system ideology," see Carol Gluck, *Japan's Modern Myths: Ideology in the Late Meiji Period* (Princeton: Princeton University Press, 1985).

6. As we shall see, I am using "universal" here in a cognitive sense: that most peoples around the planet share a cognitive disposition to order the natural world according to certain commonsense patterns. Indeed, most peoples intuitively recognize doglike animals as doglike. However, from the perspective of the cultural historian, "universal" as employed by some historians of European science is often

nothing less than a codeword for the dominant, male-centered, European classifying system thrust on much of the world in the nineteenth and early twentieth centuries, a perspective that I am quite sensitive to. For a feminist critique of the idea that European science is "universal" in nature, see the discussion in Sandra Harding, *The Science Question in Feminism* (Ithaca: Cornell University Press, 1986).

7. Scott Atran defines the "generic-specieme" as "an exhaustive and mutually exclusive partitioning of the local flora and fauna into well-bounded morpho-behavioral gestalts (which visual aspect is readily perceptible at a glance)." In turn, the "life-form" category "further assembles generic-speciemes into larger exclusive groups (tree, grass, moss, quadruped, bird, fish, insect, etc.). A salient characteristic of folk biological life-forms is that they partition the plant and animal categories into contrastive lexical fields." See Scott Atran, *Cognitive Foundations of Natural History: Toward an Anthropology of Science* (Cambridge: Cambridge University Press, 1990), 5–6, 13–14, 22.

8. Michel Foucault, *The Order of Things: An Archaeology of the Human Sciences* (New York: Vintage Books, 1970), xx. On the role of Chinese ideographs and pictographs (as well as Confucian humanism and Buddhism) in creating an East Asian civilization of which Japan was an important part, see Charles Holcombe, *The Genesis of East Asia, 221 B.C.–A.D. 907* (Honolulu: University of Hawai'i Press, 2001).

9. Confucius, *The Analects*, trans. D. C. Lau (London: Penguin Books, 1979), 118.

10. R. P. Dore, *Education in Tokugawa Japan* (London: Athlone Press, 1965; reprint, Ann Arbor: Center for Japanese Studies, University of Michigan, 1992), 52 (all page numbers cited are to the reprint edition).

11. Chu Hsi, *Learning to Be a Sage: Selections from the "Conversations of Master Chu," Arranged Topically*, trans. Daniel K. Gardner (Berkeley and Los Angeles: University of California Press, 1990), 100–101.

12. James R. Bartholomew, *The Formation of Science in Japan* (New Haven: Yale University Press, 1989), 14.

13. Atran, *Cognitive Foundations of Natural History*, 41, 213.

14. See the discussion regarding the "cumulative process of perceiving and comprehending" the notion of "principle" (*kyūri*), a protoscientific way of thinking, in Masayoshi Sugimoto and David L. Swain, *Science and Culture in Traditional Japan, A.D. 600–1854* (Cambridge: MIT Press, 1978), 303–5.

15. Hiraiwa Yonekichi, *Ōkami: Sono seitai to rekishi* (The wolf: Its ecology and history) (Tokyo: Tsukiji Shokan, 1992), 101–3.

16. For example, the *Cuon* lacks a third mandibular tooth (m3) among other anatomical differences (ibid.).

17. Fukane Sukehito, *Honzō wamyō* (Japanese names in natural studies) [922], vol. 2, ed. Masamune Atsuo and Yosano Akiko (Tokyo: Nihon Koten Zenshū Kankōkai, 1927–28), 19. See also Minamoto Shitagau, "Wamyōrui shūshō" (A collection of Japanese names) [931–37], Sapporo Agricultural College Collection, Resource Collection for Northern Studies, Hokkaido University Library, Sapporo, Japan (hereafter SACC).

18. Of course, full-blown "nationalism," as it occurs in modern nations, did not really exist in medieval or early modern Japan. However, some medieval political writers, such as Kitabatake Chikafusa in the *Jinnō shōtōki* (History of the correct succession of the gods and emperors), did explore a Japan-centered vision of imperial politics that implicitly rejected the Sinocentric order. As historian Pierre François Souyri explains, Kitabatake believed that "Japan was superior to other countries; the social hierarchy had to be respected; and the principles of good government had to be duly applied. The seed of nationalism discernible in Chikafusa's thought was the reason for his work's popularity in later centuries." See Pierre François Souyri, *The World Turned Upside Down: Medieval Japanese Society*, trans. Käthe Roth (New York: Columbia University Press, 2001), 135–360. In some respects, this "seed of nationalism" is also discernible in some of medieval Japan's taxonomy.

19. On British gastronomic taxonomies, see Harriet Ritvo, *The Platypus and the Mermaid and Other Figments of the Classifying Imagination* (Cambridge: Harvard University Press, 1997), 189–213.

20. Hitomi Hitsudai, *Honchō shokkan* (Culinary mirror of Japan) [c. 1695], vols. 1–5, transcribed and annotated by Shimada Isao (Tokyo: Tōyō Bunko, 1981).

21. Hitomi's observations remain consistent with contemporary studies of wolf and dog physiology. Barbara Lawrence and William Bossert assert that dogs can be identified "by a bend in the mid-region of the skull so that the rostrum and brain case meet at more of an angle than is usual in wild canids"; the head of the wolf, at least when compared to that of a dog, appears sharper and to taper at the muzzle. See Barbara Lawrence and William H. Bossert, "Multiple Character Analysis of *Canis lupus, latrans,* and *familiaris,* with a Discussion of the Relationships of *Canis niger,*" *American Zoologist* 7, no. 1 (February 1967): 225.

22. Hitomi, *Honchō shokkan,* 5:291–92.

23. Ono Ranzan, *Honzō kōmoku keimō* (An instructional outline of natural studies) [1803], ed. Shimonaka Hiroshi (Tokyo: Tōyō Bunko, 1992), 85–86; Hiraiwa, *Ōkami,* 101–3.

24. Marcia Yonemoto, *Mapping Early Modern Japan: Space, Place, and Culture in the Tokugawa Period (1603–1868)* (Berkeley and Los Angeles: University of California Press, 2003), 50–51. For a discussion of how *Yamato honzō* fits into the general context of Kaibara's thought, see Mary Evelyn Tucker, *Moral and Spiritual Cultivation in Japanese Neo-Confucianism: The Life and Thought of Kaibara Ekken (1630–1714)* (Albany: State University of New York Press, 1989), 47–48.

25. Kaibara Ekken, *Yamato honzō* (Natural studies in Japan) [1709], vol. 2, ed. Shirai Kōtarō (Tokyo: Ariake Shobō, 1980), 253–54.

26. Peter Steinhart concurs, stating that mountain dogs, or at least wolf-dog hybrids, "lack the caution of wolves" and therefore can be more aggressive toward people and livestock. See Peter Steinhart, *The Company of Wolves* (New York: Vintage Books, 1995), 313–14.

27. Yonemoto, *Mapping Early Modern Japan,* 105–7.

28. Terashima Ryōan, comp., *Wakan sansai zue* (The illustrated Japanese-Chinese encyclopedia of the three elements) [1713], in *Nihon shomin seikatsu shiryō shūsei* (Collected sources on the history of the daily lives of common Japanese people), vol. 10, ed. Miyamoto Tsuneichi, Haraguchi Torao, and Tanigawa Ken'ichi (Tokyo: San'ichi Shobō, 1970), 541–42.

29. Regarding tail positioning, L. David Mech, a prominent wolf biologist, has noted that one of the primary differences between wolves and dogs is that "the wolf's tail hangs, but the dog's tail usually is held high and often is curly." See L. David Mech, *The Wolf: The Ecology and Behavior of an Endangered Species* (Minneapolis: University of Minnesota Press, 1970), 26.

30. Atran, *Cognitive Foundations of Natural History*, 149.

31. For more on Itō Keisuke, see Sugimoto Isao, *Itō Keisuke* (Itō Keisuke) (Tokyo: Yoshikawa Kōbunkan, 1960).

32. Bartholomew, *Formation of Science in Japan*, 58–59; Grant K. Goodman, *Japan and the Dutch, 1600–1853* (Richmond, Surrey, UK: Curzon Press, 2000), 163.

33. Arlette Kouwenhoven and Matthi Forrer, *Siebold and Japan: His Life and Work* (Leiden: Hotei Publishing, 2000), 84–85.

34. Nakamura Kazue, "Nihon ōkami no bunrui ni kansuru seibutsu-chiri-gakuteki shiten" (A biogeographical look at the taxonomy of the Japanese wolf), *Kanagawa kenritsu hakubutsukan kenkyū hōkoku: Shizen kagaku* (Bulletin of the Kanagawa Prefectural Museum: Natural sciences) 27 (March 1998): 49. See also Philipp Franz von Siebold, *Fauna Japonica sive: Descriptie animalium, quae in itinere per Japoniam, jufsu et aufpiciis Superiorum, Qui Summum In India Batava Imperium Tenent, fuscepto, annis 1823–1830 collegit, notia, obfervationibus et adumbrationibus illuftravit*, vol. 5 (Amstelodami [Amsterdam]: Apud J. Müller et Co., 1833), 38–39.

35. Tanaka Yoshio, "Dōbutsugaku" (Zoology) [1874], SACC.

36. The quoted description comes from Tanaka Yoshio, "Dōbutsu kunmō: Honyūrui" (Instruction on animals: Mammals) [1875], SACC. A similar description (of the "wolf" or "mountain dog") can also be found in Uchida Shigeru and Ono Kyūshichi, *Futsū dōbutsu zusetsu* (Illustrated explanation of common animals) (Tokyo: Shūgakudō Shoten, 1915), 14.

37. For a discussion of wolf dispersal, see Steinhart, *Company of Wolves*, 21.

38. See Ichikawa Kōichirō, Fujita Yukinori, and Shimazu Mitsuo, *Nihon rettō chishitsu kōzō hattatsushi* (A history of the development of the geologic features of the Japanese Archipelago) (Tokyo: Tsukiji Shoten, 1970).

39. Ronald M. Nowak, "Another Look at Wolf Taxonomy," in *Ecology and Conservation of Wolves in a Changing World*, ed. Ludwig N. Carbyn, Steven H. Fritts, and Dale R. Seip (Edmonton: Canadian Circumpolar Institute, University of Alberta, 1995), 375.

40. Inukai Tetsuo, "Hokkaidō-san ōkami to sono metsubō keiro" (The Hokkaido wolf and its road to extinction), *Shokubutsu oyobi dōbutsu* (Botany and zoology) 1, no. 8 (August 1933): 14.

41. Abe Yoshio, "Nukutei ni tsuite" (On Nukutei), *Dōbutsugaku zasshi* (Zoological magazine) 35 (1923): 380–86.

42. R. I. Pocock, "The Races of Canis Lupus," *Proceedings of the Zoological Society of London*, 1935, 647–86.

43. Abe Yoshio, "Nihon ryōnai no ōkami ni tsuite Pokokku-shi ni atau [On the Korean wolf again (an answer to Mr. Pocock)]," *Dōbutsugaku zasshi* (Zoological magazine) 48 (1936): 639–44; Abe Yoshio, "On the Corean and Japanese Wolves," *Journal of Science of the Hiroshima University*, Series B, Division 1, 1 (December 1930): 33–36.

44. On Japanese policy toward Korea, see Peter Duus, *The Abacus and the Sword: The Japanese Penetration of Korea, 1895–1910* (Berkeley and Los Angeles: University of California Press, 1995), 46–49, 58; Bruce Cumings, *Korea's Place in the Sun: A Modern History* (New York: W. W. Norton and Co., 1997), 114.

45. Ritvo, *The Platypus and the Mermaid*, 42.

46. Siebold, *Fauna Japonica*, 5:38–39.

47. H. J. Snow, *Notes on the Kuril Islands* (London: John Murray, 1897), 27.

48. George Jackson Mivart, *Dogs, Jackals, Wolves, and Foxes: A Monograph of the Canidae* (London: R. H. Porter, 1890), 13–15. See also Inukai, "Hokkaidō-san ōkami to sono metsubō keiro," 13.

49. Hatta Saburō, "Hokkaidō no ōkami" (The Hokkaido wolf), *Dōbutsugaku zasshi* (Zoological magazine) 25 (January 1913): 2.

50. Inukai, "Hokkaidō-san ōkami to sono metsubō keiro," 13–14.

51. This early taxonomical classification comes from Kishida K., "Notes on the Yesso Wolf," *Lansania* 3 (1931): 72–75.

52. The following discussion is largely based on Nakamura, "Nihon ōkami no bunrui ni kansuru seibutsu-chirigakuteki shiten," 49–50.

53. Naora Nobuo, a taxonomist discussed later, pointed out that most zoologists of the early Meiji years referred to the wolf as *yamainu* (mountain dog) or *yama no inu* (dog of the mountains) in their textbooks, and it was not until about 1908, not coincidentally with the rise of Japanese ethnonationalism, that the name "Japanese wolf," *Nihon ōkami*, became commonly used by scientists. See Naora Nobuo, *Nihon-san ōkami no kenkyū* (Research on Japan's wolves) (Tokyo: Azekura Shobō, 1965), 251–52.

54. I first published these thoughts in "The History and Ecology of the Extinction of the Japanese Wolf," *Japan Foundation Newsletter* 29, no. 1 (October 2001): 10–13. For a history of zoology in Japan, including early Neo-Confucian traditions and others, see Ueno Masuzō, *Nihon dōbutsugakushi* (The history of Japan's zoology) (Tokyo: Yasaka Shobō, 1987).

55. It is interesting to note that whether one talks about the deer at Nara shrines or the potato-fed monkeys on Koshima, this notion of human-generated evolution is popular on the Japanese Archipelago; see Frans De Waal, *The Ape and the Sushi Master: Cultural Reflections of a Primatologist* (New York: Basic Books, 2001), 203.

56. Nakamura, "Nihon ōkami no bunrui ni kansuru seibutsu-chirigakuteki shiten," 53–54.

57. On the role of early hunting peoples in mass prehistoric extinctions, see Paul S. Martin and Richard G. Klein, eds., *Quaternary Extinctions: A Prehistoric Revolution* (Tucson: University of Arizona Press, 1984).

58. Nakamura, "Nihon ōkami no bunrui ni kansuru seibutsu-chirigakuteki shiten," 56–57.

59. L. David Mech, Layne G. Adams, Thomas J. Meier, John W. Burch, and Bruce W. Dale, *The Wolves of Denali* (Minneapolis: University of Minnesota Press, 1998), 47–49.

60. To illustrate what he called the chronocline of evolutionary dwarfing, Nakamura pointed out that the original Siberian wolf that migrated onto the archipelago had a mandibular carnassial (a molar in the lower jaw often used in measurements, henceforth m1) length of 29.3–34.5 mm, the wolf of the Jōmon period (10,000–300 B.C.E.) had an m1 length of 26.8–29.15 mm, and the Japanese wolf of recent centuries had an m1 length of 24.0–28.51 mm. See Nakamura, "Nihon ōkami no bunrui ni kansuru seibutsu-chirigakuteki shiten," 57–58. For all discussions of craniometrics in this chapter, I consulted Angela Von Den Driesch, *A Guide to the Measurement of Animal Bones from Archaeological Sites* (Cambridge: Peabody Museum of Archaeology and Ethnology, Harvard University, 1976).

61. The Hokkaido wolf has an m1 length of 25.8–31.0 mm.

62. Early reports tell of the Hokkaido wolf feeding on a rotten whale carcass that had washed ashore. See Kubota Shizō, *Kyōwa shieki* (Harmonious private expedition) [1856], in *Nihon shomin seikatsu shiryō shūsei* (Collected sources on the history of the daily lives of common Japanese people), vol. 4, ed. Takakura Shin'ichirō (Tokyo: San'ichi Shobō, 1969), 235. This is a frequently cited example; see Inukai, "Hokkaidō-san ōkami to sono metsubō keiro," 12–13; Inukai Tetsuo, *Hoppō dōbutsu-shi* (Northern animal journal) (Sapporo: Hokuensha, 1975), 21; Tawara Hiromi, *Hokkaidō no shizen hogo* (Protecting Hokkaido's nature) (Sapporo: Hokkaidō Daigaku Tosho Kankōkai, 1979), 103. Other reports tell of wolves feeding on herring that had washed ashore. See Captain H. C. St. John, *Notes and Sketches from the Wild Coasts of Nipon: With Chapters on Cruising After Pirates in Chinese Waters* (Edinburgh: David Douglas, 1880), 9.

63. Nakamura, "Nihon ōkami no bunrui ni kansuru seibutsu-chirigakuteki shiten," 58.

64. Conrad Totman, *A History of Japan* (Malden, MA: Blackwell Publishers, 2000), 42, 44, 146.

65. Nakamura, "Nihon ōkami no bunrui ni kansuru seibutsu-chirigakuteki shiten," 58.

66. The following discussion is largely based on Naora, *Nihon-san ōkami no kenkyū*, 236–52. For a review of this line of thinking, see Chiba Tokuji, *Shuryō den-shō* (Hunting folklore) (Tokyo: Hōsei Daigaku Shuppankyoku, 1975), 150–56.

67. Naora actually raised the possibility that two or more subspecies of wolf or of some kind of wild canine inhabited the Japanese Archipelago at the same time.

Naora explained that ancient wolf fossils recovered from the Miyata coal mine in Kuzū (Tochigi Prefecture) differ from the Japanese wolf but closely resemble the Siberian wolf. The lengths of the m_1 recovered from sites in Tochigi Prefecture range from 29.3 to 31.4 mm. Rather than illustrate a chronocline model, as they did for Nakamura, for Naora they continued to pose vexing taxonomic questions. See Naora, *Nihon-san ōkami no kenkyū*, 240–41, 244.

68. As canine taxonomists Barbara Lawrence and William Bossert have observed, wolf skulls are long and narrow, even when compared to those of coyotes, and so there is less breadth at the zygomatic arch. See Lawrence and Bossert, "Multiple Character Analysis of *Canis lupus, latrans,* and *familiaris,*" 224.

69. Naora, *Nihon-san ōkami no kenkyū*, 241–45. In a comparative osteological analysis of wolf and dog skulls (one that, I admit, yields slightly confusing results but does say a great deal about Naora's sincere belief in intraspecific homogeneity among wolves), Naora argued that the morphologic features of the Japanese wolf, in particular, actually resemble those of *Canis familiaris* more than they do those of *Canis lupus*. First, he placed a skull from a Japanese wolf alongside one from a Sakhalin Island dog, both specimens from the National Science Museum in Tokyo. The skulls are about the same size, and obvious anatomical distinctions, at a glance anyway, are hard to discern, which in itself, added Naora, warrants caution. To make his case, Naora used the ratio between total cranial length and zygomatic-arch breadth (dividing the former by the latter). The lower the ratio, the smaller the difference between skull length and cheekbone width, and hence the more the skull resembles that of a dog. One Japanese wolf skull specimen from Shizuoka Prefecture, for example, had a cranial length of 200.5 mm and a zygomatic breadth of 109.0 mm: a ratio of 1.84. Naora conceded, however, that there is a great deal of variation in Japanese wolf skull specimens. One sample from Tanzawa, for example, in Kanagawa Prefecture, had a length of 223.0 mm and a width of 139.6 mm, or a ratio of 1.60. Naora took six such samples, with the Shizuoka specimen representing the large end of the spectrum, and came up with a mean ratio of 1.72. He then compared this mean ratio to the Sakhalin Island dog, which had a ratio of 1.81 mm. This suggested that the mean of the six Japanese wolves surveyed had more in common with dogs than the one dog surveyed did. See ibid., 242. As Ron Nowak explained to me in a personal correspondence, however, Naora's conclusions are scientifically problematic because zygomatic-arch breadth can be a poor taxonomic indicator in canines.

Even Lawrence and Bossert admit that sometimes distinguishing dogs from wolves is difficult because, essentially, dogs "are small wolves." Having been manipulated by human culture for millennia, dogs lack the intraspecific homogeneity that wild canines such as wolves or coyotes supposedly possess. In other words, dogs come in many shapes, and as Lawrence and Bossert claim, a dog "often superficially resembles either of the other two [species of wild canines] more than it does other *familiaris*." Highly modified dog breeds, pugs, for example, are easily separated from wolves, of course; but less modified breeds, such as some Japanese varieties which remain extremely wolflike to this day, are distinguished from wolves by more sub-tle craniological characteristics clear only to the eye of the seasoned specialist. Lawrence and Bossert suggest that osteologically, "big dogs look rather as if they had outgrown themselves and were never meant to be that size." See Lawrence and Bossert, "Multiple Character Analysis of *Canis lupus, latrans,* and *familiaris,*" 225.

70. Taniguchi Kengo, *Inu no Nihonshi: Ningen to tomo ni ayunda ichiman nen no monogatari* (The dog in Japanese history: A tale of walking as a friend of humans for ten thousand years) (Tokyo: PHP Kenkyūjo, 2000), 16–24.

71. For more on the relationship between Native Americans and dogs, see Marion Schwartz, *A History of Dogs in the Early Americas* (New Haven: Yale University Press, 1997).

72. Michael Pollan, *The Botany of Desire: A Plant's-Eye View of the World* (New York: Random House, 2002), xviii.

73. Tsukamoto Manabu, *Edo jidaijin to dōbutsu* (The people of the Edo period and animals) (Tokyo: Nihon Editā Sukūru Shuppanbu, 1995), 235.

74. Aaron Skabelund, "Loyalty and Civilization: A Canine History of Japan, 1850–2000" (Ph.D. diss., Columbia University, 2004).

75. Tsukamoto, *Edo jidaijin to dōbutsu,* 233; Edward S. Morse, *Japan Day by Day: 1877, 1878–79, 1882–83,* vol. 2 (Boston: Houghton Mifflin, 1945; reprint, Atlanta, GA: Cherokee Publishing Co., 1990), 18 (all page numbers cited are to the reprint edi-tion). See also Aaron Skabelund, "Civilizing the Streets and Defending National Territory: The Creation of the 'Japanese' Dog, 1853–1941," paper presented at the Japanimals Symposium, Columbia University, 2001.

76. Tsukamoto, *Edo jidaijin to dōbutsu,* 233. See also Skabelund, "Civilizing the Streets and Defending National Territory."

77. Harriet Ritvo, *The Animal Estate: The English and Other Creatures in the Victorian Age* (Cambridge: Harvard University Press, 1987), 82–121. See also Kathleen Kete, *The Beast in the Boudoir: Petkeeping in Nineteenth-Century Paris* (Berkeley and Los Angeles: University of California Press, 1994).

78. Ernest Mason Satow and Lieutenant A. G. S. Hawes, *A Handbook for Travellers in Central and Northern Japan* (London: John Murray, 1884), 40. In a different (and more contemporary) sort of example, one involving foxes rather than dogs, Asahi Minoru, a leading mammalogist, told the story of a taxidermically pre-served specimen displayed at Nanao High School in Ishikawa Prefecture. Poorly pre-served, the specimen's tail had broken off and wires dangled out the tip. High school administrators had no idea when the specimen was made, but it was potentially valu-able because it might be only the fourth such whole wolf specimen in Japan. Asahi and his team took the wolf to a nearby hospital to have an x-ray photograph taken. A photograph of the skull, he reasoned, might provide evidence that the animal was a wolf. However, on returning to his lab in Osaka and conducting a thorough com-parative investigation, he determined that the skull was actually from a fox. The pelt, on the other hand, was clearly not. Asahi then came up with an interesting scenario

to explain the specimen. Some time ago, a hunter on the Noto Peninsula, on the coast of the Japan Sea, killed a wolf, skinned it, and kept the skull, which promised a high price on the open market as a charm or amulet. He decided to give the remainder of the wolf away as a gift to the high school, and so he taxidermically prepared it. But, lacking the skull, he simply could not get the head quite right, so he attached a fox skull to it instead. Even the famous specimen at Wakayama University is missing the lower parts of its legs; hunters no doubt severed them for collecting a bounty. Few specimens of the Japanese wolf exist at all, and there are even fewer of the Hokkaido wolf. Of authentic skull specimens, only twenty-seven of *C. l. hodophilax* are to be found in the entire world, and only six of *C. l. hattai*. See Asahi Minoru, *Nihon no honyū dōbutsu* (Japan's mammals) (Tokyo: Tamakawa Daigaku Shuppanbu, 1977), 127–29.

79. Regarding this "mountain dog," the condylobasal length of the skull, from the aboral border of the occipital condyles to the prosthion, was 182.5 mm, whereas the zygomatic-arch breadth was 108.2 mm. These measurements basically match midsized Japanese breeds such as the Kishū and Shikoku. See Obara Iwao and Nakamura Kazue, "Minami Ashigara shi kyōdo shiryōkan shozō, iwayuru yamainu tōkotsu ni tsuite" (Notes on the so-called *yamainu*, or wild canine, preserved in the Minami Ashigara City Folklore Museum), *Kanagawa kenritsu hakubutsukan kenkyū hōkoku: Shizen kagaku* (Bulletin of the Kanagawa Prefectural Museum: Natural sciences) 21 (March 1992): 105–10.

80. Nakamura, "Nihon ōkami no bunrui ni kansuru seibutsu-chirigakuteki shiten," 51–52.

81. See *Asahi shinbun* (Asahi newspaper), July 1, 2004, at www.asahi.com/national/update/0627/013.html. I would like to thank Aaron Skabelund for bringing this article to my attention.

82. Lawrence and Bossert, "Multiple Character Analysis of *Canis lupus, latrans,* and *familiaris*," 225.

83. Bill Devall, "The Deep Ecology Movement," in *Ecology,* ed. Carolyn Merchant, Key Concepts in Critical Theory (Atlantic Highlands, NJ: Humanities Press, 1994), 128.

2 / CULTURE AND THE CREATION
OF JAPAN'S SACRED WOLVES

1. Ichiro Hori, *Folk Religion in Japan: Continuity and Change,* ed. Joseph M. Kitagawa and Alan L. Miller (Chicago: University of Chicago Press, 1968), 143.

2. Ibid., 144, 150, 166. See also Hiromasa Ikegami, "The Significance of Mountains in the Popular Beliefs in Japan," in *Religious Studies in Japan,* ed. Japanese Association

for Religious Studies and Japanese Organizing Committee of the Ninth International Congress for the History of Religions (Tokyo: Maruzen Company, 1959), 152–60; Ichiro Hori, "Mountains and Their Importance for the Idea of the Other World in Japanese Folk Religion," *History of Religions* 6, no. 1 (August 1966): 1–23.

3. Gerald Figal, *Civilization and Monsters: Spirits of Modernity in Meiji Japan* (Durham, NC: Duke University Press, 1999), 138–40.

4. Hori, *Folk Religion in Japan*, 167–69. For one of Yanagita Kunio's more important texts related to the development of his theory of mountain worship in Japan, see *Yama no jinsei* (The lives of mountain people) [1925], in *Yanagita Kunio shū* (Yanagita Kunio collection), vol. 4 (Tokyo: Chikuma Shobō, 1968).

5. For more on the relationships between Shinto and the environment, see Sonoda Minoru, "Shinto and the Natural Environment," in *Shinto in History: Ways of the Kami*, ed. John Breen and Mark Teeuwen (Honolulu: University of Hawai'i Press, 2000), 32–46. See also my "Shinto," in *Encyclopedia of World Environmental History*, vol. 3, ed. Shepard Krech III, John R. McNeill, and Carolyn Merchant (New York: Routledge, 2004), 1110–15.

6. Later scholars, such as Ue Toshikatsu, have also explored the relationship between "mountain people" and animals. For example, see the stories contained in Ue's *Yamabito no dōbutsushi: Kishū Hatenashi sanmyaku haruaki* (The animal journal of a mountain person: Spring and fall in the Kishū and the Hatenashi Mountain Range) (Tokyo: Shinjuku Shobō, 1998).

7. Yanagita Kunio, *Tōno monogatari* (Tales of Tōno) [1910], in *Yanagita Kunio shū* (Yanagita Kunio collection), vol. 4 (Tokyo: Chikuma Shobō, 1968), 5, 11, 21–25, 38–40, 43–44, 45. For a translation, see Kunio Yanagita, *The Legends of Tōno*, trans. Ronald A. Morse (Tokyo: Japan Foundation, 1975), 5, 12, 31–39, 61–64, 71–72, 75.

8. On the relationship between Taoism and Shinto, see Tim Barrett, "Shinto and Taoism in Early Japan," in *Shinto in History: Ways of the Kami*, ed. John Breen and Mark Teeuwen (Honolulu: University of Hawai'i Press, 2000), 13–31.

9. William R. LaFleur, "Saigyō and the Buddhist Value of Nature," in *Nature in Asian Traditions of Thought: Essays in Environmental Philosophy*, ed. J. Baird Callicott and Roger T. Ames (Albany: State University of New York Press, 1989), 183, 190–91, 195, 200–208.

10. Duncan Ryūken Williams, "Animal Liberation, Death, and the State: Rites to Release Animals in Medieval Japan," in *Buddhism and Ecology: The Interconnection of Dharma and Deeds*, ed. Mary Evelyn Tucker and Duncan Ryūken Williams (Cambridge: Harvard University Press, 1997), 149.

11. Jane Marie Law, "Violence, Ritual Reenactment, and Ideology: The *Hōjō-e* (Rite for Release of Sentient Beings) of the Usa Hachiman Shrine in Japan," *History of Religions* 33, no. 4 (1994): 325–26.

12. Williams, "Animal Liberation, Death, and the State," 149–57; Law, "Violence, Ritual Reenactment, and Ideology," 325–57.

13. Karen A. Smyers, *The Fox and the Jewel: Shared and Private Meanings in Contemporary Japanese Inari Worship* (Honolulu: University of Hawai'i Press, 1999), 15–22, 43, 47.

14. *Nihongi: Chronicles of Japan from the Earliest Times to A.D. 697*, 2 vols. in 1, trans. W. G. Aston (Rutland, VT: Charles E. Tuttle Co., 1972), 2:36–37. See also Okada Akio, *Dōbutsu: Nihonshi shōhyakka* (Animals: An Encyclopedia of Japanese history) (Tokyo: Kondō Shuppansha, 1979), 98. Some of this ground is also covered in *Kokushi daijiten*, vol. 2, ed. Kokushi Daijiten Henshū Iinkai (The comprehensive dictionary of Japanese history) (Tokyo: Yoshikawa Kōbunkan, 1980), 523.

15. Ihara Saikaku, *Honchō nijū fukō* (Twenty stories of unfilial behavior in the realm) [1686], in *Taiyaku Saikaku zenshū* (The complete works of Saikaku with transliteration), vol. 10, ed. Asō Isoji and Fuji Akio (Tokyo: Meiji Shoin, 1976), 13–21. See also Okada, *Dōbutsu*, 99.

16. For more on Shinto, see Sokyo Ono, *Shinto: The Kami Way* (Rutland, VT: Charles E. Tuttle Co., 1962); Helen Hardacre, *Shintō and the State, 1868–1988* (Princeton: Princeton University Press, 1989); John K. Nelson, *A Year in the Life of a Shinto Shrine* (Seattle: University of Washington Press, 1996); John Breen and Mark Teeuwen, eds., *Shinto in History: Ways of the Kami* (Honolulu: University of Hawai'i Press, 2000).

17. *Kofudoki itsubun* (Lost writings on ancient customs) [713], ed. Kurita Hiroshi (Tokyo: Ō Okayama Shoten, 1927), 13.

18. On Ōguchi no Magami, or Large-Mouthed Pure God, as a literary metaphor, see Asahi Minoru, *Nihon no honyū dōbutsu* (Japan's mammals) (Tokyo: Tamagawa Daigaku Shuppanbu, 1977), 126.

19. *Man'yōshū*, vol. 2, in *Shinchō Nihon koten shūsei*, vol. 21, ed. Aoki Takako, Ide Itaru, Itō Haku, Shimizu Katsuhiko, and Hashimoto Shirō (Tokyo: Shinchōsha, 1978), 360.

20. *The Man'yōshū* (Collection of ten thousand leaves) [c. 759], vol. 13, trans. Nippon Gakujutsu Shinkōkai (New York: Columbia University Press, 1965), 3268–69. See also *Man'yōshū*, vol. 4, in *Shinchō Nihon koten shūsei*, vol. 55, ed. Aoki Takako, Ide Itaru, Itō Haku, Shimizu Katsuhiko, and Hashimoto Shirō (Tokyo: Shinchōsha, 1982), 45. I have modified this translation slightly.

21. *Man'yōshū*, trans. Nippon Gakujutsu Shinkōkai, 39. See also *Man'yōshū*, vol. 1, in *Shinchō Nihon koten shūsei*, vol. 6, ed. Aoki Takako, Ide Itaru, Itō Haku, Shimizu Katsuhiko, and Hashimoto Shirō (Tokyo: Shinchōsha, 1976), 138–39. On the establishment of the Kiyomihara Palace at Asuka, see Joan R. Piggott, *The Emergence of Japanese Kingship* (Stanford: Stanford University Press, 1997), 131–35.

22. Barry Holstun Lopez, *Of Wolves and Men* (New York: Simon and Schuster, 1978), 204.

23. Pamela J. Asquith and Arne Kalland, "Japanese Perceptions of Nature: Ideals and Illusions," in *Japanese Images of Nature: Cultural Perspectives*, ed. Pamela J. Asquith and Arne Kalland (London: Curzon Press, 1997), 2–3, 21, 11–16.

24. Nonoguchi Takamasa, comp., *Ōō hitsugo* (Stories of Ōō) [1842], vol. 2, in

Nihon zuihitsu taisei (Survey of Japan's literary miscellany), vol. 9, ed. Nihon Zuihitsu Taisei Henshūbu (Tokyo: Yoshikawa Kōbunkan, 1975), 220–23.

25. Ibid., 221. See also Hiraiwa Yonekichi, *Ōkami: Sono seitai to rekishi* (The wolf: Its ecology and history) (Tokyo: Tsukiji Shokan, 1992), 89.

26. Hiraiwa, *Ōkami*, 89.

27. Legends regarding Yamato Takeru no Mikoto and the Mitsumine Shrine are sometimes discussed in the local journal published by the shrine; for example, see Machida Hiroshi, "Hikage no kazura no mabushii midori" (The radiant green of the shade vine), *Mitsumine-san* (Mount Mitsumine) 168 (April 1992): 6. See also *Nihongi*, 1:208.

28. Ōga Tetsuo, ed., *Nihon daihyakka zensho* (The complete encyclopedia of Japan), vol. 22 (Tokyo: Shōgakukan, 1988), 372; "Saitama-ken" (Saitama Prefecture), in *Kadokawa Nihon chimei daijiten* (The Kadokawa dictionary of Japanese place-names), vol. 11, ed. Takeuchi Rizō (Tokyo: Kadokawa Shoten, 1980), 812; Hiraiwa, *Ōkami*, 90.

29. From the sixteenth century, the Kumano shrines also fostered close ties to the Shōgo'in sect; see Miyake Hitoshi, *Shugendō: Essays on the Structure of Japanese Folk Religion* (Ann Arbor: Center for Japanese Studies, University of Michigan, 2001), 25–27. For more specific information related to Mitsumine Shrine and Shugendō traditions, see Yokoyama Haruo, "Honzanha shugen Mitsumine-san no kōryū" (The rise of Honzan school mountain asceticism at Mount Mitsumine), *Kokugakuin zasshi* (Journal of Kokugakuin University) 80, no. 10 (October 1979): 284–96; Miyake Hitoshi, "Mitsumine-san no shugen" (The mountain asceticism of Mount Mitsumine), *Mitsumine-san* (Mount Mitsumine) 144 (April 1994): 6–7. On Shugendō and Kōfukuji, see Royall Tyler, "Kōfuku-ji and Shugendō," *Japanese Journal of Religious Studies* 16, nos. 2–3 (June–September 1989): 143–80.

30. *Kokushi daijiten*, 13:380–81; *Kadokawa Nihon chimei daijiten*, 11:812.

31. Hiraiwa, *Ōkami*, 89–90.

32. *The Upanishads*, trans. Juan Mascaró (New York: Penguin Books, 1965), 83–84.

33. For more on the raiding of mushroom crops by macaques, see John Knight, *Waiting for Wolves in Japan: An Anthropological Study of People-Wildlife Relations* (Oxford: Oxford University Press, 2003), 87–92.

34. *Nanpō zuihitsu* (Literary miscellany on southern regions), cited in Okada, *Dōbutsu*, 98–99.

35. Negishi Yasumori, *Mimibukuro* (Ear bag) [1782–1814], vol. 1, ed. Suzuki Tōzō (Tokyo: Heibonsha, 2000), 242–43.

36. U. A. Casal, "The Goblin Fox and Badger and Other Witch Animals of Japan," *Folklore Studies* 18 (1959): 82–84.

37. Anne Walthall, *The Weak Body of a Useless Woman: Matsuo Taseko and the Meiji Restoration* (Chicago: University of Chicago Press, 1998), 136.

38. John Knight, "On the Extinction of the Japanese Wolf," *Asian Folklore Studies*, 56 (1997): 140.

39. Sutō Isao, *Yama no hyōteki: Inoshishi to yamabito no seikatsushi* (Target in the mountains: A history of the daily lives of wild boars and mountain people) (Tokyo: Miraisha, 1991). See also Knight, *Waiting for Wolves in Japan*, 51–68.

40. Constantine Vaporis, *Breaking Barriers: Travel and the State in Early Modern Japan* (Cambridge: Harvard University Press, 1994).

41. For more on the sending wolf, see Knight, "On the Extinction of the Japanese Wolf," 136. For Okuriōkami lore from the Higashi Yoshino region, see Higashi Yoshino-mura Kyōiku Iinkai, ed., *Higashi Yoshino no minwa* (The folklore of Higashi Yoshino) (Osaka: Gyōsei, 1992), 245–47.

42. Emiko Ohnuki-Tierney, *The Monkey as Mirror: Symbolic Transformations in Japanese History and Ritual* (Princeton: Princeton University Press, 1987).

43. Knight, "On the Extinction of the Japanese Wolf," 136.

44. Chiba Tokuji, *Shuryō denshō* (Hunting folklore) (Tokyo: Hōsei Daigaku Shuppankyoku, 1975), 155; Okada, *Dōbutsu*, 99; Hiraiwa, *Ōkami*, 88. See also Terashima Ryōan, comp., *Wakan sansai zue* (The illustrated Japanese-Chinese encyclopedia of the three elements) [1713], in *Nihon shomin seikatsu shiryō shūsei* (Collected sources on the history of the daily lives of common Japanese people), vol. 10, ed. Miyamoto Tsuneichi, Haraguchi Torao, and Tanigawa Ken'ichi (Tokyo: San'ichi Shobō, 1970), 541–42.

45. Negishi, *Mimibukuro*, 2:87–88.

46. For more on the travels and maps of Furukawa Koshōken, see Marcia Yonemoto, *Mapping Early Modern Japan: Space, Place, and Culture in the Tokugawa Period (1603–1868)* (Berkeley and Los Angeles: University of California Press, 2003).

47. "Miyagi-ken" (Miyagi Prefecture), in *Kadokawa Nihon chimei daijiten* (The Kadokawa dictionary of Japanese place-names), vol. 4, ed. Takeuchi Rizō (Tokyo: Kadokawa Shoten, 1979), 105.

48. Furukawa Koshōken, *Tōyū zakki* (Miscellaneous notes on travels in the east) [1788], ed. Ōtō Tokihiko (Tokyo: Tōyō Bunko, 1964), 223.

49. Ibid. See also Chiba, *Shuryō denshō*, 157; Hiraiwa, *Ōkami*, 89.

50. Nishimura Hakū, *Enka kidan* (Strange stories from the mist) [1773], in *Nihon zuihitsu taisei* (Survey of Japan's literary miscellany), vol. 4, ed. Nihon Zuihitsu Taisei Henshūbu (Tokyo: Yoshikawa Kōbunkan, 1975), 210.

51. On the medieval Hōjō-e Ceremony, see Williams, "Animal Liberation, Death, and the State," 149–57. On the Buddha-nature of plants and trees, see LaFleur, "Saigyō and the Buddhist Value of Nature."

52. Harold Bolitho, "The Dog Shogun," in *Self and Biography: Essays on the Individual and Society in Asia*, ed. Wang Gungwu (Sydney: Sydney University Press, 1975), 127–28. See also Donald H. Shively, "Tokugawa Tsunayoshi, the Genroku Shogun," in *Personality in Japanese History*, ed. Albert M. Craig and Donald H. Shively (Berkeley and Los Angeles: University of California Press, 1970; reprint, Ann Arbor: Center for Japanese Studies, University of Michigan, 1995), 85–126 (all page num-

bers cited are to the reprint edition); Beatrice Bodart Bailey, "The Laws of Compassion," *Monumenta Nipponica* 40, no. 2 (1985): 163–89.

53. *Tokugawa jikki* (The true chronicle of the Tokugawa) [1809–43], vol. 6, in *Shintei zōho kokushi taikei* (Newly revised and enlarged survey of Japanese history), vol. 43, ed. Kuroita Katsumi (Tokyo: Yoshikawa Kōbunkan, 1931), 47. See also Hiraiwa, *Ōkami*, 104.

54. *Tokugawa jikki*, 71.

55. Ibid., 169.

56. Ibid., 230.

57. Hiraiwa, *Ōkami*, 104–6.

58. Ibid., 106–7.

59. On the Ainu, see my *The Conquest of Ainu Lands: Ecology and Culture in Japanese Expansion, 1590–1800* (Berkeley and Los Angeles: University of California Press, 2001).

60. Inukai Tetsuo, "Hokkaidō-san ōkami to sono metsubō keiro" (The Hokkaido wolf and its road to extinction), *Shokubutsu oyobi dōbutsu* (Botany and zoology) 1, no. 8 (August 1933): 12.

61. I am relying on the version of this origin myth contained in Sarashina Genzō and Sarashina Kō, *Kotan seibutsuki: Yajū-kaijū-gyozoku hen* (A biological chronicle of Ainu villages: Volume on animals, marine mammals, and fishes), vol. 2 (Tokyo: Hōsei Daigaku Shuppankyoku, 1976), 291–92. A variant, English version of this story can be found in Carl Etter, *Ainu Folklore: Traditions and Culture of the Vanishing Aborigines of Japan* (Chicago: Wilcox and Follett Co., 1949), 20–21.

62. Sarashina and Sarashina, *Kotan seibutsuki*, 2:292. Regional variations of the name for the Hokkaido wolf expose the respect Ainu had for its hunting ability. In the Kushiro region, for example, the wolf was sometimes called Onrupp Kamuy, the God Who Hunts, while Ainu from the Tokachi region called the wolf Yukkoiki Kamuy, the God Who Takes Deer. According to Chiri Mashiho, who studied Ainu folklore, regional variations of the name for wolf include Horkew or Horkew Kamuy (in the regions of Iburi, Hidaka, Sorachi, Ishikari, Teshio, and Taraika); Horokew or Horokew Kamuy (in Maoka, Shiraura, Ochiho, and Niitoi); Orkew (in Ashoro and Biboro); and Onrupus Kamuy (in Biboro, Kutcharo, Harutori, Shiranuka, and Fushiko). Other, lesser-known variations include Nupuripaunkur, the God Hidden in the Mountains, and Nupuripakor Kamuy, a term that also refers to the presence of the wolf god in the mountains. Chiri wrote that some Ainu elders insisted that the name Ose Kamuy, sometimes used to describe the wolf, actually referred to a mountain dog, raising some speculation of a possible second subspecies on Hokkaido or large numbers of feral dogs. Ainu actually called feral dogs Kimuy Seta, or the Dog That Lives in the Mountains (of course, reminding one of the mountain dogs of mainland Japan), and such canines apparently thrived around Shiraoi and Horobetsu. See Chiri Mashiho, *Chiri Mashiho chosakushū: Bunrui Ainugo jiten—*

shokubutsu dōbutsu (The collected works of Chiri Mashiho: Ainu language by classification—plants and animals) (Tokyo: Heibonsha, 1976), appendix 1:141–42.

63. Sarashina and Sarashina, *Kotan seibutsuki*, 2:289.

64. Walter McClintock, *The Old North Trail, or Life, Legends, and Religions of the Blackfeet Indians* (London: Macmillan and Co., 1910; reprint, Lincoln: University of Nebraska, 1999), 473–78 (all page numbers cited are to the reprint edition). See also Rick McIntyre, ed., *War against the Wolf: America's Campaign to Exterminate the Wolf* (Stillwater, MN: Voyageur Press, 1995), 267–68.

65. See Nagata Hōsei, *Hokkaidō Ezogo chimei kai* (Understanding Hokkaido's Ainu-language place-names) (Sapporo: Hokkaidō Chō, 1891), 1:25 and 2:72; Inukai, "Hokkaidō-san ōkami to sono metsubō keiro," 12; Chiri, *Chiri Mashiho chosakushū*, appendix 1:142; Sarashina and Sarashina, *Kotan seibutsuki*, 2:290.

66. Sarashina and Sarashina, *Kotan seibutsuki*, 2:291.

67. Uchida Kiyoshi, Tauchi Suteroku, and Fujita Kusaburō, "Hidaka Tokachi Kushiro Kitami Nemuro junkai fukumeisho" (Documentary report of tour of Hidaka, Tokachi, Kushiro, Kitami, and Nemuro) [1881], Resource Collection for Northern Studies, Hokkaido University Library, Sapporo, Japan. This text is mentioned in Inukai, "Hokkaidō-san ōkami to sono metsubō keiro," 12; and in Sarashina and Sarashina, *Kotan seibutsuki*, 2:291.

68. Sarashina and Sarashina, *Kotan seibutsuki*, 2:291. See also Tawara Hiromi, *Hokkaidō no shizen hogo* (Protecting Hokkaido's nature) (Sapporo: Hokkaidō Daigaku Tosho Kankōkai, 1979), 104.

69. Hatakeyama Saburōta, "Hokkaidō no inu ni tsuite no oboegaki: Senshi jidai kaizuka-ken to Ainu-ken no hikaku" (A memorandum on the Hokkaido dog: A comparison of the prehistoric shell mound dog to the Ainu dog), *Hokkaidōshi no kenkyū* (Research on Hokkaido history) 1 (December 1973): 41. For information related to Hokkaido and Sakhalin dogs, see Kushihara Seihō, *Igen zokuwa* (Ainu proverbs and gossip) [1792], in *Nihon shomin seikatsu shiryō shūsei* (Collected sources on the history of the daily lives of common Japanese people), vol. 4, ed. Takakura Shin'ichirō (Tokyo: San'ichi Shobō, 1969), 492–93.

70. Matsuura Takeshirō, *Shinpan Ezo nisshi: Higashi Ezo nisshi* (Newly published Ezo diary: Eastern Ezo diary), vol. 1, ed. Yoshida Tsunekichi (Tokyo: Jiji Tsūshinsha, 1984), 159. Actually, understanding the origins of the Ainu dog is important because some scholars assume that the canine was a locally bred, domesticated wolf. Others argue the opposite, that it was a northern variation on such Japanese breeds as the Shiba or Kishū. Saitō Hirokichi, a canine specialist, held that Ainu dogs were closely related to southern Japanese breeds and not related to, say, Sakhalin dogs or local wolves. Saitō argued that these dogs had lived with earlier residents of Hokkaido (who originally had contact with Honshu) and, after becoming isolated, evolved into a distinct northern breed. In other words, the evolution of the Ainu dog needs to be viewed in the same light as the development of Ainu culture: the product of interaction between the Epi-Jōmon and later Satsumon and Okhostk cultures. Ainu

dogs, like the Ainu themselves, probably included traces of these three early cultures. The Moyoro shell mounds, for example, which contain relics of the Okhostk culture, also contain the remnants of dogs, suggesting that Okhotsk people had their own breeds, and it remains likely that these dogs bred with Epi-Jōmon canines, creating the Ainu dog. Japanese and Ainu dogs may have had similar distant ancestors in the Jōmon world; but Ainu dogs later became distinct as they bred with northern Okhostk breeds and, as Kaitakushi officials observed, occasionally even the Hokkaido wolf. See Hatakeyama, "Hokkaidō no inu ni tsuite no oboegaki," 42–46.

71. Tessa Morris-Suzuki, "Creating the Frontier: Border, Identity, and History in Japan's Far North," *East Asian History* 7 (June 1994): 18–23.

72. Inukai Tetsuo, *Hoppō dōbutsushi* (Northern animal journal) (Sapporo: Hokuensha, 1975), 20–21.

73. Hiraga Satano, "Horkew kotan kor kur" (Japanese: Ōkami-kami to mura osa; English: The wolf god and the village man) [October 13, 1967], in *Kayano Shigeru no Ainu shinwa shūsei: Kamuy yukar* (Kayano Shigeru's collection of Ainu divine tales: Kamuy-yukar), vol. 2, ed. Kayano Shigeru (Tokyo: Heibonsha, 1998), 59.

74. Ibid., 61.

75. Ibid., 63.

76. Ibid., 65.

77. Ibid., 67.

78. Ibid., 81.

79. Ibid., 83.

80. Ibid., 85.

81. Ibid., 87.

82. Ibid., 89.

83. Kayano Shigeru, *Honoo no uma: Ainu minwashū* (Horse of fire: A collection of Ainu folklore) (Tokyo: Suzusawa Shoten, 1977), 72–82.

84. Kayano Shigeru and Saitō Hiroyuki, *Kibori no ōkami* (The carved wooden wolf) (Tokyo: Shōhō Shoten, 1998).

85. Inukai, "Hokkaidō-san ōkami to sono metsubō keiro," 12.

86. Toyohara Shōji, "Unmemke ni okeru senkō ichi ni tsuite" (On the positioning of the puncture in the *unmemke*), *Kan Ohōtsuku* (Circle Okhotsk) 3 (1995): 63–69.

3. THE CONFLICTS BETWEEN WOLF HUNTERS AND RABID MAN-KILLERS IN EARLY MODERN JAPAN

1. Shiga Naoya, "Takibi" (Night fires) [1920], in *Gendai Nihon bungaku taikei: Shiga Naoya shū* (Survey of Japan's modern literature: Shiga Naoya collection), vol. 34 (Tokyo: Chikuma Shobō, 1968), 287–92. For two English translations, see William F. Sibley, *The Shiga Hero* (Chicago: University of Chicago Press, 1979),

186–97; Theodore W. Goossen, ed., *The Oxford Book of Japanese Short Stories* (Oxford: Oxford University Press, 1997), 52–61.

2. The following discussion is largely based on Hiraiwa Yonekichi, *Ōkami: Sono seitai to rekishi* (The wolf: Its ecology and history) (Tokyo: Tsukiji Shokan, 1992), 85–86.

3. Ivan Morris, *The World of the Shining Prince: Court Life in Ancient Japan* (New York: Kōdansha International, 1964), 19.

4. Ronald P. Toby, "Why Leave Nara? Kanmu and the Transfer of the Capital," *Monumenta Nipponica* 40, no. 3 (Autumn 1985): 331–47.

5. Conrad Totman, *The Green Archipelago: Forestry in Pre-industrial Japan* (Berkeley and Los Angeles: University of California Press, 1989; reprint, Athens: Ohio University Press, 1998), 12–16 (all page numbers cited are to the reprint edition).

6. "Kamo no Chōmei: An Account of My Hut," trans. Donald Keene, in *Anthology of Japanese Literature from the Earliest Era to the Mid–Nineteenth Century*, ed. Donald Keene (New York: Grove Press, 1999), 202–3.

7. Pierre François Souyri, *The World Turned Upside Down: Medieval Japanese Society*, trans. Käthe Roth (New York: Columbia University Press, 2001), 177–78.

8. Hiraiwa, *Ōkami*, 135–36.

9. *Kofudoki itsubun* (Lost writings on ancient customs) [713], ed. Kurita Hiroshi (Tokyo: Ō Okayama Shoten, 1927), 13. See also Hiraiwa, *Ōkami*, 84.

10. *Nihon kōki* (Japan's postscript) [796–833], in *Shintei zōho kokushi taikei* (Newly revised and enlarged survey of Japanese history), vol. 3, ed. Kuroita Katsumi (Tokyo: Yoshikawa Kōbunkan, 1961), 104. See also Hiraiwa, *Ōkami*, 85.

11. See Hiraiwa, *Ōkami*, 85, 100.

12. Ella E. Clark, *Indian Legends from the Northern Rockies* (Norman: University of Oklahoma Press, 1966), 310–12. See also Denise Casey and Tim W. Clark, comps., *Tales of the Wolf: Fifty-one Stories of Wolf Encounters in the Wild* (Moose, WY: Homestead Publishing, 1996), 4–6.

13. See Hiraiwa, *Ōkami*, 85.

14. Fujiwara Mototsune, *Nihon Montoku tennō jitsuroku* (The true record of Japan's Emperor Montoku) [878], in *Shintei zōho kokushi taikei* (Newly revised and enlarged survey of Japanese history), vol. 3, ed. Kuroita Katsumi (Tokyo: Yoshikawa Kōbunkan, 1981), 75. See also Hiraiwa, *Ōkami*, 85.

15. Both accounts from Fujiwara Tokihira, *Nihon sandai jitsuroku* (The true record of Japan's three eras) [858–87], vol. 2, in *Shintei zōho kokushi taikei* (Newly revised and enlarged survey of Japanese history), vol. 4, ed. Kuroita Katsumi (Tokyo: Yoshikawa Kōbunkan, 1973), 507, 618. See also Hiraiwa, *Ōkami*, 85–86.

16. Hiraiwa, *Ōkami*, 85.

17. Ibid., 85–86.

18. Ibid., 128.

19. Ibid., 132.

20. The economy of Morioka domain is discussed in Susan B. Hanley and Kozo

Yamamura, *Economic and Demographic Change in Preindustrial Japan, 1600–1868* (Princeton: Princeton University Press, 1977).

21. Tsumura Masayuki, *Tankai* (Sea of conversations) [1795] (Tokyo: Kokusho Kankōkai, 1917), 125.

22. Hiraiwa, *Ōkami*, 128–29.

23. Tsukamoto Manabu, *Shōrui o meguru seiji* (The politics of the clemency for living creatures) (Tokyo: Heibonsha, 1993), 9–95.

24. Hiraiwa, *Ōkami*, 129.

25. Ibid., 129–30.

26. Ibid., 130.

27. Ibid.

28. Negishi Yasumori, *Mimibukuro* (Ear bag) [1782–1814], vol. 2, ed. Suzuki Tōzō (Tokyo: Heibonsha, 2000), 417.

29. Hiraiwa, *Ōkami*, 130.

30. Ernest Thompson Seton, *Life-Histories of Northern Animals: An Account of the Mammals of Manitoba*, vol. 2 (New York: Doubleday and Co., 1909; reprint, New York: Arno Press, 1974), 765 (all page numbers cited are to the reprint edition).

31. Hiraiwa, *Ōkami*, 131.

32. Tsumura, *Tankai*, 253–54. See also Hiraiwa, *Ōkami*, 131.

33. Tanamori Fusaaki, *Tanamori Fusaaki shuki* (The notes of Tanamori Fusaaki) [1580], in *Zokuzoku gunsho ruijū*, vol. 4, ed. Kokusho Kankōkai (Tokyo: Zoku Gunsho Ruijū Kanseikai, 1970), 167; Hiraiwa, *Ōkami*, 131–32.

34. Hiraiwa, *Ōkami*, 132.

35. Ibid., 132–33.

36. T. Fujitani, *Splendid Monarchy: Power and Pageantry in Modern Japan* (Berkeley and Los Angeles: University of California Press, 1996).

37. Kishida Ginkō, "Tōhoku gojunkōki" (Northeastern imperial procession chronicle) [1876], in *Meiji bunka zenshū* (The complete works of Meiji culture), vol. 17, ed. Meiji Bunka Kenkyūkai (Tokyo: Nihon Hyōronsha, 1967), 383. See also Hiraiwa, *Ōkami*, 133.

38. Tadano Makuzu, *Ōshū banashi* (Tales from Ōshū) [1832], in *Tadano Makuzu shū* (Tadano Makuzu collection), revised by Suzuki Yoneko (Tokyo: Kabushikigaisha Kokusho Kankōkai, 1994), 233–36. See also Hiraiwa, *Ōkami*, 113–14.

39. C. H. D. Clarke, "The Beast of Gévaudan," *Natural History* 80, no. 4 (April 1971): 46. See also Barry Holstun Lopez, *Of Wolves and Men* (New York: Simon and Schuster, 1978), 69–71.

40. Clarke, "Beast of Gévaudan," 44–73.

41. Charles Larpenteur, *Forty Years a Fur Trader on the Upper Missouri: The Personal Narrative of Charles Larpenteur, 1833–72*, vol. 1, ed. Elliot Cous (New York: Francis P. Harper, 1898), 36–41; W. A. Ferris, *Life in the Rocky Mountains: A Diary of Wandering on the Sources of the Rivers Missouri, Columbia, and Colorado from February, 1830, to November, 1835*, ed. Paul C. Phillips (Denver: Fred A. Rosenstock,

Old West Publishing Co., 1940), 265–66; Washington Irving, *Adventures of Captain Bonneville* (Portland, OR: Binfords and Mort, Publishers, 1963), 145–46. For excerpts, see Casey and Clark, *Tales of the Wolf*, 102–4.

42. John Richardson, *Fauna Boreali-Americana; or the Zoology of the Northern Parts of British America*, vol. 1 (London: John Murray, 1829), 64. See also Casey and Clark, *Tales of the Wolf*, 75.

43. Richard Irving Dodge, *The Plains of North America and Their Inhabitants* (Newark: University of Delaware Press, 1989), 124–25. See also Casey and Clark, *Tales of the Wolf*, 129–30.

44. Ernest S. Tierkel, "Canine Rabies," in *The Natural History of Rabies*, vol. 2, ed. George M. Baer (New York: Academic Press, 1975), 123–37; K. M. Charlton, "The Pathogenesis of Rabies and Other Lyssaviral Infections: Recent Studies," in *Lyssaviruses*, ed. C. E. Rupprecht, B. Dietzschold, and H. Koprowski (Berlin: Springer-Verlag, 1994), 95–119.

45. On the introduction of smallpox to Japan, see William Wayne Farris, *Population, Disease, and Land in Early Japan, 645–900* (Cambridge: Harvard University Press, 1985), 50–73. Whether smallpox came from Korean fishermen in the eighth century and whether new strains of tuberculosis originated with Western visitors in the nineteenth century, as currently portrayed in mainstream Japanese historiography, disease has always been imported to Japan. For thoughts on foreigners and tuberculosis in Japan, see William Johnston, *The Modern Epidemic: A History of Tuberculosis in Japan* (Cambridge: Harvard University Press, 1995), 50. Alfred Crosby observes that throughout the world syphilis in particular has been associated with foreign countries. In Japan it was sometimes called the "Tang sore," a reference to China. See Alfred W. Crosby, *The Columbian Exchange: Biological and Cultural Consequences of 1492* (Westport, CT: Greenwood Press, 1972), 124–25.

46. Hiraiwa, *Ōkami*, 108. Noro also compiled a botanical and zoological taxonomy, entitled *Oranda kinjūchūgyozu wage* (Japanese version of Dutch explanations of drawings of birds, beasts, insects, and fish; 1741), which gave Dutch, Latin, and Japanese names for a variety of creatures. Later, he focused on herbal medicines in *Oranda honzō wage* (Japanese version of Dutch natural studies; 1750). See Masayoshi Sugimoto and David L. Swain, *Science and Culture in Traditional Japan*, A.D. 600–1854 (Cambridge: MIT Press, 1978), 315–16.

47. Kanzawa Teikan, *Okinagusa* (Pasqueflower) [1851], vol. 5, in *Nihon zuihitsu taisei* (Survey of Japan's literary miscellany), 3rd ed., vol. 23, ed. Nihon Zuihitsu Taisei Henshūbu (Tokyo: Yoshikawa Kōbunkan, 1978), 254–55. See also Hiraiwa, *Ōkami*, 108–9.

48. Nishimura Hakū, *Enka kidan* (Strange stories from the mist) [1773], in *Nihon zuihitsu taisei* (Survey of Japan's literary miscellany), vol. 4, ed. Nihon Zuihitsu Taisei Henshūbu (Tokyo: Yoshikawa Kōbunkan, 1975), 210. See also Hiraiwa, *Ōkami*, 109.

49. Hiraiwa, *Ōkami*, 109.

50. "Ōkami Yuatsumi o arashi jūkyū nin o kuitaru koto" (Incident of wolf vio-

lently eating nineteen people in Yuatsumi), in *Shōnai shiryō* (Shōnai sources), ed. Tetsuya Shigeta (Tsuruoka: Tōhoku Shuppan Kikaku, 1912), 12. See also Hiraiwa, *Ōkami*, 109–10.

51. Tachibana Nankei, *Hokusō sadan* (Brief conversations through a northern window) [1825], vol. 2, in *Nihon zuihitsu taisei* (Survey of Japan's literary miscellany), 2nd ed., vol. 15, ed. Nihon Zuihitsu Taisei Henshūbu (Tokyo: Nihon Zuihitsu Taisei Kankōkai, 1974), 202. Evidence of people not feeling any pain has led Hiraiwa Yonekichi, in his survey of wolf-bite accounts, to speculate that some sort of paralysis may have set in. Indeed, as discussed earlier, the third and final stage of rabies is the paralytic stage. See Hiraiwa, *Ōkami*, 110.

52. Hiraiwa, *Ōkami*, 110.

53. Ibid., 110–11.

54. Tachibana Nankei, *Tōyūki—Seiyūki* (Lyrical record of a journey to the east—Lyrical record of a journey to the west) [1795], ed. Kurimoto Chikara (Tokyo: Kenkyūsha Gakusei Bunko 307, 1940), 91–114. See also Hiraiwa, *Ōkami*, 111–13.

55. Furukawa Koshōken, *Tōyū zakki* (Miscellaneous notes on travels in the east) [1788], ed. Ōtō Tokihiko (Tokyo: Tōyō Bunko, 1964), 223.

56. Komiyama Masahide, *Fūken gūki* (Whimsical chronicle of maple eaves) [1807–10], in *Nihon zuihitsu taisei* (Survey of Japan's literary miscellany), vol. 10, ed. Nihon Zuihitsu Taisei Henshūbu (Tokyo: Nihon Zuihitsu Taisei Kankōkai, 1929); Hiraiwa, *Ōkami*, 116.

57. Tōdō Takasawa, *Tōdō Takasawa kōdenryaku* (Abridged public biography of Tōdō Takasawa) (Tokyo: Tōdō-ke, 1918), 33. See also Hiraiwa, *Ōkami*, 116–17.

58. Hiraiwa, *Ōkami*, 117.

59. Hokkaidō Keisatsu Henshū Iinkai, ed., *Hokkaidō keisatsushi* (History of the Hokkaido police), vol. 1 (Sapporo: Hokkaidō Keisatsu Honbu, 1968), 188.

60. Harriet Ritvo, *The Animal Estate: The English and Other Creatures in the Victorian Age* (Cambridge: Harvard University Press, 1987).

61. For more on the international influence on the development of rules related to dog ownership on Hokkaido, see "Chikken no kisoku" (Regulations regarding domesticated dogs), *Hokkai tsūshi* (Northern sea interpreter) 6 (June 1880): 6–9.

62. Hokkaidō Keisatsu Henshū Iinkai, *Hokkaidō keisatsushi*, 188–89.

63. For examples of such language related to wild and rabid dogs, see "Inu no ryūkōbyō no ken" (Matter of canine epizootics) [1873.6.2], in Kaitakushi kōbun roku (Record of official Kaitakushi documents) (5750); "Yaken bokusatsu no ken" (Matter of killing wild dogs) [1878.8.19], in Kakugun bun iroku (Transferred records of written documents from all districts) (A4-49); "Akuken karitori no ken" (Matter of hunting and killing bad dogs) [1877.6.19], in Shoka bun iroku (Transferred records of written documents from various divisions) (A4-24); and "Men'yō kyōken no tame daigai o uke ni tsuki arawa ni kansatsu fuyo nai inu no bunshu no ken" (Matter of openly conferring licenses to hunt dogs because of pervasive damage to sheep caused by mad dogs) [1878.6.18], in Kachū bun iroku (Transferred records

of divisional written documents) (A4-27); all in Hokkaido Prefectural Archives, Sapporo, Japan.

64. Rick McIntyre, ed., *War against the Wolf: America's Campaign to Exterminate the Wolf* (Stillwater, MN: Voyageur Press, 1995), 43–47.

65. Takebe Ayatari, *Suzumigusa* (Cool grass) [1771], in *Takebe Ayatari zenshū* (The complete works of Takebe Ayatari), vol. 6 (Tokyo: Kokusho Kankōkai, 1987), 287–88. I am indebted to Lawrence Marceau for not only bringing this source to my attention but also providing a draft translation. For more on the life and writings of Takebe, see Lawrence E. Marceau, *Takebe Ayatari: A Bunjin Bohemian in Early Modern Japan* (Ann Arbor: Center for Japanese Studies, University of Michigan, 2002).

66. Hiraiwa, *Ōkami*, 131–35.

67. "Satake-ke gonikki" (Chronicle of the Satake family) [1673–1854]. I obtained copies held in the private collection of Tozawa Shirō of the Kakunodate Rekishi Annainin Kumiai. For selections from the "Satake-ke gonikki" pertaining to wolf hunts near Kakunodate, see Kakunodateshi Henshū Iinkai, ed., *Kakunodateshi: Kitake jidai 2* (The history of Kakunodate: The northern household period 2), vol. 4 (Tokyo: Daiichi Hōki Shuppan Kabushikigaisha, 1970), 151–55. For a transliteration and description of the 1810 hunt, see Hiraiwa, *Ōkami*, 133–35.

68. Santō Kyōzan, *Oshiegusa nyōbō katagi* (The character of moral wives) [1853], in *Musume setsuyō-Oshiegusa nyōbō katagi* (The frugality of daughters—The character of moral wives), ed. Tsukamoto Tetsuzō (Tokyo: Yūhodo Shobō, 1927), 500–588. In particular, see 503. Also see Hiraiwa, *Ōkami*, 134.

69. Hiraiwa, *Ōkami*, 134.

4 / MEIJI MODERNIZATION, SCIENTIFIC AGRICULTURE, AND DESTROYING THE HOKKAIDO WOLF

1. The classic study of early modern Japanese diplomacy is Ronald P. Toby, *State and Diplomacy in Early Modern Japan: Asia in the Development of the Tokugawa Bakufu* (Princeton: Princeton University Press, 1984; reprint, Stanford: Stanford University Press, 1991). For a historiographical assessment of English-language scholarship on early modern Japan's foreign relations, see my "Foreign Affairs and Frontiers in Early Modern Japan: A Historiographical Essay of the Field," *Early Modern Japan: An Interdisciplinary Journal* 10, no. 2 (Fall 2002): 44–62.

2. For English-language studies of the politics of the early Meiji period, see Joseph Pittau, *Political Thought in Early Meiji Japan, 1868–1889* (Cambridge: Harvard University Press, 1967); W. G. Beasley, *The Meiji Restoration* (Stanford: Stanford University Press, 1972); Marius B. Jansen and Gilbert Rozman, eds., *Japan in Transition: From Tokugawa to Meiji* (Princeton: Princeton University Press, 1986).

For a pioneering discussion of "nature" in Japanese political discourse, including in the early Meiji years, see Julia Adeney Thomas, *Reconfiguring Modernity: Concepts of Nature in Japanese Political Ideology* (Berkeley and Los Angeles: University of California Press, 2001). For fresh thoughts on the Satchō alliance and the character of early Meiji governance, see L. M. Cullen, *A History of Japan, 1582–1941: Internal and External Worlds* (Cambridge: Cambridge University Press, 2003), 218–24. For specific information on American experts and the colonization of Hokkaido, see Fumiko Fujita, *American Pioneers and the Japanese Frontier: American Experts in Nineteenth-Century Japan* (Westport, CT: Greenwood Press, 1994). On early Japanese embassies that brought back information regarding Europe and the United States, see Masao Miyoshi, *As We Saw Them: The First Japanese Embassy to the United States* (New York: Kōdansha International, 1979); Peter Duus, *The Japanese Discovery of America: A Brief History with Documents* (Boston: Bedford Books, 1997); Ian Nish, ed., *The Iwakura Mission in America and Europe: A New Assessment* (Richmond, Surrey, UK: Japan Library, 1998).

3. The Meiji government established the Kaitakushi in July 1869 to oversee the economic development and military defense of Hokkaido. Likewise, between 1871 and 1876 the Kaitakushi oversaw the development and defense of Sakhalin Island and, after 1876, the Kuril Islands as well. Primarily, the Kaitakushi tried to entice Japanese to settle Hokkaido and to develop the island's many natural resources. However, one year after the Hokkaido colonization "assets scandal" of 1881, the Meiji government abolished the Kaitakushi and Hokkaido officially lost its colonial status. On the activities of the Kaitakushi, see John A. Harrison, *Japan's Northern Frontier: A Preliminary Study in Colonization and Expansion with Special Reference to the Relations of Japan and Russia* (Gainesville: University of Florida Press, 1953), 59–139.

4. On the 1871 Iwakura mission, see Beasley, *Meiji Restoration*, 366–74. On the use of the Morrill Act as a model for the establishment of the Sapporo Agricultural College (Hokkaido University), see John M. Maki, *William Smith Clark: A Yankee in Hokkaido* (Sapporo: Hokkaido University Press, 1996), 77–145. In 1872, while viewing military drills at the Massachusetts Agricultural College, a land-grant institution, Mori Arinori, chargé d'affaires at the Japanese legation (1870–72) and later minister of education (1885), exclaimed, "That is the kind of an institution Japan must have, that is what we need, an institution that shall teach young men to feed themselves and to defend themselves" (Ibid., 124).

5. For a history of early modern trade and other forms of cultural and ecological exchange between Japanese and Ainu, see my *The Conquest of Ainu Lands: Ecology and Culture in Japanese Expansion, 1590–1800* (Berkeley and Los Angeles: University of California Press, 2001). For a more general history of Ainu-Japanese relations, see Richard Siddle, *Race, Resistance and the Ainu of Japan* (London: Routledge, 1996).

6. "Mōjū ryōsatsusha e teatekin kyūyo no ken" (Matter of bounty allowance paid to wild animal hunters) [1878.2.15], in Shūsairoku (Collected jurisdictional

records) (A4-54-49), Hokkaido Prefectural Archives, Sapporo, Japan (hereafter HPA). For a history of Russo-Japanese relations in the North Pacific, see George Alexander Lensen, *The Russian Push toward Japan: Russo-Japanese Relations, 1697–1875* (Princeton: Princeton University Press, 1959).

7. On the "live machines" of Meiji Japan, see Hazel J. Jones, *Live Machines: Hired Foreigners in Meiji Japan* (Vancouver: University of British Columbia Press, 1980).

8. Tessa Morris-Suzuki, *A History of Japanese Economic Thought* (London: Routledge, 1989), 17.

9. Nativist (*kokugaku*) scholars such as Hirata Atsutane valorized agriculture in Japan's early modern period as the "ancient way" (*kodō*). According to H. D. Harootunian, Hirata believed that "Japanese rice was superior to the rice of all countries, and those who consumed it took in a divine food that guaranteed their uniqueness and superiority over all others." See H. D. Harootunian, *Things Seen and Unseen: Discourse and Ideology in Tokugawa Nativism* (Chicago: University of Chicago Press, 1988), 23, 212. For more on rice in early modern Japan, see Emiko Ohnuki-Tierney, *Rice as Self: Japanese Identities through Time* (Princeton: Princeton University Press, 1993), 86–88.

10. Of course, there are many exceptions to this Japanese disdain for meat eating. The widespread consumption of wild game is perhaps the most important. To begin with, Japan boasts an ancient tradition of hunting wild game, and the inhabitants of mountain communities often ate the flesh of many wild animals. In some instances, Japanese attributed medicinal qualities to certain wild game, which seems to have alleviated some of the Buddhist constraints against its consumption; see my "Commercial Growth and Environmental Change in Early Modern Japan: Hachinohe's Wild Boar Famine of 1749," *Journal of Asian Studies* 60, no. 2 (May 2001): 329–51. In the sixteenth century, some European missionaries described butcher shops in Edo (present-day Tokyo) where urban consumers could buy the meat from a variety of wild game; see Rodrigo de Vivero y Velasco's remarks in Michael Cooper, comp., *They Came to Japan: An Anthology of European Reports on Japan, 1543–1640* (Berkeley and Los Angeles: University of California Press, 1965), 284–85.

11. Shibusawa Keizō, comp., *Japanese Life and Culture in the Meiji Era*, trans. Charles S. Terry (Tokyo: Ōbunsha, 1958), 65–66.

12. Chiba Tokuji, *Shuryō denshō* (Hunting folklore) (Tokyo: Hōsei Daigaku Shuppankyoku, 1975), 157; Hiraiwa Yonekichi, *Ōkami: Sono seitai to rekishi* (The wolf: Its ecology and history) (Tokyo: Tsukiji Shokan, 1992), 89.

13. On the impact of Western notions of "civilization and enlightenment" on Meiji Japan, see Irokawa Daikichi, *The Culture of the Meiji Period*, trans. Marius B. Jansen (Princeton: Princeton University Press, 1985), 51–75.

14. Describing the role of European fables, myths, and biblical interpretations in creating the Western hatred for wolves, a highly astute (but sadly anonymous) European writer observed from British colonial India in 1927: "As a good European, I inherit a whole huddle of dark neolithic fears which the poets and magicians and

schoolmasters of my tribe have sedulously kept alive through the safe, comfortable centuries. I am not to blame. From my cradle have I been bidden, enjoined, commanded to fear the wolf. He tears you to pieces alive and digs you up when you are dead, and before the maid has time to run to your frantic ringing he pulls you down on your own threshold; between the pillarbox and the front-door he pulls you down, in the dark, after tea. No, I am not to blame" (C. G. C. T. cited in Denise Casey and Tim W. Clark, comps., *Tales of the Wolf: Fifty-one Stories of Wolf Encounters in the Wild* [Moose, WY: Homestead Publishing, 1996], 134).

15. Several histories have been written that cover the reintroduction of the wolf to Yellowstone. Michael K. Phillips and Douglas W. Smith, the wolf restoration project leaders, wrote the "inside story" of wolf reintroduction: *The Wolves of Yellowstone* (Stillwater, MN: Voyageur Press, 1996). On the fight leading up to the wolf reintroduction and the reintroduction itself, see Hank Fischer, *Wolf Wars* (Helena, MT: Falcon Press Publishing Co., 1995); Rick McIntyre, *A Society of Wolves: National Parks and the Battle over the Wolf* (Stillwater, MN: Voyageur Press, 1993); Thomas McNamee, *The Return of the Wolf to Yellowstone* (New York: Henry Holt and Co., 1997). For a history and guide with documents, see Paul Schullery, ed., *The Yellowstone Wolf: A Guide and Sourcebook* (Worland, WY: High Plains Publishing Co., 1996). On the recovery and management of wolf populations after reintroduction, see Martin A. Nie, *Beyond Wolves: The Politics of Wolf Recovery and Management* (Minneapolis: University of Minnesota Press, 2003).

16. Gananath Obeyesekere, *Medusa's Hair: An Essay on Personal Symbols and Religious Experience* (Chicago: University of Chicago Press, 1984), 1–2, 8–9.

17. See, e.g., Donald Worster, *Nature's Economy: A History of Ecological Ideas* (Cambridge: Cambridge University Press, 1977), 258–90; Barry Holstun Lopez, *Of Wolves and Men* (New York: Simon and Schuster, 1978); Bruce Hampton, *The Great American Wolf* (New York: Henry Holt and Co., 1997); Karen R. Jones, *Wolf Mountains: A History of Wolves along the Great Divide* (Calgary: University of Calgary Press, 2002); Jon Thomas Coleman, "Wolves in American History," (Ph.D. diss., Yale University, 2003).

18. Ben Corbin is cited in Rick McIntyre, ed., *War against the Wolf: America's Campaign to Exterminate the Wolf* (Stillwater, MN: Voyageur Press, 1995), 123–24.

19. Ibid., 124, 128–29.

20. Hokkaido Prefectural Government, ed., *Foreign Pioneers: A Short History of the Contribution of Foreigners to the Development of Hokkaido* (Sapporo: Hokkaido Prefectural Government, 1968), 23, 24.

21. Funayama Kaoru, *Zoku Otōsei* (Otōsei continued), in *Funayama Kaoru shōsetsu zenshū* (The complete collection of novels by Funayama Kaoru), vol. 9 (Tokyo: Kawade Shobō Shinsha, 1975). The section of Funayama's novel dealing with wolves (134–50) includes quotations from Dun's unpublished memoirs describing the Hokkaido wolf (146). In Togawa Yukio and Honjō Kei's work, by contrast, the comic (*manga*) style affords these authors greater flexibility. Although the authors portray

Dun heroically, they portray the Hokkaido wolves sympathetically as well. The authors even give the wolves voices in the narrative. Wolves excitedly exclaim, "Meat!" when they first discover the strychnine-laced bait, and then cry out, "Oh, the pain!" and "Water!" as the strychnine begins to kill them (160–205). A young lad called "Boy," who, on one occasion, embraces a wolf as it dies from strychnine, protects the wolves. In many ways, this comic is one of the first environmental histories on the disappearance of either subspecies of Japanese wolf. See Togawa Yukio and Honjō Kei, *Ōkami no hi: Ezo ōkami no zetsumetsuki* (Memorial to the wolf: A record of the extinction of the Hokkaido wolf) (Tokyo: Shōnen Champion Comics, 1994).

22. For biographic information on Edwin Dun, see Hokkaido Prefectural Government, *Foreign Pioneers*, 15–27. See also Fujita, *American Pioneers and the Japanese Frontier*, 69–87. Dun's daughter, Dun [Dan] Michiko, also wrote about her father, but she offers little information, at least related to Niikappu and wolves, that is not contained in his unpublished memoirs; see Dan Michiko, *Meiji no bokusaku* (Fenced pastures of Meiji) (Tokyo: Sumire Gakuennai, 1968), 43–53.

23. Edwin Dun, "Reminiscences of Nearly a Half Century in Japan" [n.d.], Resource Collection for Northern Studies, Hokkaido University Library, Sapporo, Japan (hereafter RCNS).

24. On the industrialization and modernization of American agriculture, see Peter D. McClelland, *Sowing Modernity: America's First Agricultural Revolution* (Ithaca: Cornell University Press, 1997); Deborah Kay Fitzgerald, *Every Farm a Factory: The Industrial Ideal in American Agriculture* (New Haven: Yale University Press, 2003). On the roots of the industrial ideal in American agriculture and literary traditions, see William Conlogue, *Working the Garden: American Writers and the Industrialization of Agriculture* (Chapel Hill: University of North Carolina Press, 2002).

25. Dun, "Reminiscences," 6.

26. For more in English on Kuroda Kiyotaka, see David Forsyth Anthony, "The Administration of Hokkaido under Kuroda Kiyotaka—1870–1882: An Early Example of Japanese-American Cooperation" (Ph.D. diss., Yale University, 1968).

27. Dun, "Reminiscences," 12–13.

28. On the revolution in Japanese meat-eating habits during the Meiji period, see Shibusawa, *Japanese Life and Culture in the Meiji Era*, 65–69. For more on Japanese food culture, see Michael Ashkenazi and Jeanne Jacob, *The Essence of Japanese Cuisine: An Essay on Food and Culture* (Philadelphia: University of Pennsylvania Press, 2000).

29. On Japanese and wild boars, for example, see my "Commercial Growth and Environmental Change in Early Modern Japan."

30. At the same time in the United States, after an expansion of the cattle industry in the American West, "Americans came to think that they were living in the 'Golden Age of American Beef'"; see Richard White, "Animals and Enterprise," in *The Oxford History of the American West*, ed. Clyde A. Milner II, Carol A. O'Connor, and Martha A. Sandweiss (New York: Oxford University Press, 1994), 256. For more

on the historical development of the North American cattle industry, see Terry G. Jordon, *North American Cattle-Ranching Frontiers: Origins, Diffusion, and Differentiation* (Albuquerque: University of New Mexico Press, 2000).

31. Department of Agriculture and Commerce, ed., *Japan in the Beginning of the 20th Century* (Tokyo: Shoin, 1904), 184–200.

32. Dun, "Reminiscences," 13–14.

33. On the ideology of the Meiji period, see Carol Gluck, *Japan's Modern Myths: Ideology in the Late Meiji Period* (Princeton: Princeton University Press, 1985); T. Fujitani, *Splendid Monarchy: Power and Pageantry in Modern Japan* (Berkeley and Los Angeles: University of California Press, 1996).

34. Dun, "Reminiscences," 15–16, 30, 74–75.

35. On the importance of zootechny in the context of the colonization of the Philippines, see Greg Bankoff, "A Question of Breeding: Zootechny and Colonial Attitudes towards the Tropical Environment in Late Nineteenth-Century Philippines," *Journal of Asian Studies* 60, no. 2 (Spring 2001): 413–37.

36. Dun, "Reminiscences," 32–34.

37. Ibid., 35.

38. Edwin Dun, "His Excellency, Matsumoto" [Nanai, 1875.10.4], RCNS.

39. Edwin Dun, "Mr Dzushio Hirotake, Acting Daishioke Kwan of Kaitakushi" [Sapporo, 1877.7.9], RCNS.

40. Edwin Dun, "His Excellency, Dzushio Hirotaki, Kaitaku, Daishioki Kwan" [Sapporo, 1881.5.28], RCNS.

41. Even without Dun's advice, Kaitakushi officials, increasingly knowledgeable about the American experience, probably became well aware of the potential problems that predators posed to raising sheep on Hokkaido. E.g., see "Beikoku kuma ōkami no men'yō o gai suru keikyō" (The situation of sheep depredation by bears and wolves in the United States), *Hokkai tsūshi* (Northern sea interpreter) 14 (October 1880): 8–11.

42. Edwin Dun, "Shiriuchi" [Sapporo, 1881.7.1], RCNS.

43. Edwin Dun, "His Excellency, Governor Hori, Kaitaku, Daishioki Kwan" [Sapporo, 1877.10.18], RCNS.

44. Edwin Dun, "Mr Satow, Gon shio shioki Kwan of Kaitakushi" [Sapporo, 1878.3.25], RCNS.

45. Edwin Dun, "His Excellency, Matsumoto" [Nanai, 1875.10.4], RCNS.

46. For some slightly later reportage on the predator situation at the Niikappu ranch, see "Niikappu bokujō no keikyō" (The situation at the Niikappu ranch), *Hokkai tsūshi* (Northern sea interpreter) 7 (July 1880): 4–9.

47. Dun, "Reminiscences," 36.

48. "Yachū bokuba o oiireru saku chikuzō hoka yonkado no ken" (Matter of four reasons to construct a fence to herd horses into at night) [1878.5.11], in Shūsairoku (Collected jurisdictional records) (A4-54), HPA.

49. Ibid.

50. "Niikappu bokujō sanjiba kuma ōkami nado no higai bōjo no tame yōjū dan sōchi no ken" (Matter of sending Western firearms and ammunition to prevent bear and wolf damage to newborn horses at the Niikappu ranch) [1878.5.23], in *Honka todokeroku* (Records of home division reports) (A4-51-72), HPA.

51. For some comparative statistics of roughly the same time period regarding wolf and coyote predation in the American West, see Theodore S. Palmer, "Extermination of Noxious Animals by Bounties," in *Yearbook of the United States Department of Agriculture* (Washington, DC: Government Printing Office, 1896), 55–56.

52. Inukai Tetsuo, "Hokkaidō-san ōkami to sono metsubō keiro" (The Hokkaido wolf and its road to extinction), *Shokubutsu oyobi dōbutsu* (Botany and zoology) 1, no. 8 (August 1933): 17.

53. For more on the history of Hokkaido's deer population, see Inukai Tetsuo, "Hokkaidō no shika to sono kōbō" (The Hokkaido deer and its rise and fall), *Hoppō bunka kenkyū hōkoku* (Research reports on northern culture) 7 (March 1952): 1–22.

54. Inukai Tetsuo, *Hoppō dōbutsushi* (Northern animal journal) (Sapporo: Hokuensha, 1975), 23.

55. Dun, "Reminiscences," 52–53.

56. Inukai Tetsuo, *Waga dōbutsuki* (My animal chronicle) (Tokyo: Kurashi no Techōsha, 1970), 117–18.

57. Stanley Paul Young, *The Wolf in North American History* (Caldwell, ID: Caxton Printers, 1946), 100, 104–5. See also Andrew C. Isenberg, *The Destruction of the Bison* (Cambridge: Cambridge University Press, 2000).

58. For more on this bitter war between wolves and humans, and a discussion of the changing attitudes toward wolves in the United States, see Thomas R. Dunlap, *Saving America's Wildlife: Ecology and the American Mind, 1850–1990* (Princeton: Princeton University Press, 1988).

59. Benjamin S. Lyman, "Journal of a trip from Nemoro by Soya and Hakodate to Yedo" [1874], RCNS.

60. Dun, "Reminiscences," 36–37.

61. Ibid.

62. Stanley Paul Young, *The Last of the Loners* (New York: Macmillan Co., 1970), 40–44.

63. William Edward Webb, *Buffalo Land: An Authentic Account of the Discoveries, Adventures, and Mishaps of a Scientific and Sporting Party in the Wild West with Graphic Descriptions of the Country; the Red Man, Savage and Civilized; Hunting the Buffalo, Antelope, Elk and Wild Turkey; Etc; Etc* (Chicago and Cincinnati: E. Hannaford and Co., 1872), 290–93. See also McIntyre, *War against the Wolf*, 54–55.

64. Granville Stuart, *Forty Years on the Frontier* (Glendale, CA: Arthur H. Clark Co., 1925). See also McIntyre, *War against the Wolf*, 59.

65. Joseph H. Batty, *How to Hunt and Trap* (New York: Orange Judd Co., 1884). See also McIntyre, *War against the Wolf*, 60.

66. L. S. Kelly, "Wolves and Coyotes," *Forest and Stream* 48, no. 8 (February 20, 1897): 144–45. See also Casey and Clark, *Tales of the Wolf*, 162–64. The sporting magazine *Forest and Stream* often ran articles on how to poison wolves with strychnine, some going into grotesque detail on what poisoning techniques offered the wolfer the greatest killing power. E.g., see Orin Belknap, "Poisoning Wolves," *Forest and Stream* 48, no. 9 (February 27, 1897): 168.

67. "Niikappu bokujō nai ni oite yajū dokusatsu no gi ukagai no ken" (Matter of inquiry into poisoning wild animals at the Niikappu ranch) [1878.6.20], in Shūsairoku (Collected jurisdictional records) (A4-54), HPA.

68. Dun, "Reminiscences," 37–38.

69. In the nineteenth-century modern order, Westerners and Japanese came to view wolves not only as ruthless "murderers" but as trespassers and violators of property rights as well. This fact helps to explain the brutality of wolf extermination, as well as the extermination of other less "murderous" creatures such as prairie dogs. See Susan Jones, "Becoming a Pest: Prairie Dog Ecology and the Human Economy in the Euroamerican West," *Environmental History* 4, no. 4 (October 1999): 531–52.

70. "Hondō rōgai zankoku ni tsuki satsukaku no tame sutorikiniine kōkyū no ue sōchi kata no ken" (Matter of the brutal damage caused by wolves throughout Hokkaido and the purchasing and sending of strychnine to kill them) [1880.3.16], in Tōkyō bun iroku (Transferred records of Tokyo written documents) (03774), HPA.

71. For an overview of the "underside" of Japan's modernization, see Mikiso Hane, *Peasants, Rebels, and Outcastes: The Underside of Modern Japan* (New York: Pantheon Books, 1982). On "factory girls," see E. Patricia Tsurumi, *Factory Girls: Women in the Thread Mills of Meiji Japan* (Princeton: Princeton University Press, 1990). On tuberculosis in Meiji Japan's thread mills, see William Johnston, *The Modern Epidemic: A History of Tuberculosis in Japan* (Cambridge: Harvard University Press, 1995), 69–90. For a personal account of Ainu being forced to work in forestry and at fisheries, see the memoirs of Kayano Shigeru, *Our Land Was a Forest: An Ainu Memoir* (Boulder: Westview Press, 1980). For a fictional account of life on a crab "factory ship," see Kobayashi Takiji, *"The Factory Ship" and "The Absentee Landlord,"* trans. Frank Motofuji (Seattle: University of Washington Press, 1973).

5 / WOLF BOUNTIES AND THE ECOLOGIES OF PROGRESS

1. Quoted from Tessa Morris-Suzuki, *Re-inventing Japan: Time, Space, Nation* (New York: M. E. Sharpe, 1998), 55. See also Kenneth Strong, *Ox against the Storm: A Biography of Tanaka Shozo—Japan's Conservationist Pioneer* (Richmond, Surrey, UK: Japan Library, 1977). On the Ashio copper mine disaster, see F. G. Notehelfer, "Japan's First Pollution Incident," *Journal of Japanese Studies* 1, no. 2 (1975): 351–83.

2. "Kamiiso-gun Sakkari Kikonai ryōson nite ōkami kujo no tame ryōshi yatoi

ire kata no ken" (Matter of hiring hunters to exterminate wolves from the two villages of Sakkari and Kikonai in Kamiiso District) [1880.3.22], in Kannai gunku yakusho bun iroku (Transferred records of written documents from district-ward jurisdictional government offices) (4051), Hokkaido Prefectural Archives, Sapporo, Japan (hereafter HPA).

3. It is not surprising that Enomoto Takeaki had such a painting in his office, considering his diplomatic career. After fighting as a Tokugawa loyalist against the imperial restoration of 1868, he was pardoned by the Meiji government in 1872 and subsequently worked for the Kaitakushi. In 1874, he was appointed vice admiral and, that same year, served as a special envoy to Russia, where he negotiated a treaty that placed the Kuril Islands under Japanese control. He may have received the painting at this time.

4. See Paul Schach, "Russian Wolves in Folktales and Literature of the Plains: A Question of Origins," *Great Plains Quarterly* 3, no. 2 (Spring 1983): 67–78.

5. "Roryō Shiberia chihō ōkami ryō keikyō gakumenzu utsushitori sashi mawashi no ken" (Matter of sending around copy of picture depicting wolf hunt in area of Russian Siberia) [1879.7.22], in Tōkyō bun iroku (Transferred records of Tokyo written documents) (3031), HPA.

6. Jonathan Weiner, *The Beak of the Finch* (New York: Vintage Books, 1994), 142, 144–45.

7. Fukuzawa Yukichi, *An Outline of a Theory of Civilization*, trans. David A. Dilworth and G. Cameron Hurst (Tokyo: Sophia University, 1973), 20. See also Fukuzawa Yukichi, *Bunmeiron no gairyaku* (An outline of a theory of civilization) [1875], in *Fukuzawa Yukichi zenshū* (The complete works of Fukuzawa Yukichi), vol. 4, ed. Keiō Gijuku (Tokyo: Iwanami Shoten, 1959), 23.

8. Julia Adeney Thomas, *Reconfiguring Modernity: Concepts of Nature in Japanese Political Ideology* (Berkeley and Los Angeles: University of California Press, 2001).

9. Fukuzawa, *Outline of a Theory of Civilization*, 13–15. In Japanese, Fukuzawa, *Bunmeiron no gairyaku*, 16–18.

10. Fukuzawa, *Outline of a Theory of Civilization*, 30.

11. Theodore S. Palmer, "Extermination of Noxious Animals by Bounties," in *Yearbook of the United States Department of Agriculture* (Washington, DC: Government Printing Office, 1896), 55–56. For an excellent chronology of significant game and hunting laws passed in the United States that also includes some information on predator bounties, see Theodore S. Palmer, "Chronology and Index of the More Important Events in American Game Protection, 1776–1911," *Bulletin of the United States Bureau of Biological Survey* 41 (1912): 7–46.

12. Bruce Hampton, *The Great American Wolf* (New York: Henry Holt and Co., 1997), 118.

13. Inukai Tetsuo, "Hokkaidō-san ōkami to sono metsubō keiro" (The Hokkaido wolf and its road to extinction), *Shokubutsu oyobi dōbutsu* (Botany and zoology) 1, no. 8 (August 1933): 13.

14. Kaitakushi, ed., *Hokkaidōshi* (Hokkaido records) [1892] (Tokyo: Rekishi Toshosha, 1973), 397–99.

15. Stanley Paul Young, *The Wolf in North American History* (Caldwell, ID: Caxton Printers, 1946), 19.

16. On the creation of the Neo-Europes, see Alfred W. Crosby, *Ecological Imperialism: The Biological Expansion of Europe, 900–1900* (Cambridge: Cambridge University Press, 1986), 2–3.

17. Hampton, *Great American Wolf*, 70.

18. For documents related to early wolf killing in North America, see Rick McIntyre, ed., *War against the Wolf: America's Campaign to Exterminate the Wolf* (Stillwater, MN: Voyageur Press, 1995), 29–52.

19. Hampton, *Great American Wolf*, 114–20.

20. Palmer, "Extermination of Noxious Animals by Bounties," 55, 58.

21. Ibid., 55, 59–60. Palmer believed that the bounty system in the United States had certainly contributed to keeping "noxious animal" numbers down, but he also believed that the drain on county and state finances was too much and that private means, such as shooting competitions organized by gun clubs, should be used to exterminate certain animals.

22. Young, *Wolf in North American History*, 141.

23. Ibid., 146.

24. Ibid., 118.

25. Hampton, *Great American Wolf*, 120–40.

26. "Mōjū ryōsatsusha e teatekin kyūyo no ken" (Matter of bounty allowance paid to wild animal hunters) [1878.2.15], in Shūsairoku (Collected jurisdictional records) (A4-54-49), HPA.

27. "Kuma ōkami nado mōjū ryōsatsu no ken" (Matter of hunting and killing bears, wolves, and other animals) [1878.1.21 and 1878.5.31], in Shūsairoku (Collected jurisdictional records) (A4-54-48), HPA.

28. "Kuma ōkami narabi ni karasu hokaku todoke tori matome hōkoku sashidasase kata no ken" (Matter of the final settling of notification regarding the capture of bears, wolves, and crows) [1880.1.21], in Shūsairoku (Collected jurisdictional records) (04061), HPA.

29. See Ōta Shinya, *Karasu wa machi no ōsama da* (Crows are kings of the city streets) (Fukuoka: Ashishobō Yūgengaisha, 1999).

30. Frans De Waal, *The Ape and the Sushi Master: Cultural Reflections of a Primatologist* (New York: Basic Books, 2001), 201.

31. Eytan Avital and Eva Jablonka, *Animal Traditions: Behavioral Inheritance in Evolution* (Cambridge: Cambridge University Press, 2000), 2.

32. Bernd Heinrich, *Mind of the Raven* (New York: Cliff Street Books, 1999), 49, 68, 130.

33. Ibid., 141, 226–35.

34. "Jūsan nen jūni gatsu yori jūyon nen rokugatsu made yūgai chōjū kakusatsu

teate shiharai no ken" (Matter of paying bounties between December 1880 and June 1881 for killing damaging animals) [1881.12.14], in Shūsairoku (Collected jurisdictional records) (05036), HPA.

35. "Kuma ōkami bōgai no tame Noboribetsu bokujō e jūhō narabi ni dan'yaku watashi kata no ken" (Matter of sending firearms and ammunition to the Noboribetsu ranch to prevent damage caused by bears and wolves) [1874.1.14], in Kaitakushi kōbun roku (Record of official Kaitakushi documents) (06001), HPA.

36. "Hidaka-koku Mitsuishi-gun fukutōchō Ōshima Kunitarō yori yūgai no ōkami jūsatsu negaide ni tsuki shobun sai no ken" (Matter of Assistant Head Ōshima Kunitarō of Hidaka Province, Mitsuishi District, requesting permission to deal with damaging wild wolves by shooting them) [1878.7.7], in Shoka bun iroku (Transferred records of written documents from various divisions) (A4-54-71), HPA.

37. "Usu-gun ekitei toriatsukai nindai Iida Ichigorō hoka yonjūni mei ōkami ko hokaku ni tsuki teatekin kashi no ken" (Matter of imperial grant bounty allowance paid to Iida Ichigorō and forty-two others for captured wolf pups) [1878.8.5], in Shūsairoku (Collected jurisdictional records) (A4-54), HPA.

38. "Oshamanbe-mura Ega Jūjirō ōkami ko shukaku ni kan shi kensa no tame zenku sashidashi kata no ken" (Matter of Ega Jūjirō from Oshamanbe Village submitting for inspection entire corpse of captured wolf pup) [1880.5.13], in Kannai gunku yakusho bun iroku (Transferred records of written documents from district-ward jurisdictional government offices) (4051), HPA.

39. "Ōkami no ko tōchaku no gi narabi ni hokaku shonyūhi shōkai sono hoka no ken" (Matter of inquiries into various expenses regarding the capture and arrival of wolf pups) [1878.7], in Tōkyō shoka bun iroku (Transferred records of various Tokyo divisional written documents) (02641), HPA.

40. "Komagadake kinbō ni ōkami no ko kore aru yoshi Tōkyō hakubutsujo e sashidashi taku ikedori kata no ken" (Matter of capturing live wolves to send to Tokyo museum because of wolf pups in the Komagadake area) [1878.1.15], in [file name not given] (2647-3), HPA.

41. "Komagatake nite ikedori no ōkami no ko chōekijo nite shiiki kata no ken" (Matter of caring for captive wolf pups captured at Komagatake) [1878.5.10], in Shūsairoku (Collected jurisdictional records) (02650), HPA.

42. "Nemuro gunnai ni oite shukaku no ōkami no kawa sōchi no ken" (Matter of sending skins from wolves killed within Nemuro District) [1881.3.1], in Honshichō ōrai bun iroku (Transferred records of written documents between home and branch offices) (05329), HPA.

43. Donna Haraway, "Teddy Bear Patriarchy: Taxidermy in the Garden of Eden, New York City, 1908–1936," *Social Text: Theory, Culture, Ideology,* Winter 1984–85, 30, 34.

44. "Esashi nite ikedori no ōkami no ko Tōkyō kari hakubutsukan e shuppin ni tsuki sashiokurikata no ken" (Matter of sending wolf pups captured alive at Esashi

for display at Tokyo provisional museum) [1877.6], in Kaka bun iroku (Transferred records of divisional written documents) (02136), HPA.

45. "Hiyama-gun Otobe-mura sanchū ni oite shūkaku no ōkami no ko Hakodate-maru e tsumiiremawashi itashi no ken" (Matter of shipping on the *Hakodate-maru* wolf pups caught in the mountains of Hiyama District, Otobe Village) [1877.6.16], in Tōkyō bun iroku (Transferred records of Tokyo written documents) (02122), HPA.

46. "Tōkyō e sashiokuri no ōkami no ko no tsukisoi toshite Miyamura Ichitarō sashitate no ken" (Matter of sending Miyamura Ichitarō to accompany wolf pups being shipped to Tokyo) [1877.6.18], in Shūsairoku (Collected jurisdictional records) (02259), HPA; "Tōkyō e sashiokuri ōkami nitō kaiage keihi torishirabe no ken" (Matter of examining how to pay the expenses of shipping two wolves to Tokyo) [1877.7.29], in Esashi bunsho ōraisho (Documents that came and went from Esashi branch office) (02256), HPA.

47. On European menageries, see Harriet Ritvo, *The Animal Estate: The English and Other Creatures in the Victorian Age* (Cambridge: Harvard University Press, 1987), 205–42.

48. For a brief mention of the Japanese wolf that was exhibited in 1878, see Nibbashi Kazuaki, "Nihon no dōbutsuen no rekishi" (The history of Japanese zoos), in *Dōbutsuen to iu media* (The media called zoos), ed. Watanabe Morio et al. (Tokyo: Seikyūsha, 2000), 154–55.

49. Tōkyō To, ed., *Ueno dōbutsuen hyakunenshi* (One hundred year history of the Ueno zoo), vol. 1 (Tokyo: Daiichi Hōki Shuppan Kabushikigaisha, 1982), 1–21, 527, 529. See also Nishimura Saburō, *Bunmei no naka no hakubutsugaku: Seiō to Nihon* (Natural history in civilizations, west and east), vol. 2 (Tokyo: Kinokuniya Shoten, 2000), 515–33.

50. "Kameda-gun Komagatake ni oite ikedori no ōkami sōchi kata no ken" (Matter of sending live wolves from Komagatake in Kameda District) [1878.6.8], in Tōkyō shoka bun iroku (Transferred records of various Tokyo divisional written documents) (02641), HPA.

51. "Kaitakushi chichū kari hakubutsujo nite kaitate no mesuōkami niwaka ni hanmon kutsū ni tsuki gaikoku iin jushin kata torihakarai korearitaku no ken" (Matter of arranging physical examination by foreign doctor of female wolf raised at the Kaitakushi provisional museum with sudden severe pain) [1877.10.17], in Hokkaidō shobuppin ōfukusho oyobi shoshōken tsūshin shorui (Varieties of documents regarding Hokkaido goods and ministry and prefecture correspondences) (02380), HPA; "Kaitakushi kari hakubutsujo nite kaioki no yamainu kitoku ni tsuki heishin no sai wa gaikokujin ni takushi kaibō itashitaki ken" (Matter of entrusting foreigner to perform dissection when mountain dog that was raised at Kaitakushi provisional museum and that is in critical condition dies) [1877.10.18], in Hokkaidō shobuppin ōfukusho oyobi shoshōken tsūshin shorui (Varieties of documents regarding Hokkaido goods and ministry and prefecture correspondences) (02380),

HPA; "Kaitakushi chikuyō no mesuōkami kyūbyō ni tsuki yatoi jūi shindan no gi do-nichiyō igai no hi tōka made hikitsuke kudasare shinryō no ken" (Matter of request for consultation with hired veterinarian at [agricultural sciences] division on any day other than Saturday or Sunday regarding diagnosis of sick female wolf raised by Kaitakushi) [1877.10.18], in Hokkaidō shobuppin ōfukusho oyobi shoshōken tsūshin shorui (Varieties of documents regarding Hokkaido goods and ministry and prefecture correspondences) (02380), HPA; "Enchū chikuyō no ōkami ribyō jūtai no ken" (Matter of wolf raised in garden contracting disease and falling into critical condition) [1879.7.20], in Bun iroku (Transferred records of written documents) (03736), HPA.

52. Haraway, "Teddy Bear Patriarchy," 26.

53. Stanley Paul Young, *The Last of the Loners* (New York: Macmillan Co., 1970), 222.

54. For more on wolf behavior and ecology, see L. David Mech, *The Wolves of Isle Royale* (Washington, DC: U.S. Government Printing Office, 1966); Richard Fiennes, *The Order of Wolves* (Indianapolis: Bobbs-Merrill Co., 1976); Durward L. Allen, *Wolves of Minong: Isle Royale's Wild Community* (Ann Arbor: University of Michigan Press, 1979); Michael W. Fox, *The Soul of the Wolf* (Boston: Little, Brown, and Co., 1980); Sylvia A. Johnson and Alice Aamodt, *Wolf Pack: Tracking Wolves in the Wild* (Minneapolis: Lerner Publications Co., 1985); L. David Mech, *The Arctic Wolf: Ten Years with the Pack* (Stillwater, MN: Voyageur Press, 1988); L. David Mech, *Wolves of the High Arctic* (Stillwater, MN: Voyageur Press, 1992); L. David Mech, *The Way of the Wolf* (Stillwater, MN: Voyageur Press, 1991); Robert H. Busch, *The Wolf Almanac* (New York: Lyons Press, 1995); Ludwig N. Carbyn, Steven H. Fritts, and Dale R. Seip, eds., *Ecology and Conservation of Wolves in a Changing World* (Edmonton: Canadian Circumpolar Institute, University of Alberta, 1995); Ken L. Jenkins, *Wolf Reflections: Reflections of the Wilderness Series* (Merrillville, IN: ICS Books, 1996); L. David Mech, Layne G. Adams, Thomas J. Meier, John W. Burch, and Bruce W. Dale, *The Wolves of Denali* (Minneapolis: University of Minnesota Press, 1998).

55. Kadosaki Masaaki and Seki Hideshi, "Ezochi ni okeru dōbutsusō no bunkengakuteki kenkyū" (Philological research on the fauna in the Yezo-Island), in *Hokkaidō kaitaku kinenkan chōsa hōkoku* (Bulletin of the Hokkaido Colonization Museum) 38 (1999): 96–108.

56. Matsumae Hironaga, *Matsumaeshi* (Matsumae record) [1781], in *Hokumon sōsho* (Northern gate library), vol. 2, ed. Ōtomo Kisaku (Tokyo: Hokkō Shobō, 1943), 197. See also Tawara Hiromi, *Hokkaidō no shizen hogo* (Protecting Hokkaido's nature) (Sapporo: Hokkaidō Daigaku Tosho Kankōkai, 1979), 103.

57. Kubota Shizō, *Kyōwa shieki* (Harmonious private expedition) [1856], in *Nihon shomin seikatsu shiryō shūsei* (Collected sources on the history of the daily lives of common Japanese people), vol. 4, ed. Takakura Shin'ichirō (Tokyo: San'ichi Shobō, 1969), 235. This is a frequently cited example; see Inukai, "Hokkaidō-san ōkami to

sono metsubō keiro," 12–13; Inukai Tetsuo, *Hoppō dōbutsushi* (Northern animal journal) (Sapporo: Hokuensha, 1975), 21; Tawara, *Hokkaidō no shizen hogo*, 103.

58. Captain H. C. St. John, *Notes and Sketches from the Wild Coasts of Nipon: With Chapters on Cruising After Pirates in Chinese Waters* (Edinburgh: David Douglas, 1880), 9. See also Inukai, "Hokkaidō-san ōkami to sono metsubō keiro," 12–13; Inukai, *Hoppō dōbutsushi*, 21.

59. David L. Howell, *Capitalism from Within: Economy, Society, and the State in a Japanese Fishery* (Berkeley and Los Angeles: University of California Press, 1995), 117.

60. Peter Steinhart, *The Company of Wolves* (New York: Vintage Books, 1995), 127, 130.

61. By far the most up-to-date treatment of wolf behavior and biology can be found in L. David Mech and Luigi Boitani, eds., *Wolves: Behavior, Ecology, and Conservation* (Chicago: University of Chicago Press, 2003); chapter 1 deals with the intricacies of wolf social ecology, and chapter 2 explores wolf behavior.

62. L. David Mech, *The Wolf: The Ecology and Behavior of an Endangered Species* (Minneapolis: University of Minnesota Press, 1970), 38–41, 46–47, 50; Barry Holstun Lopez, *Of Wolves and Men* (New York: Simon and Schuster, 1978), 26. See also Steinhart, *Company of Wolves*, 13–16.

63. Adolph Murie, *The Wolves of Mount McKinley* (Seattle: University of Washington Press, 1985), 43.

64. Mech, *Wolf*, 68.

65. Murie, *Wolves of Mount McKinley*, 21.

66. Lopez, *Of Wolves and Men*, 29.

67. Mech, *Wolf*, 68, 69, 102. See also Lopez, *Of Wolves and Men*, 39. For a general status report on wolf senses, see Cheryl S. Asa and L. David Mech, "A Review of the Sensory Organs in Wolves and Their Importance to Life History," in *Ecology and Conservation of Wolves in a Changing World*, ed. Ludwig N. Carbyn, Steven H. Fritts, and Dale R. Seip (Edmonton: Canadian Circumpolar Institute, University of Alberta, 1995), 287–91.

68. Jeffrey Moussaieff Masson and Susan McCarthy, *When Elephants Weep: The Emotional Lives of Animals* (New York: Delacorte Press, 1995), 15–16, 31.

69. Mech, *Wolf*, 105. See also Steinhart, *Company of Wolves*, 16.

6 / WOLF EXTINCTION THEORIES AND THE BIRTH
OF JAPAN'S DISCIPLINE OF ECOLOGY

1. For more on the origins of ecology, see Garland E. Allen, *Life Science in the Twentieth Century* (Cambridge: Cambridge University Press, 1975), 1–19.

2. On the origins of the word "ecology," see Charles E. Bessey, "The Word 'Ecology,'" *Science* 15 (April 11, 1902): 593; W. F. Ganong, "The Word 'Ecology,'"

Science 15 (April 11, 1902): 593–94; F. A. Bather, "Discussion and Correspondence: Scientific Terminology," *Science* 15 (May 9, 1902): 747–49; W. F. Ganong, "The Word 'Ecology,'" *Science* 15 (May 16, 1902): 792–93.

3. Bessey, "The Word 'Ecology,'" 593.

4. Ganong, "The Word 'Ecology,'" 594.

5. Ernst Haeckel cited in W. C. Allee, Alfred E. Emerson, Orlando Park, Thomas Park, and Karl P. Schmidt, *Principles of Animal Ecology* (Philadelphia and London: W. B. Saunders Co., 1949), v.

6. William Morton Wheeler, "'Natural History,' 'Oecology,' or 'Ethology,'" *Science* 15 (June 20, 1902): 973, 974.

7. Robert P. McIntosh, *The Background of Ecology: Concept and Theory* (Cambridge: Cambridge University Press, 1985), 43.

8. Chapters 10–12 in Kitarō Nishida, *An Inquiry into the Good*, trans. Masao Abe and Christopher Ives (New Haven: Yale University Press, 1990), most influenced the ecological theories of Imanishi Kinji.

9. McIntosh, *Background of Ecology*, 27.

10. Ibid., 52.

11. Karl Semper, *Animal Life as Affected by the Natural Conditions of Existence* (New York: D. Appleton and Co., 1881; reprint, New York: Arno Press, 1977), 29, 32–33 (all page numbers cited are to the reprint edition). See also McIntosh, *Background of Ecology*, 63, 73.

12. McIntosh, *Background of Ecology*, 87.

13. Donald Worster, *Nature's Economy: A History of Ecological Ideas* (Cambridge: Cambridge University Press, 1977), 204.

14. Victor E. Shelford, *Animal Communities in Temperate America as Illustrated in the Chicago Region* (Chicago: University of Chicago Press, 1913), 35.

15. Charles S. Elton, *Animal Ecology*, intro. Mathew A. Leibold and J. Timothy Wootton (London: Sidgwick and Jackson, 1927; reprint, Chicago: University of Chicago Press, 1966), 1 (all page numbers cited are to the reprint edition).

16. Ibid., 1–4, 190. See also McIntosh, *Background of Ecology*, 89, 302.

17. Charles S. Elton, *The Ecology of Invasions by Animals and Plants*, foreword by Daniel Simberloff (London: Methuen and Co., 1958; reprint, Chicago: University of Chicago Press, 2000), 125–26 (all page numbers cited are to the reprint edition).

18. Gregg Mitman, *The State of Nature: Ecology, Community, and American Social Thought, 1900–1950* (Chicago: University of Chicago Press, 1992), 1.

19. Ibid., 5.

20. Imanishi Kinji, *Seibutsu shakai no ronri* (The logic of biological societies) [1948], in *Imanishi Kinji zenshū* (The complete works of Imanishi Kinji), vol. 4, ed. Noma Shōichi (Tokyo: Kōdansha, 1974), 3–217. Imanishi included a discussion of Frederic Clements and his "succession" and "climax" theories (32–33); contrasted the writings of Warder Clyde Allee with the writings of William Morton Wheeler (60); referenced Victor Shelford's scholarship on the geographical distribution of

tiger beetles (151), animal geographies (155, 200), and ecological succession (159); referenced Allee's writings on animal societies (244); and explored Charles Elton's theories on animal ecology (248).

21. Mitman, *State of Nature*, 74.

22. Ibid., 133.

23. Allee et al., *Principles of Animal Ecology*, 6.

24. Mitman, *State of Nature*, 143.

25. Julia Adeney Thomas, *Reconfiguring Modernity: Concepts of Nature in Japanese Political Ideology* (Berkeley and Los Angeles: University of California Press, 2001), 179.

26. Ibid., 180–81.

27. Imanishi Kinji, *Dōbutsuki* (Animal chronicle) [1946], in *Imanishi Kinji zenshū* (The complete works of Imanishi Kinji), vol. 2, ed. Noma Shōichi (Tokyo: Kōdansha, 1974), 358–81.

28. Henry David Thoreau, "The Natural History of Massachusetts" [1842], in *Henry David Thoreau: The Natural History Essays* (Salt Lake City: Peregrine Smith, 1980), 29. See also Worster, *Nature's Economy*, 96.

29. Imanishi, *Dōbutsuki*, 358–59.

30. Ibid., 358.

31. Ibid., 359.

32. Worster, *Nature's Economy*, 183.

33. Imanishi, *Dōbutsuki*, 359–61.

34. For more on Yanagita Kunio, see Ronald A. Morse, *Yanagita Kunio and the Folklore Movement: The Search for Japan's National Character and Distinctiveness* (New York: Garland Publishing, 1990); Minoru Kawada, *The Origin of Ethnography in Japan: Yanagita Kunio and His Times*, trans. Toshiko Kishida-Ellis (London: Kegan Paul International, 1993). For a guide to the folklore of Yanagita, see Fanny Hagin Mayer, trans. and ed., *The Yanagita Kunio Guide to the Japanese Folk Tale* (Bloomington: Indiana University Press, 1948); an unremarkable reference to a wolf can be found on pp. 285–86. For a collection of essays highlighting international perspectives on Yanagita's work, see J. Victor Koschmann, Ōiwa Keibō, and Yamashita Shinji, eds., *International Perspectives on Yanagita Kunio and Japanese Folklore Studies* (Ithaca: Cornell University East Asia Program, 1985).

35. Imanishi, *Dōbutsuki*, 362–64. Yanagita explored the ethnology and ecology of wolf extinction in *Koen zuihitsu* (Miscellaneous writings on the lone monkey) [1939], in *Yanagita Kunio shū* (Yanagita Kunio collection), vol. 22 (Tokyo: Chikuma Shobō, 1970), 426–60. Tsukamoto Manabu covers the debate between Imanishi and Yanagita briefly in *Edo jidaijin to dōbutsu* (The people of the Edo period and animals) (Tokyo: Nihon Editā Sukūru Shuppanbu, 1995), 156–61.

36. Yanagita, *Koen zuihitsu*, 435.

37. Ibid., 430–33.

38. Ibid., 437. See also Imanishi, *Dōbutsuki*, 363–64.

39. Yanagita, *Koen zuihitsu*, 437.

40. See Imanishi Kinji, *Seibutsu no sekai* (The world of living things) [1941], in *Imanishi Kinji zenshū* (The complete works of Imanishi Kinji), vol. 1, ed. Noma Shōichi (Tokyo: Kōdansha, 1974), 1–164. For an English translation, see Kinji Imanishi, *A Japanese View of Nature: "The World of Living Things" by Imanishi Kinji*, trans. Pamela J. Asquith, Heita Kawakatsu, Shusuke Yagi, and Hiroyuki Takasaki (London: RoutledgeCurzon, 2002).

41. Imanishi, *Japanese View of Nature*, 41–42, 63.

42. Ibid., 41–42.

43. For a discussion of the intellectual relationship between Imanishi Kinji and Nishida Kitarō, see ibid., xxxiv–xxxvi. See also Nishida, *Inquiry into the Good*, 63. For more on Imanishi's thoughts on the "specia," see Pamela J. Asquith, "Primate Research Groups in Japan: Orientations and East-West Differences," in *The Monkeys of Arashiyama: Thirty-five Years of Research in Japan and the West*, ed. L. M. Fedigan and P. J. Asquith (Albany: State University of New York Press, 1991).

44. Itō Hirobumi, "From an Address on the Constitution to the Conference of Presidents of Prefectural Assemblies, February 15, 1889," in *Sources of Japanese Tradition*, comp. Ryusaku Tsunoda, W. Theodore de Bary, and Donald Keene (New York: Columbia University Press, 1958), 667–68.

45. Imanishi, *Dōbutsuki*, 366–67.

46. Ibid., 369–70.

47. Roy Chapman Andrews, Walter Granger, Clifford Hillhouse Pope, and Nels Christian Nelson, *The New Conquest of Central Asia: A Narrative of the Explorations of the Central Asiatic Expeditions in Mongolia and China, 1921–1930* (New York: American Museum of Natural History, 1932).

48. Imanishi, *Dōbutsuki*, 371.

49. Yanagita Kunio, *Tōno monogatari* (Tales of Tōno) [1910], in *Yanagita Kunio shū* (Yanagita Kunio collection), vol. 4 (Tokyo: Chikuma Shobō, 1968), 22.

50. Imanishi, *Dōbutsuki*, 378–79.

51. To make this point, Imanishi refers to Walter Heape and F. H. A. Marshall, *Emigration, Migration and Nomadism* (Cambridge: W. Heffer and Sons, 1931).

52. Inukai Tetsuo, "Hokkaidō-san ōkami to sono metsubō keiro" (The Hokkaido wolf and its road to extinction), *Shokubutsu oyobi dōbutsu* (Botany and zoology) 1, no. 8 (August 1933): 11–18.

53. See Thomas R. Dunlap, *Saving America's Wildlife: Ecology and the American Mind, 1850–1990* (Princeton: Princeton University Press, 1988).

54. Inukai, "Hokkaidō-san ōkami to sono metsubō keiro," 11.

55. Ibid.

56. Ibid., 12.

57. Other ecologists, such as Tawara Hiromi, have also emphasized Edwin Dun's role in the extinction of the Hokkaido wolf. Tawara even goes so far as to suggest that had someone other than Dun been in charge of the Niikappu ranch,

the wolf might not have been wiped out. He points to the long-standing place of the wolf in premodern Japanese culture, implying that the Western hatred for the wolf, in the person of Dun, led to the extinction of the Hokkaido wolf. See Tawara Hiromi, *Hokkaidō no shizen hogo* (Protecting Hokkaido's nature) (Sapporo: Hokkaidō Daigaku Tosho Kankōkai, 1979), 105.

58. Inukai, "Hokkaidō-san ōkami to sono metsubō keiro," 14–16.

59. Inukai Tetsuo covers the impact of Kaitakushi deer hunting in *Hoppō dōbutsu-shi* (Northern animal journal) (Sapporo: Hokuensha, 1975), 23; and *Waga dōbutsu-ki* (My animal chronicle) (Tokyo: Kurashi no Techōsha, 1970), 117–21.

60. Inukai, "Hokkaidō-san ōkami to sono metsubō keiro," 17.

61. Ibid.

62. Chiba Tokuji, *Ōkami wa naze kietaka* (Why did the wolf disappear?) (Tokyo: Shin Jinbutsu Ōraisha, 1995).

63. Ibid., 166–73.

64. Ibid., 174–75.

65. Ibid., 175–76.

66. Ibid., 176–77.

67. Ibid., 177.

68. Ibid., 177–78.

69. Ibid., 179–82.

70. Ibid., 183–85.

71. Ibid., 186–89.

72. Ibid., 190.

73. Ibid., 192–93.

74. Ibid., 193–94.

EPILOGUE

1. Yoshida Noriyuki, "Robo People: And Robo Cats and Dogs," *Look Japan* 47, no. 547 (October 2001): 10.

2. See, e.g., Clive Ponting, *A Green History of the World: The Environment and the Collapse of Great Civilizations* (New York: Penguin Books, 1991).

3. J. R. McNeill, *Something New under the Sun: An Environmental History of the Twentieth-Century World* (New York: W. W. Norton and Co., 2000), 262–64.

4. Stephen R. Kellert, "Attitudes, Knowledge, and Behavior toward Wildlife among the Industrial Superpowers: United States, Japan, and Germany," *Journal of Social Issues* 49, no. 1 (1993): 53–69.

5. John D. Varley, Wayne G. Brewster, Sarah E. Broadbent, and Renee Evanoff, eds., *Wolves for Yellowstone? A Report to the United States Congress*, vol. 4, *Research and Analysis* (Billings, MT: National Park Service, Yellowstone National Park, 1992).

6. Fish and Wildlife Service, U.S. Department of the Interior, *The Reintroduction of Gray Wolves to Yellowstone National Park and Central Idaho: Summary Environmental Impact Statement* (Washington, DC: U.S. Government Printing Office, 1993).

7. For a summary of the early phase of wolf reintroduction, see Edward E. Bangs et al., "Status of Gray Wolf Restoration in Montana, Idaho, and Wyoming," *Wildlife Society Bulletin* 26, no. 4 (1998): 787.

8. Brett L. Walker, "Commercial Growth and Environmental Change in Early Modern Japan: Hachinohe's Wild Boar Famine of 1749," *Journal of Asian Studies* 60, no. 2 (May 2001): 329–51.

9. Kikuchi Isao, *Kinsei no kikin* (Early modern famines) (Tokyo: Yoshikawa Kōbunkan, 1997), 122, 125.

10. Koganezawa Masaaki, "Nikkō ni okeru shika no zōka to shinrin seitaikei e no eikyō, soshite ōkami dōnyū no hitsuyōsei" (The link between the rise in deer numbers and forest ecology in Nikkō and the necessity of wolf introduction), *Foresuto kōru* (Forest call) 6 (May 1999): 4–5.

11. See Wajiro Suzuki, Hiroshi Tanaka, and Tohru Nakashizuka, "Long Term Ecological Research at Senju-ga-hara Forest Reserve in Nikko, Central Japan," at http://ss.ffpri.affrc.go.jp/labs/femnet/nikko/nikko.htm.

12. See, e.g., Ue Toshikatsu, *Yamabito no dōbutsushi: Kishū Hatenashi sanmyaku haruaki* (The animal journal of a mountain person: Spring and fall in the Kishū and the Hatenashi Mountain Range) (Tokyo: Shinjuku Shobō, 1998).

13. E.g., see Yanai Kenji, *Maboroshi no Nihon ōkami* (The elusive Japanese wolf) (Urawashi: Sakitama Shuppankai, 1993).

14. John Knight, "On the Extinction of the Japanese Wolf," *Asian Folklore Studies* 56 (1997): 145–49. For a more thorough treatment of the reintroduction debate, see John Knight, *Waiting for Wolves in Japan: An Anthropological Study of People-Wildlife Relations* (Oxford: Oxford University Press, 2003), 216–34.

WORKS CITED

UNPUBLISHED DOCUMENTARY SOURCES

"Akuken karitori no ken" (Matter of hunting and killing bad dogs) [1877.6.19].
In Shoka bun iroku (Transferred records of written documents from various
divisions) (A4-24). Hokkaido Prefectural Archives, Sapporo, Japan.

Dun, Edwin. "His Excellency, Dzushio Hirotaki, Kaitaku, Daishioki Kwan"
[Sapporo, 1881.5.28]. Resource Collection for Northern Studies, Hokkaido
University Library, Sapporo, Japan.

————. "His Excellency, Governor Hori, Kaitaku, Daishioki Kwan" [Sapporo,
1877.10.18]. Resource Collection for Northern Studies, Hokkaido University
Library, Sapporo, Japan.

————. "His Excellency, Matsumoto" [Nanai, 1875.10.4]. Resource Collection
for Northern Studies, Hokkaido University Library, Sapporo, Japan.

————. "Mr Dzushio Hirotake, Acting Daishioke Kwan of Kaitakushi"
[Sapporo, 1877.7.9]. Resource Collection for Northern Studies, Hokkaido
University Library, Sapporo, Japan.

————. "Mr Satow, Gon shio shioki Kwan of Kaitakushi" [Sapporo, 1878.3.25].
Resource Collection for Northern Studies, Hokkaido University Library,
Sapporo, Japan.

————. "Reminiscences of Nearly a Half Century in Japan" [n.d.]. Resource
Collection for Northern Studies, Hokkaido University Library, Sapporo, Japan.

————. "Shiriuchi" [Sapporo, 1881.7.1]. Resource Collection for Northern
Studies, Hokkaido University Library, Sapporo, Japan.

"Enchū chikuyō no ōkami ribyō jūtai no ken" (Matter of wolf raised in garden
contracting disease and falling into critical condition) [1879.7.20]. In Bun
iroku (Transferred records of written documents) (03736). Hokkaido
Prefectural Archives, Sapporo, Japan.

"Esashi nite ikedori no ōkami no ko Tōkyō kari hakubutsukan e shuppin ni tsuki sashiokurikata no ken" (Matter of sending wolf pups captured alive at Esashi for display at Tokyo provisional museum) [1877.6]. In Kaka bun iroku (Transferred records of divisional written documents) (02136). Hokkaido Prefectural Archives, Sapporo, Japan.

"Hidaka-koku Mitsuishi-gun fukutochō Ōshima Kunitarō yori yūgai no ōkami jūsatsu negaide ni tsuki shobun sai no ken" (Matter of Assistant Head Ōshima Kunitarō of Hidaka Province, Mitsuishi District, requesting permission to deal with damaging wild wolves by shooting them) [1878.7.7]. In Shoka bun iroku (Transferred records of written documents from various divisions) (A4-54-71). Hokkaido Prefectural Archives, Sapporo, Japan.

"Hiyama-gun Otobe-mura sanchū ni oite shūkaku no ōkami no ko Hakodate-maru e tsumiiremawashi itashi no ken" (Matter of shipping on the *Hakodate-maru* wolf pups caught in the mountains of Hiyama District, Otobe Village) [1877.6.16]. In Tōkyō bun iroku (Transferred records of Tokyo written documents) (02122). Hokkaido Prefectural Archives, Sapporo, Japan.

"Hondō rōgai zankoku ni tsuki satsukaku no tame sutorikiniine kōkyū no ue sōchi kata no ken" (Matter of the brutal damage caused by wolves throughout Hokkaido and the purchasing and sending of strychnine to kill them) [1880. 3.16]. In Tōkyō bun iroku (Transferred records of Tokyo written documents) (03774). Hokkaido Prefectural Archives, Sapporo, Japan.

"Inu no ryūkōbyō no ken" (Matter of canine epizootics) [1873.6.2]. In Kaitakushi kōbun roku (Record of official Kaitakushi documents) (5750). Hokkaido Prefectural Archives, Sapporo, Japan.

"Jūsan nen jūni gatsu yori jūyon nen rokugatsu made yūgai chōjū kakusatsu teate shiharai no ken" (Matter of paying bounties between December 1880 and June 1881 for killing damaging animals) [1881.12.14]. In Shūsairoku (Collected jurisdictional records) (05036). Hokkaido Prefectural Archives, Sapporo, Japan.

"Kaitakushi chichū kari hakubutsujo nite kaitate no mesuōkami niwaka ni han-mon kutsū ni tsuki gaikoku iin jushin kata torihakarai korearitaku no ken" (Matter of arranging physical examination by foreign doctor of female wolf raised at the Kaitakushi provisional museum with sudden severe pain) [1877. 10.17]. In Hokkaidō shobuppin ōfukusho oyobi shoshōken tsūshin shorui (Varieties of documents regarding Hokkaido goods and ministry and prefecture correspondences) (02380). Hokkaido Prefectural Archives, Sapporo, Japan.

"Kaitakushi chikuyō no mesuōkami kyūbyō ni tsuki yatoi jūi shindan no gi do-nichiyō igai no hi tōka made hikitsuke kudasare shinryō no ken" (Matter of request for consultation with hired veterinarian at [agricultural sciences] division on any day other than Saturday or Sunday regarding diagnosis of sick female wolf raised by Kaitakushi) [1877.10.18]. In Hokkaidō shobuppin ōfukusho oyobi shoshōken tsūshin shorui (Varieties of documents regarding Hokkaido goods and ministry and prefecture correspondences) (02380). Hokkaido Prefectural Archives, Sapporo, Japan.

"Kaitakushi kari hakubutsujo nite kaioki no yamainu kitoku ni tsuki heishin no sai wa gaikokujin ni takushi kaibō itashitaki ken" (Matter of entrusting foreigner to perform dissection when mountain dog that was raised at Kaitakushi provisional museum and that is in critical condition dies) [1877.10.18]. In Hokkaidō shobuppin ōfukusho oyobi shoshōken tsūshin shorui (Varieties of documents regarding Hokkaido goods and ministry and prefecture correspondences) (02380). Hokkaido Prefectural Archives, Sapporo, Japan.

"Kameda-gun Komagatake ni oite ikedori no ōkami sōchi kata no ken" (Matter of sending live wolves from Komagatake in Kameda District) [1878.6.8]. In Tōkyō shoka bun iroku (Transferred records of various Tokyo divisional written documents) (02641). Hokkaido Prefectural Archives, Sapporo, Japan.

"Kamiiso-gun Sakkari Kikonai ryōson nite ōkami kujo no tame ryōshi yatoi ire kata no ken" (Matter of hiring hunters to exterminate wolves from the two villages of Sakkari and Kikonai in Kamiiso District) [1880.3.22]. In Kannai gunku yakusho bun iroku (Transferred records of written documents from district-ward jurisdictional government offices) (4051). Hokkaido Prefectural Archives, Sapporo, Japan.

"Komagatake kinbō ni ōkami no ko kore aru yoshi Tōkyō hakubutsujo e sashidashi taku ikedori kata no ken" (Matter of capturing live wolves to send to Tokyo museum because of wolf pups in the Komagatake area) [1878.1.15]. In [file name not given] (2647-3). Hokkaido Prefectural Archives, Sapporo, Japan.

"Komagatake nite ikedori no ōkami no ko chōekijo nite shiiki kata no ken" (Matter of caring for captive wolf pups captured at Komagatake) [1878.5.10]. In Shūsairoku (Collected jurisdictional records) (02650). Hokkaido Prefectural Archives, Sapporo, Japan.

"Kuma ōkami bōgai no tame Noboribetsu bokujō e jūhō narabi ni dan'yaku watashi kata no ken" (Matter of sending firearms and ammunition to the Noboribetsu ranch to prevent damage caused by bears and wolves) [1874.1.14]. In Kaitakushi kōbun roku (Record of official Kaitakushi documents) (06001). Hokkaido Prefectural Archives, Sapporo, Japan.

"Kuma ōkami nado mōjū ryōsatsu no ken" (Matter of hunting and killing bears, wolves, and other animals) [1878.1.21 and 1878.5.31]. In Shūsairoku (Collected jurisdictional records) (A4-54-48). Hokkaido Prefectural Archives, Sapporo, Japan.

"Kuma ōkami narabi ni karasu hokaku todoke tori matome hōkoku sashidasase kata no ken" (Matter of the final settling of notification regarding the capture of bears, wolves, and crows) [1880.1.21]. In Shūsairoku (Collected jurisdictional records) (04061). Hokkaido Prefectural Archives, Sapporo, Japan.

Lyman, Benjamin S. "Journal of a trip from Nemoro by Soya and Hakodate to

Yedo" [1874]. Resource Collection for Northern Studies, Hokkaido University Library, Sapporo, Japan.

"Men'yō kyōken no tame daigai o uke ni tsuki arawa ni kansatsu fuyo nai inu no bunshu no ken" (Matter of openly conferring licenses to hunt dogs because of pervasive damage to sheep caused by mad dogs) [1878.6.18]. In Kachū bun iroku (Transferred records of divisional written documents) (A4-27). Hokkaido Prefectural Archives, Sapporo, Japan.

Minamoto Shitagau. "Wamyōrui shūshō" (A collection of Japanese names) [931–37]. Sapporo Agricultural College Collection, Resource Collection for Northern Studies, Hokkaido University Library, Sapporo, Japan.

"Mōjū ryōsatsusha e teatekin kyūyo no ken" (Matter of bounty allowance paid to wild animal hunters) [1878.2.15]. In Shūsairoku (Collected jurisdictional records) (A4-54-49). Hokkaido Prefectural Archives, Sapporo, Japan.

"Nemuro gunnai ni oite shukaku no ōkami no kawa sōchi no ken" (Matter of sending skins from wolves killed within Nemuro District) [1881.3.1]. In Honshichō ōrai bun iroku (Transferred records of written documents between home and branch offices) (05329). Hokkaido Prefectural Archives, Sapporo, Japan.

"Niikappu bokujō nai ni oite yajū dokusatsu no gi ukagai no ken" (Matter of inquiry into poisoning wild animals at the Niikappu ranch) [1878.6.20]. In Shūsairoku (Collected jurisdictional records) (A4-54). Hokkaido Prefectural Archives, Sapporo, Japan.

"Niikappu bokujō sanjiba kuma ōkami nado no higai bōjo no tame yōjū dan sōchi no ken" (Matter of sending Western firearms and ammunition to prevent bear and wolf damage to newborn horses at the Niikappu ranch). In Honka todokeroku (Records of home division reports) (A4-51-72). Hokkaido Prefectural Archives, Sapporo, Japan.

"Ōkami no ko tōchaku no gi narabi ni hokaku shonyūhi shōkai sono hoka no ken" (Matter of inquiries into various expenses regarding the capture and arrival of wolf pups) [1878.7]. In Tōkyō shoka bun iroku (Transferred records of various Tokyo divisional written documents) (02641). Hokkaido Prefectural Archives, Sapporo, Japan.

"Oshamanbe-mura Ega Jūjirō ōkami ko shukaku ni kan shi kensa no tame zenku sashidashi kata no ken" (Matter of Ega Jūjirō from Oshamanbe Village submitting for inspection entire corpse of captured wolf pup) [1880.5.13]. In Kannai gunku yakusho bun iroku (Transferred records of written documents from district-ward jurisdictional government offices) (4051). Hokkaido Prefectural Archives, Sapporo, Japan.

"Roryō Shiberia chihō ōkami ryō keikyō gakumenzu utsushitori sashi mawashi no ken" (Matter of sending around copy of picture depicting wolf hunt in area of Russian Siberia) [1879.7.22]. In Tōkyō bun iroku (Transferred records of Tokyo written documents) (3031). Hokkaido Prefectural Archives, Sapporo, Japan.

"Satake-ke gonikki" (Chronicle of the Satake family) [1673–1854]. Copy from the personal collection of Mr. Tozawa Shirō of the Kakunodate Rekishi Annainin Kumiai, Kakunodate, Akita Prefecture.

Tanaka Yoshio. "Dōbutsugaku" (Zoology) [1874]. Sapporo Agricultural College Collection, Resource Collection for Northern Studies, Hokkaido University Library, Sapporo, Japan.

———. "Dōbutsu kunmō: Honyūrui" (Instruction on animals: Mammals) [1875]. Sapporo Agricultural College Collection, Resource Collection for Northern Studies, Hokkaido University Library, Sapporo, Japan.

"Tōkyō e sashiokuri no ōkami no ko no tsukisoi toshite Miyamura Ichitarō sashitate no ken" (Matter of sending Miyamura Ichitarō to accompany wolf pups being shipped to Tokyo) [1877.6.18]. In Shūsairoku (Collected jurisdictional records) (02259). Hokkaido Prefectural Archives, Sapporo, Japan.

"Tōkyō e sashiokuri ōkami nitō kaiage keihi torishirabe no ken" (Matter of examining how to pay the expenses of shipping two wolves to Tokyo) [1877.7.29]. In Esashi bunsho ōraisho (Documents that came and went from Esashi branch office) (02256). Hokkaido Prefectural Archives, Sapporo, Japan.

Uchida Kiyoshi, Tauchi Suteroku, and Fujita Kusaburō. "Hidaka Tokachi Kushiro Kitami Nemuro junkai fukumeisho" (Documentary report of tour of Hidaka, Tokachi, Kushiro, Kitami, and Nemuro) [1881]. Resource Collection for Northern Studies, Hokkaido University Library, Sapporo, Japan.

"Usu Shizunai ryōgun nōji keikyō narabi ni imin gaikyō chōsho no ken" (Matter of memorandum regarding agricultural conditions and general outlook of immigrants in Usu and Shizunai Districts) [1878.5.23]. In Honka todokeroku (Records of home division reports) (A4-53-73). Hokkaido Prefectural Archives, Sapporo, Japan.

"Usu-gun ekitei toriatsukai nindai Iida Ichigorō hoka yonjūni mei ōkami ko hokaku ni tsuki teatekin kashi no ken" (Matter of imperial grant bounty allowance paid to Iida Ichigorō and forty-two others for captured wolf pups) [1878.8.5]. In Shūsairoku (Collected jurisdictional records) (A4-54). Hokkaido Prefectural Archives, Sapporo, Japan.

"Yachū bokuba o oiireru saku chikuzō hoka yonkado no ken" (Matter of four reasons to construct a fence to herd horses into at night) [1878.5.11]. In Shūsairoku (Collected jurisdictional records) (A4-54). Hokkaido Prefectural Archives, Sapporo, Japan.

"Yaken bokusatsu no ken" (Matter of killing wild dogs) [1878.8.19]. In Kakugun bun iroku (Transferred records of written documents from all districts) (A4-49). Hokkaido Prefectural Archives, Sapporo, Japan.

PUBLISHED DOCUMENTARY SOURCES

Abe Yoshio. "Nihon ryōnai no ōkami ni tsuite Pokokku-shi ni atau [On the Korean wolf again (and answer to Mr. Pocock)]." *Dōbutsugaku zasshi* (Zoological magazine) 48 (1936): 639–44.

———. "Nukutei ni tsuite" (On Nukutei). *Dōbutsugaku zasshi* (Zoological magazine) 35 (1923): 380–86.

———. "On the Corean and Japanese Wolves." *Journal of Science of the Hiroshima University,* Series B, Division 1, 1 (December 1930): 33–36.

Allee, W. C., Alfred E. Emerson, Orlando Park, Thomas Park, and Karl P. Schmidt. *Principles of Animal Ecology.* Philadelphia and London: W. B. Saunders Co., 1949.

Andrews, Roy Chapman, Walter Granger, Clifford Hillhouse Pope, and Nels Christian Nelson. *The New Conquest of Central Asia: A Narrative of the Explorations of the Central Asiatic Expeditions in Mongolia and China, 1921–1930.* New York: American Museum of Natural History, 1932.

Bather, F. A. "Discussion and Correspondence: Scientific Terminology." *Science* 15 (May 9, 1902): 747–49.

Batty, Joseph H. *How to Hunt and Trap.* New York: Orange Judd Co., 1884.

Belknap, Orin. "Poisoning Wolves." *Forest and Stream* 48, no. 9 (February 27, 1897): 168.

Bessey, Charles E. "The Word 'Ecology.'" *Science* 15 (April 11, 1902): 593.

Brehms, Alfred Edmund, Walther Kahle, and Walter Rammner. *Brehms Tierleben: Kleine Ausgabe für Volk und Schule.* Leipzig: Bibliographisches Institut, 1929.

Chiba Tokuji. *Ōkami wa naze kietaka* (Why did the wolf disappear?). Tokyo: Shin Jinbutsu Ōraisha, 1995.

———. *Shuryō denshō* (Hunting folklore). Tokyo: Hōsei Daigaku Shuppankyoku, 1975.

Chu Hsi. *Learning to Be a Sage: Selections from the "Conversations of Master Chu," Arranged Topically.* Translated by Daniel K. Gardner. Berkeley and Los Angeles: University of California Press, 1990.

Clark, Ella E. *Indian Legends from the Northern Rockies.* Norman: University of Oklahoma Press, 1966.

Confucius. *The Analects.* Translated by D. C. Lau. London: Penguin Books, 1979.

Corbin, Ben. *Corbin's Advice, or the Wolfer's Guide.* North Dakota: Tribune Co., 1900.

Darwin, Charles. *The Expression of the Emotions in Man and Animals.* New York: D. Appleton and Co., 1872. Reprint, Chicago: University of Chicago Press, 1965.

Department of Agriculture and Commerce, ed. *Japan in the Beginning of the 20th Century.* Tokyo: Shoin, 1904.

Dodge, Richard Irving. *The Plains of North America and Their Inhabitants.* Newark: University of Delaware Press, 1989.

Elton, Charles S. *Animal Ecology*. Introduction by Mathew A. Leibold and J. Timothy Wootton. London: Sidgwick and Jackson, 1927. Reprint, Chicago: University of Chicago Press, 1966.

———. *The Ecology of Invasions by Animals and Plants*. Foreword by Daniel Simberloff. London: Methuen and Co., 1958. Reprint, Chicago: University of Chicago Press, 2000.

Essays in Idleness: The Tsurezuregusa of Kenkō. Translated by Donald Keene. New York: Columbia University Press, 1967.

Ferris, W. A. *Life in the Rocky Mountains: A Diary of Wandering on the Sources of the Rivers Missouri, Columbia, and Colorado from February, 1830, to November, 1835*. Edited by Paul C. Phillips. Denver: Fred A. Rosenstock, Old West Publishing Co., 1940.

Fujiwara Mototsune. *Nihon Montoku tennō jitsuroku* (The true record of Japan's Emperor Montoku) [878]. In *Shintei zōho kokushi taikei* (Newly revised and enlarged survey of Japanese history), vol. 3, edited by Kuroita Katsumi. Tokyo: Yoshikawa Kōbunkan, 1981.

Fujiwara Tokihira. *Nihon sandai jitsuroku* (The true record of Japan's three eras) [858–87]. Vol. 2. In *Shintei zōho kokushi taikei* (Newly revised and enlarged survey of Japanese history), vol. 4, edited by Kuroita Katsumi. Tokyo: Yoshikawa Kōbunkan, 1973.

Fukane Sukehito. *Honzō wamyō* (Japanese names in natural studies) [922]. Vol. 2, edited by Yosano Tekkan, Masamune Atsuo, and Yosano Akiko. Tokyo: Nihon Koten Zenshū Kankōkai, 1927–28.

Fukuzawa Yukichi. *An Outline of a Theory of Civilization* [1875]. Translated by David A. Dilworth and G. Cameron Hurst. Tokyo: Sophia University, 1973.

———. *Bunmeiron no gairyaku* (An outline of a theory of civilization) [1875]. In *Fukuzawa Yukichi zenshū* (The complete works of Fukuzawa Yukichi), vol. 4, edited by Keiō Gijuku. Tokyo: Iwanami Shoten, 1959.

Funayama Kaoru. *Zoku Otōsei* (Otōsei continued). In *Funayama Kaoru shōsetsu zenshū* (The complete collection of novels by Funayama Kaoru), vol. 9. Tokyo: Kawade Shobō Shinsha, 1975.

Furukawa Koshōken. *Tōyū zakki* (Miscellaneous notes on travels in the east) [1788]. Edited by Ōtō Tokihiko. Tokyo: Tōyō Bunko, 1964.

Ganong, W. F. "The Word 'Ecology.'" *Science* 15 (April 11, 1902): 593.

———. "The Word 'Ecology.'" *Science* 15 (May 16, 1902): 792–93.

Hatta Saburō. "Hokkaidō no ōkami" (The Hokkaido wolf). *Dōbutsugaku zasshi* (Zoological magazine) 25 (January 1913).

Heape, Walter, and F. H. A. Marshall. *Emigration, Migration and Nomadism*. Cambridge: W. Heffer and Sons, 1931.

Hiraga Satano. "Horkew kotan kor kur" (Japanese: Ōkami-kami to mura osa; English: The wolf god and the village man) [October 13, 1967]. In *Kayano Shigeru no Ainu shinwa shūsei: Kamuy-yukar* (Kayano Shigeru's collection of Ainu divine tales: Kamuy-yukar), vol. 2, edited by Kayano Shigeru. Tokyo: Heibonsha, 1998.

Hitomi Hitsudai. *Honchō shokkan* (Culinary mirror of Japan) [c. 1695]. Vols. 1–5. Transcribed and annotated by Shimada Isao. Tokyo: Tōyō Bunko, 1981.

Hokkaidō Keisatsu Henshū Iinkai, ed. *Hokkaidō keisatsushi* (History of the Hokkaido police). Vol 1. Sapporo: Hokkaidō Keisatsu Honbu, 1968.

Ihara Saikaku. *Honchō nijū fukō* (Twenty stories of unfilial behavior in the realm) [1686]. In *Taiyaku Saikaku zenshū* (The complete works of Saikaku with transliteration), vol. 10, edited by Asō Isoji and Fuji Akio. Tokyo: Meiji Shoin, 1976.

Imanishi, Kinji. *A Japanese View of Nature: "The World of Living Things" by Imanishi Kinji.* Translated by Pamela J. Asquith, Heita Kawakatsu, Shusuke Yagi, and Hiroyuki Takasaki. London: RoutledgeCurzon, 2002.

Imanishi Kinji. *Dōbutsuki* (Animal chronicle) [1946]. In *Imanishi Kinji zenshū* (The complete works of Imanishi Kinji), vol. 2, edited by Noma Shōichi. Tokyo: Kōdansha, 1974.

―――. *Seibutsu no sekai* (The world of living things) [1941]. In *Imanishi Kinji zenshū* (The complete works of Imanishi Kinji), vol. 1, edited by Noma Shōichi. Tokyo: Kōdansha, 1974.

―――. *Seibutsu shakai no ronri* (The logic of biological societies) [1948]. In *Imanishi Kinji zenshū* (The complete works of Imanishi Kinji), vol. 4, edited by Noma Shōichi. Tokyo: Kōdansha, 1974.

Inukai Tetsuo. "Hokkaidō no shika to sono kōbō" (The Hokkaido deer and its rise and fall). *Hoppō bunka kenkyū hōkoku* (Research reports on northern culture) 7 (March 1952): 1–45.

―――. "Hokkaidō-san ōkami to sono metsubō keiro" (The Hokkaido wolf and its road to extinction). *Shokubutsu oyobi dōbutsu* (Botany and zoology) 1, no. 8 (August 1933): 11–18.

―――. *Hoppō dōbutsushi* (Northern animal journal). Sapporo: Hokuensha, 1975.

―――. *Waga dōbutsuki* (My animal chronicle). Tokyo: Kurashi no Techōsha, 1970.

Inukai Tetsuo and Kadosaki Masaaki. *Higuma: Hokkaidō no shizen* (Brown bear: Hokkaido's nature). Sapporo: Hokkaidō Shinbunsha, 1993.

Irving, Washington. *Adventures of Captain Bonneville.* Portland, OR: Binfords and Mort, Publishers, 1963.

Itō Hirobumi. "From an Address on the Constitution to the Conference of Presidents of Prefectural Assemblies, February 15, 1889." In *Sources of Japanese Tradition,* compiled by Ryusaku Tsunoda, W. Theodore de Bary, and Donald Keene. New York: Columbia University Press, 1958.

Kaibara Ekken. *Yamato honzō* (Natural studies in Japan) [1709]. 2 vols. Edited by Shirai Kōtarō. Tokyo: Ariake Shobō, 1980.

Kaitakushi, ed. *Hokkaidōshi* (Hokkaido records) [1892]. Tokyo: Rekishi Toshosha, 1973.

"Kamo no Chōmei: An Account of My Hut." Translated by Donald Keene. In

Anthology of Japanese Literature from the Earliest Era to the Mid–Nineteenth Century. Edited by Donald Keene. New York: Grove Press, 1999.

Kanzawa Teikan. *Okinagusa* (Pasqueflower) [1851]. Vol 5. In *Nihon zuihitsu taisei* (Survey of Japan's literary miscellany), 3rd ed., vol. 23, edited by Nihon Zuihitsu Taisei Henshūbu. Tokyo: Yoshikawa Kōbunkan, 1978.

Kayano Shigeru. *Honoo no uma: Ainu minwashū* (Horse of fire: A collection of Ainu folklore). Tokyo: Suzusawa Shoten, 1977.

Kayano Shigeru and Saitō Hiroyuki. *Kibori no ōkami* (The carved wooden wolf). Tokyo: Shōhō Shoten, 1998.

Kelly, L. S. "Wolves and Coyotes." *Forest and Stream* 48, no. 8 (February 20, 1897): 144–45.

Kishida Ginkō. "Tōhoku gojunkōki" (Northeastern imperial procession chronicle) [1876]. In *Meiji bunka zenshū* (The complete works of Meiji culture), vol. 17, edited by Meiji Bunka Kenkyūkai. Tokyo: Nihon Hyōronsha, 1967.

Kishida K. "Notes on the Yesso Wolf." *Lansania* 3 (1931): 72–75.

Kofudoki itsubun (Lost writings on ancient customs) [713]. Edited by Kurita Hiroshi. Tokyo: Ō Okayama Shoten, 1927.

Komiyama Masahide. *Fūken gūki* (Whimsical chronicle of maple eaves) [1807–10]. In *Nihon zuihitsu taisei* (Survey of Japan's literary miscellany), vol. 10, edited by Nihon Zuihitsu Taisei Henshūbu. Tokyo: Nihon Zuihitsu Taisei Kankōkai, 1929.

Kubota Shizō. *Kyōwa shieki* (Harmonious private expedition) [1856]. In *Nihon shomin seikatsu shiryō shūsei* (Collected sources on the history of the daily lives of common Japanese people), vol. 4, edited by Takakura Shin'ichirō. Tokyo: San'ichi Shobō, 1969.

Kushihara Seihō. *Igen zokuwa* (Ainu proverbs and gossip) [1792]. In *Nihon shomin seikatsu shiryō shūsei* (Collected sources on the history of the daily lives of common Japanese people), vol. 4, edited by Takakura Shin'ichirō. Tokyo: San'ichi Shobō, 1969.

Larpenteur, Charles. *Forty Years a Fur Trader on the Upper Missouri: The Personal Narrative of Charles Larpenteur, 1833–72.* 2 vols. Edited by Elliot Cous. New York: Francis P. Harper, 1898.

Lorenz, Konrad Z. *King Solomon's Ring: New Light on Animal Ways.* New York: Harper and Row Publishers, 1952. Reprint, New York: Time-Life Books, 1962.

The Man'yōshū (Collection of ten thousand leaves) [c. 759]. Translated by Nippon Gakujutsu Shinkōkai. New York: Columbia University Press, 1965.

Man'yōshū. Vol. 1. In *Shinchō Nihon koten shūsei*, vol. 6, edited by Aoki Takako, Ide Itaru, Itō Haku, Shimizu Katsuhiko, and Hashimoto Shirō. Tokyo: Shinchōsha, 1976.

Man'yōshū. Vol. 2. In *Shinchō Nihon koten shūsei*, vol. 21, edited by Aoki Takako, Ide Itaru, Itō Haku, Shimizu Katsuhiko, and Hashimoto Shirō. Tokyo: Shinchōsha, 1978.

Man'yōshū. Vol. 4. In *Shinchō Nihon koten shūsei*, vol. 55, edited by Aoki Takako, Ide Itaru, Itō Haku, Shimizu Katsuhiko, and Hashimoto Shirō. Tokyo: Shinchōsha, 1982.

Matsumae Hironaga. *Matsumaeshi* (Matsumae record) [1781]. In *Hokumon sōsho* (Northern gate library), vol. 2, edited by Ōtomo Kisaku. Tokyo: Hokkō Shobō, 1943.

Matsuura Takeshirō. *Shinpan Ezo nisshi: Higashi Ezo nisshi* (Newly published Ezo diary: Eastern Ezo diary). Vol. 1. Edited by Yoshida Tsunekichi. Tokyo: Jiji Tsūshinsha, 1984.

McClintock, Walter. *The Old North Trail, or Life, Legends, and Religions of the Blackfeet Indians*. London: Macmillan and Co., 1910. Reprint, Lincoln: University of Nebraska Press, 1999.

Mivart, George Jackson. *Dogs, Jackals, Wolves, and Foxes: A Monograph of the Canidae*. London: R. H. Porter, 1890.

Morse, Edward S. *Japan Day by Day: 1877, 1878–79, 1882–83*. Vol. 2. Boston: Houghton Mifflin Co., 1945. Reprint, Atlanta, GA: Cherokee Publishing Co., 1990.

Nakamura Kazue. "Nihon ōkami no bunrui ni kansuru seibutsu-chirigakuteki shiten" (A biogeographical look at the taxonomy of the Japanese wolf). *Kanagawa kenritsu hakubutsukan kenkyū hōkoku: Shizen kagaku* (Bulletin of the Kanagawa Prefectural Museum: Natural sciences) 27 (March 1998): 49–60.

Naora Nobuo. *Nihon-san ōkami no kenkyū* (Research on Japan's wolves). Tokyo: Azekura Shobō, 1965.

Negishi Yasumori. *Mimibukuro* (Ear bag) [1782–1814]. 2 vols. Edited by Suzuki Tōzō. Tokyo: Heibonsha, 2000.

Nihon kiryaku (Abridged annals of Japan) [897–1036]. 2 vols. In *Shintei zōho kokushi taikei* (Newly revised and enlarged survey of Japanese history), vols. 10–11, edited by Kuroita Katsumi. Tokyo: Yoshikawa Kōbunkan, 1985–88.

Nihon kōki (Japan's postscript) [796–833]. In *Shintei zōho kokushi taikei* (Newly revised and enlarged survey of Japanese history), vol. 3, edited by Kuroita Katsumi. Tokyo: Yoshikawa Kōbunkan, 1961.

"Nihon ōkami ka? Yaken ka? Kyūshū no sanchū deno mokugeki sōdō" (Japanese wolf? Feral dog? Debate over an observation on Kyushu). *Asahi shinbun* (Asashi newspaper), November 22, 2000, evening edition.

Nihongi: Chronicles of Japan from the Earliest Times to A.D. 697. 2 vols. in 1. Translated by W. G. Aston. Rutland, VT: Charles E. Tuttle Co., 1972.

Nishida, Kitarō. *An Inquiry into the Good*. Translated by Masao Abe and Christopher Ives. New Haven: Yale University Press, 1990.

Nishimura Hakū. *Enka kidan* (Strange stories from the mist) [1773]. In *Nihon zuihitsu taisei* (Survey of Japan's literary miscellany), vol. 4, edited by Nihon Zuihitsu Taisei Henshūbu. Tokyo: Yoshikawa Kōbunkan, 1975.

Nitobe, Inazo. *The Works of Inazo Nitobe*. Vol. 1. Edited by Yasaka Takagi, Shigeharu Matsumoto, Yoichi Maeda, Kiyoko Takeda, Makoto Saito, and Yoshihiko Kobayashi. Tokyo: University of Tokyo Press, 1972.

Nonoguchi Takamasa, comp. *Ōō hitsugo* (Stories of Ōō) [1842]. Vol. 2. In *Nihon zuihitsu taisei* (Survey of Japan's literary miscellany), vol. 9, edited by Nihon Zuihitsu Taisei Henshūbu. Tokyo: Yoshikawa Kōbunkan, 1975.

Obara Iwao and Nakamura Kazue. "Minami Ashigara-shi kyōdo shiryōkan shozō no, iwayuru yamainu tōkotsu ni tsuite" (Notes on the so-called *yamainu*, or wild canine, preserved in the Minami Ashigara Folklore Museum). *Kanagawa kenritsu hakubutsukan kenkyū hōkoku: Shizen kagaku* (Bulletin of the Kanagawa Prefectural Museum: Natural sciences) 21 (March 1992): 105–10.

"Ōkami Yuatsumi o arashi jūkyū nin o kuitaru koto" (Incident of wolf violently eating nineteen people in Yuatsumi). In *Shōnai shiryō* (Shōnai sources), edited by Tetsuya Shigeta. Tsuruoka: Tōhoku Shuppan Kikaku, 1912.

Ōkurashō, ed. *Kaitakushi jigyō hōkoku* (Report on Kaitakushi industries). Vol. 2. Sapporo: Hokkaidō Shuppan Kikaku Sentā, 1885.

Ono Ranzan. *Honzō kōmoku keimō* (An instructional outline of natural studies) [1803]. Edited by Shimonaka Hiroshi. Tokyo: Tōyō Bunko, 1992.

Ōuchi Yoan. *Tōkai yawa* (Night talk from eastern Ezo) [1854–59]. In *Hokumon sōsho* (Northern gate library), vol. 5, edited by Ōtomo Kisaku. Tokyo: Hokkō Shobō, 1943–44.

Palmer, Theodore S. "Chronology and Index of the More Important Events in American Game Protection, 1776–1911." *Bulletin of the United States Bureau of Biological Survey* 41 (1912): 7–46.

———. "Extermination of Noxious Animals by Bounties." In *Yearbook of the United States Department of Agriculture*. Washington, DC: Government Printing Office, 1896.

Pocock, R. I. "The Races of Canis Lupus." *Proceedings of the Zoological Society of London*, 1935, 647–86.

Richardson, John. *Fauna Boreali-Americana; or the Zoology of the Northern Parts of British America*. 4 vols. London: John Murray, 1829.

Santō Kyōzan. *Oshiegusa nyōbō katagi* (The character of moral wives) [1853]. In *Musume setsuyō—Oshiegusa nyōbō katagi* (The frugality of daughters—The character of moral wives). Edited by Tsukamoto Tetsuzō. Tokyo: Yūhodo Shobō, 1927.

Satow, Ernest Mason, and Lieutenant A. G. S. Hawes. *A Handbook for Travellers in Central and Northern Japan*. London: John Murray, 1884.

Semper, Karl. *Animal Life as Affected by the Natural Conditions of Existence*. New York: D. Appleton and Co., 1881. Reprint, New York: Arno Press, 1977.

Seton, Ernest Thompson. *Life-Histories of Northern Animals: An Account of the Mammals of Manitoba*. 2 vols. New York: Doubleday and Co., 1909. Reprint, New York: Arno Press, 1974.

Shelford, Victor E. *Animal Communities in Temperate America as Illustrated in the Chicago Region*. Chicago: University of Chicago Press, 1913.

Shiga Naoya. "Takibi" (Night fires) [1920]. In *Gendai Nihon bungaku taikei: Shiga Naoya shū* (Survey of Japan's modern literature: Shiga Naoya collection), vol. 34. Tokyo: Chikuma Shobō, 1968.

Snow, H. J. *Notes on the Kuril Islands*. London: John Murray, 1897.

St. John, Captain H. C. *Notes and Sketches from the Wild Coasts of Nipon: With Chapters on Cruising After Pirates in Chinese Waters*. Edinburgh: David Douglas, 1880.

Stuart, Granville. *Forty Years on the Frontier*. Glendale, CA: Arthur H. Clark Co., 1925.

Tachibana Nankei. *Hokusō sadan* (Brief conversations through a northern window) [1825]. 2 vols. In *Nihon zuihitsu taisei* (Survey of Japan's literary miscellany), 2nd ed., vol. 15, edited by Nihon Zuihitsu Taisei Henshūbu. Tokyo: Nihon Zuihitsu Taisei Kankōkai, 1974.

———. *Tōyūki—Saiyūki* (Lyrical record of a journey to the east—Lyrical record of a journey to the west) [1795]. Edited by Kurimoto Chikara. Tokyo: Kenkyūsha Gakusei Bunko 307, 1940.

Tadano Makuzu. *Ōshū banashi* (Tales from Ōshū) [1832]. In *Tadano Makuzu shū* (Tadano Makuzu collection), revised by Suzuki Yoneko. Tokyo: Kabushikigaisha Kokusho Kankōkai, 1994.

Takebe Ayatari. *Suzumigusa* (Cool grass) [1771]. In *Takebe Ayatari zenshū* (The complete works of Takebe Ayatari), vol. 6. Tokyo: Kokusho Kankōkai, 1987.

Tanamori Fusaaki. *Tanamori Fusaaki shuki* (The notes of Tanamori Fusaaki) [1580]. In *Zokuzoku gunsho ruijū*, vol. 4, ed. Kokusho Kankōkai. Tokyo: Zoku Gunsho Ruijū Kanseikai, 1970.

Terashima Ryōan, comp. *Wakan sansai zue* (The illustrated Japanese-Chinese encyclopedia of the three elements) [1713]. In *Nihon shomin seikatsu shiryō shūsei* (Collected sources on the history of the daily lives of common Japanese people), vol. 10, edited by Miyamoto Tsuneichi, Haraguchi Torao, and Tanigawa Ken'ichi. Tokyo: San'ichi Shobō, 1970.

Thoreau, Henry David. "The Natural History of Massachusetts" [1842]. In *Henry David Thoreau: The Natural History Essays*. Salt Lake City: Peregrine Smith, 1980.

Tōdō Takasawa. *Tōdō Takasawa kōdenryaku* (Abridged public biography of Tōdō Takasawa). Tokyo: Tōdō-ke, 1918.

Togawa Yukio and Honjō Kei. *Ōkami no hi: Ezo ōkami no zetsumetsuki* (Memorial to the wolf: A record of the extinction of the Hokkaido wolf). Tokyo: Shōnen Champion Comics, 1994.

Tokugawa jikki (The true chronicle of the Tokugawa) [1809–43]. Vol. 6. In *Shintei zōho kokushi taikei* (Newly revised and enlarged survey of Japanese history), vol. 43, edited by Kuroita Katsumi. Tokyo: Yoshikawa Kōbunkan, 1931.

Tsumura Masayuki. *Tankai* (Sea of conversations) [1795]. Tokyo: Kokusho Kankōkai, 1917.

Uchida Shigeru and Ono Kyūshichi. *Futsū dōbutsu zusetsu* (Illustrated explanation of common animals). Tokyo: Shūgakudō Shoten, 1915.

The Upanishads. Translated by Juan Mascaró. New York: Penguin Books, 1965.

von Siebold, Philipp Franz. *Fauna Japonica sive: Descriptie animalium, quae in itinere per Japoniam, jufsu et aufpiciis Superiorum, Qui Summum In India Batava Imperium Tenent, fuscepto, annis 1823–1830 collegit, notia, obfervationibus et adumbrationibus illuftravit.* Amstelodami (Amsterdam): Apud J. Müller et Co., 1833.

Webb, William Edward. *Buffalo Land: An Authentic Account of the Discoveries, Adventures, and Mishaps of a Scientific and Sporting Party in the Wild West with Graphic Descriptions of the Country; the Red Man, Savage and Civilized; Hunting the Buffalo, Antelope, Elk and Wild Turkey; Etc; Etc.* Chicago and Cincinnati: E. Hannaford and Co., 1872.

Wheeler, William Morton. "'Natural History,' 'Oecology,' or 'Ethology.'" *Science* 15 (June 20, 1902): 971–76.

Yanagita, Kunio. *The Legends of Tōno.* Translated by Ronald A. Morse. Tokyo: Japan Foundation, 1975.

Yanagita Kunio. *Koen zuihitsu* (Miscellaneous writings on the lone monkey) [1939]. In *Yanagita Kunio shū* (Yanagita Kunio collection), vol. 22. Tokyo: Chikuma Shobō, 1970.

———. *Tōno monogatari* (Tales of Tōno) [1910]. In *Yanagita Kunio shū* (Yanagita Kunio collection), vol. 4. Tokyo: Chikuma Shobō, 1968.

———. *Yama no jinsei* (The lives of mountain people) [1925]. In *Yanagita Kunio shū* (Yanagita Kunio collection), vol. 4. Tokyo: Chikuma Shobō, 1968.

Young, Stanley Paul. *The Last of the Loners.* New York: Macmillan Co., 1970.

———. *The Wolf in North American History.* Caldwell, ID: Caxton Printers, 1946.

OTHER SOURCES

Abbey, Edward. *Desert Solitaire: A Season in the Wilderness.* New York: Touchstone Book, 1968.

Allen, Durward L. *Wolves of Minong: Isle Royale's Wild Community.* Ann Arbor: University of Michigan Press, 1979.

Allen, Garland E. *Life Science in the Twentieth Century.* Cambridge: Cambridge University Press, 1975.

Anthony, David Forsyth. "The Administration of Hokkaido under Kuroda Kiyotaka—1870–1882: An Early Example of Japanese-American Cooperation." Ph.D. diss., Yale University, 1968.

Asa, Cheryl S., and L. David Mech. "A Review of the Sensory Organs in Wolves and

Their Importance to Life History." In *Ecology and Conservation of Wolves in a Changing World*, edited by Ludwig N. Carbyn, Steven H. Fritts, and Dale R. Seip. Edmonton: Canadian Circumpolar Institute, University of Alberta, 1995.

Asahi Minoru. *Nihon no honyū dōbutsu* (Japan's mammals). Tokyo: Tamakawa Daigaku Shuppanbu, 1977.

Asahi shinbun (Asahi newspaper), July 1, 2004. See www.asahi.com/national/update/0627/013.html.

Ashkenazi, Michael, and Jeanne Jacob. *The Essence of Japanese Cuisine: An Essay on Food and Culture*. Philadelphia: University of Pennsylvania Press, 2000.

Asquith, Pamela J. "Primate Research Groups in Japan: Orientations and East-West Differences." In *The Monkeys of Arashiyama: Thirty-five Years of Research in Japan and the West*, edited by L. M. Fedigan and P. J. Asquith. Albany: State University of New York Press, 1991.

Asquith, Pamela J., and Arne Kalland. "Japanese Perceptions of Nature: Ideals and Illusions." In *Japanese Images of Nature: Cultural Perspectives*, edited by Pamela J. Asquith and Arne Kalland. London: Curzon Press, 1997.

Atran, Scott. *Cognitive Foundations of Natural History: Toward an Anthropology of Science*. Cambridge: Cambridge University Press, 1990.

Avital, Eytan, and Eva Jablonka. *Animal Traditions: Behavioral Inheritance in Evolution*. Cambridge: Cambridge University Press, 2000.

Bailey, Beatrice Bodart. "The Laws of Compassion." *Monumenta Nipponica* 40, no. 2 (1985): 163–89.

Bangs, Edward E., Steven H. Fritts, Joseph A. Fontaine, Douglas W. Smith, Kerry M. Murphy, Curtis M. Mack, and Carter C. Niemeyer. "Status of Gray Wolf Restoration in Montana, Idaho, and Wyoming." *Wildlife Society Bulletin* 26, no. 4 (1998): 785–98.

Bankoff, Greg. "A Question of Breeding: Zootechny and Colonial Attitudes towards the Tropical Environment in Late Nineteenth-Century Philippines." *Journal of Asian Studies* 60, no. 2 (Spring 2001): 413–37.

Barrett, Tim. "Shinto and Taoism in Early Japan." In *Shinto in History: Ways of the Kami*, ed. John Breen and Mark Teeuwen. Honolulu: University of Hawai'i Press, 2000.

Bartholomew, James R. *The Formation of Science in Japan*. New Haven: Yale University Press, 1989.

Bass, Rick. *The New Wolves: The Return of the Mexican Wolf to the American Southwest*. New York: Lyons Press, 1998.

———. *The Ninemile Wolves*. New York: Ballantine Books, 1992.

Beasley, W. G. *The Meiji Restoration*. Stanford: Stanford University Press, 1972.

"Beikoku kuma ōkami no men'yō o gai suru keikyō" (The situation of sheep depredation by bears and wolves in the United States). *Hokkai tsūshi* (Northern sea interpreter) 14 (October 1880): 8–11.

Bolitho, Harold. "The Dog Shogun." In *Self and Biography: Essays on the*

Individual and Society in Asia, edited by Wang Gungwu. Sydney: Sydney University Press, 1975.

Breen, John, and Mark Teeuwen, eds. *Shinto in History: Ways of the Kami.* Honolulu: University of Hawaiʻi Press, 2000.

Brittan, Gordon. "The Mental Life of Other Animals." *Evolution and Cognition* 5, no. 2 (1999): 105–13.

————. "The Secrets of Antelope." *Erkenntnis* 51 (1999): 59–77.

Burbank, James C. *Vanishing Lobo: The Mexican Wolf and the Southwest.* Boulder: Johnson Books, 1990.

Busch, Robert H. *The Wolf Almanac.* New York: Lyons Press, 1995.

————, ed. *Wolf Songs.* San Francisco: Sierra Club Books, 1994.

Caras, Roger A. *The Custer Wolf: Biography of an American Renegade.* Boston: Little, Brown and Co., 1966.

Carbyn, Ludwig N., Steven H. Fritts, and Dale R. Seip, eds. *Ecology and Conservation of Wolves in a Changing World.* Edmonton: Canadian Circumpolar Institute, University of Alberta, 1995.

Casal, U. A. "The Goblin Fox and Badger and Other Witch Animals of Japan." *Folklore Studies* 18 (1959): 1–93.

Casey, Denise, and Tim W. Clark, comps. *Tales of the Wolf: Fifty-one Stories of Wolf Encounters in the Wild.* Moose, WY: Homestead Publishing, 1996.

Charlton, K. M. "The Pathogenesis of Rabies and Other Lyssaviral Infections: Recent Studies." In *Lyssaviruses,* edited by C. E. Rupprecht, B. Dietzschold, and H. Koprowski. Berlin: Springer-Verlag, 1994.

"Chikken no kisoku" (Regulations regarding domesticated dogs). *Hokkai tsūshi* (Northern sea interpreter) 6 (June 1880): 6–9.

Chiri Mashiho. *Chiri Mashiho chosakushū: Bunri Ainugo jiten—shokubutsu dōbutsu* (The collected works of Chiri Mashiho: Ainu language by classification—plants and animals). Appendix 1. Tokyo: Heibonsha, 1976.

Clarke, C. H. D. "The Beast of Gévaudan." *Natural History* 80, no. 4 (April 1971): 44–73.

Coleman, Jon Thomas. "Wolves in American History." Ph.D. diss., Yale University, 2003.

Conlogue, William. *Working the Garden: American Writers and the Industrialization of Agriculture.* Chapel Hill: University of North Carolina Press, 2002.

Cooper, Michael, comp. *They Came to Japan: An Anthology of European Reports on Japan, 1543–1640.* Berkeley and Los Angeles: University of California Press, 1965.

Coppinger, Raymond, and Lorna Coppinger. *Dogs: A Startling New Understanding of Canine Origin, Behavior, and Evolution.* New York: Scribner, 2001.

Cullen, L. M. *A History of Japan, 1582–1941: Internal and External Worlds.* Cambridge: Cambridge University Press, 2003.

Cumings, Bruce. *Korea's Place in the Sun: A Modern History*. New York: W. W. Norton and Co., 1997.

Crosby, Alfred W. *The Columbian Exchange: Biological and Cultural Consequences of 1492*. Westport, CT: Greenwood Press, 1972.

———. *Ecological Imperialism: The Biological Expansion of Europe, 900–1900*. Cambridge: Cambridge University Press, 1986.

The Daily Yomiuri. February 29, 2004. Online edition at www.yomiuri.co.jp/newse/20040229w071.htm.

Dan Michiko. *Meiji no bokusaku* (Fenced pastures of Meiji). Tokyo: Sumire Gakuennai, 1968.

Davis, Mike. *Ecology of Fear: Los Angeles and the Imagination of Disaster*. New York: Metropolitan Books, 1998.

Devall, Bill. "The Deep Ecology Movement." In *Ecology*, edited by Carolyn Merchant. Key Concepts in Critical Theory. Atlantic Highlands, NJ: Humanities Press, 1994.

De Waal, Frans. *The Ape and the Sushi Master: Cultural Reflections of a Primatologist*. New York: Basic Books, 2001.

Doak, Kevin M. "What Is a Nation and Who Belongs? National Narratives and the Ethnic Imagination in Twentieth-Century Japan." *American Historical Review* 102, no. 2 (1997): 283–309.

Dore, R. P. *Education in Tokugawa Japan*. London: Athlone Press, 1965. Reprint, Ann Arbor: Center for Japanese Studies, University of Michigan, 1992.

Dunlap, Thomas R. *Saving America's Wildlife: Ecology and the American Mind, 1850–1990*. Princeton: Princeton University Press, 1988.

Duus, Peter. *The Abacus and the Sword: The Japanese Penetration of Korea, 1895–1910*. Berkeley and Los Angeles: University of California Press, 1995.

———. *The Japanese Discovery of America: A Brief History with Documents*. Boston: Bedford Books, 1997.

Etter, Carl. *Ainu Folklore: Traditions and Culture of the Vanishing Aborigines of Japan*. Chicago: Wilcox and Follett Co., 1949.

Farris, William Wayne. *Population, Disease, and Land in Early Japan, 645–900*. Cambridge: Harvard University Press, 1985.

Fiennes, Richard. *The Order of Wolves*. Indianapolis: Bobbs-Merrill Co., 1976.

Figal, Gerald. *Civilization and Monsters: Spirits of Modernity in Meiji Japan*. Durham, NC: Duke University Press, 1999.

Fischer, Hank. *Wolf Wars*. Helena, MT: Falcon Press Publishing Co., 1995.

Fish and Wildlife Service, U.S. Department of the Interior. *The Reintroduction of Gray Wolves to Yellowstone National Park and Central Idaho: Summary Environmental Impact Statement*. Washington, DC: U.S. Government Printing Office, 1993.

Fitzgerald, Deborah Kay. *Every Farm a Factory: The Industrial Ideal in American Agriculture*. New Haven: Yale University Press, 2003.

Flader, Susan L. *Thinking like a Mountain: Aldo Leopold and the Evolution of an Ecological Attitude toward Deer, Wolves, and Forests.* Madison: University of Wisconsin Press, 1974.

Foresuto kōru (Forest call) 1 (August 1994): 1–14.

Foucault, Michel. *The Order of Things: An Archaeology of the Human Sciences.* New York: Vintage Books, 1970.

Fox, Michael W. *The Soul of the Wolf.* Boston: Little, Brown, and Co., 1980.

Fujita, Fumiko. *American Pioneers and the Japanese Frontier: American Experts in Nineteenth-Century Japan.* Westport, CT: Greenwood Press, 1994.

Fujitani, T. *Splendid Monarchy: Power and Pageantry in Modern Japan.* Berkeley and Los Angeles: University of California Press, 1996.

Fujiwara Hitoshi. *Maboroshi no Nihon ōkami: Fukushima-ken no seisoku kiroku* (The elusive Japanese wolf: A record of life in Fukushima Prefecture). Wakamatsu: Rekishi Shunjū Shuppan Kabushikigaisha, 1994.

Gluck, Carol. *Japan's Modern Myths: Ideology in the Late Meiji Period.* Princeton: Princeton University Press, 1985.

Goodman, Grant K. *Japan and the Dutch, 1600–1853.* Richmond, Surrey, UK: Curzon Press, 2000.

Goossen, Theodore W. *The Oxford Book of Japanese Short Stories.* Oxford: Oxford University Press, 1997.

Hampton, Bruce. *The Great American Wolf.* New York: Henry Holt and Co., 1997.

Hane, Mikiso. *Peasants, Rebels, and Outcastes: The Underside of Modern Japan.* New York: Pantheon Books, 1982.

Hanley, Susan B., and Kozo Yamamura. *Economic and Demographic Change in Preindustrial Japan, 1600–1868.* Princeton: Princeton University Press, 1977.

Haraway, Donna. "Teddy Bear Patriarchy: Taxidermy in the Garden of Eden, New York City, 1908–1936." *Social Text: Theory, Culture, Ideology,* Winter 1984–85, 20–64.

Hardacre, Helen. *Shintō and the State, 1868–1988.* Princeton: Princeton University Press, 1989.

Harding, Sandra. *The Science Question in Feminism.* Ithaca: Cornell University Press, 1986.

Harootunian, H. D. *Things Seen and Unseen: Discourse and Ideology in Tokugawa Nativism.* Chicago: University of Chicago Press, 1988.

Harrison, John A. *Japan's Northern Frontier: A Preliminary Study in Colonization and Expansion with Special Reference to the Relations of Japan and Russia.* Gainesville: University of Florida Press, 1953.

Hatakeyama Saburōta. "Hokkaidō no inu ni tsuite no oboegaki: Senshi jidai kaizuka-ken to Ainu-ken no hikaku" (A memorandum on the Hokkaido dog: A comparison of the prehistoric shell mound dog to the Ainu dog). *Hokkai-dōshi no kenkyū* (Research on Hokkaido history) 1 (December 1973): 41–63.

Heinrich, Bernd. *Mind of the Raven.* New York: Cliff Street Books, 1999.

Higashi Yoshino-mura Kyōiku Iinkai, ed. *Higashi Yoshino no minwa* (The folklore of Higashi Yoshino). Osaka: Gyōsei, 1992.

Hiraiwa Yonekichi. *Inu no kōdō to shinri* (The actions and mentality of dogs). Tokyo: Tsukiji Shokan, 1991.

———. *Inu no seitai* (The ecology of dogs). Tokyo: Tsukiji Shokan, 1989.

———. *Inu o kau chie* (Wisdom for raising dogs). Tokyo: Tsukiji Shokan, 1999.

———. *Inu to ōkami* (Dogs and wolves). Tokyo: Tsukiji Shokan, 1990.

———. *Neko no rekishi to kiwa* (History and strange stories regarding cats). Tokyo: Tsukiji Shokan, 1992.

———. *Ōkami: Sono seitai to rekishi* (The wolf: Its ecology and history). Tokyo: Tsukiji Shokan, 1992.

———. *Watashi no inu* (My dog). Tokyo: Tsukiji Shokan, 1991.

Hokkaido Prefectural Government, ed. *Foreign Pioneers: A Short History of the Contribution of Foreigners to the Development of Hokkaido*. Sapporo: Hokkaido Prefectural Government, 1968.

Holcombe, Charles. *The Genesis of East Asia, 221 B.C.–A.D. 907*. Honolulu: University of Hawai'i Press, 2001.

Hori, Ichiro. *Folk Religion in Japan: Continuity and Change*. Edited by Joseph M. Kitagawa and Alan L. Miller. Chicago: University of Chicago Press, 1968.

———. "Mountains and Their Importance for the Idea of the Other World in Japanese Folk Religion." *History of Religions* 6, no. 1 (August 1966): 1–23.

Howell, David L. *Capitalism from Within: Economy, Society, and the State in a Japanese Fishery*. Berkeley and Los Angeles: University of California Press, 1995.

Ichikawa Kōichirō, Fujita Yukinori, and Shimazu Mitsuo, eds. *Nihon rettō chishitsu kōzō hattatsushi* (A history of the development of the geologic features of the Japanese Archipelago). Tokyo: Tsukiji Shoten, 1970.

Ikegami, Hiromasa. "The Significance of Mountains in the Popular Beliefs in Japan." In *Religious Studies in Japan*, edited by Japan Association for Religious Studies and Japanese Organizing Committee of the Ninth International Congress for the History of Religions. Tokyo: Maruzen Co., 1959.

Imagawa Isao. *Inu no gendaishi* (A modern history of dogs). Tokyo: Gendai Shokan, 1996.

Irokawa Daikichi. *The Culture of the Meiji Period*. Translated by Marius B. Jansen. Princeton: Princeton University Press, 1985.

Isenberg, Andrew C. *The Destruction of the Bison*. Cambridge: Cambridge University Press, 2000.

Ivy, Marilyn. *Discourses of the Vanishing: Modernity, Phantasm, Japan*. Chicago: University of Chicago, 1995.

Jansen, Marius B., and Gilbert Rozman, eds. *Japan in Transition: From Tokugawa to Meiji*. Princeton: Princeton University Press, 1986.

Jenkins, Ken L. *Wolf Reflections: Reflections of the Wilderness Series*. Merrillville, IN: ICS Books, 1996.

Johnson, Sylvia A., and Alice Aamodt. *Wolf Pack: Tracking Wolves in the Wild.* Minneapolis: Lerner Publications Co., 1985.

Johnston, William. *The Modern Epidemic: A History of Tuberculosis in Japan.* Cambridge: Harvard University Press, 1995.

Jones, Hazel J. *Live Machines: Hired Foreigners in Meiji Japan.* Vancouver: University of British Columbia Press, 1980.

Jones, Karen R. *Wolf Mountains: A History of Wolves along the Great Divide.* Calgary: University of Calgary Press, 2002.

Jones, Susan. "Becoming a Pest: Prairie Dog Ecology and the Human Economy in the Euroamerican West." *Environmental History* 4, no. 4 (October 1999): 531–52.

Jordon, Terry G. *North American Cattle-Ranching Frontiers: Origins, Diffusion, and Differentiation.* Albuquerque: University of New Mexico Press, 2000.

Kadosaki Masaaki and Seki Hideshi. "Ezochi ni okeru dōbutsusō no bunkengakuteki kenkyū" (Philological research on the fauna on Yezo Island). *Hokkaidō kaitaku kinenkan chōsa hōkoku* (Bulletin of the Hokkaido Colonization Museum) 38 (1999): 96–108.

Kakunodateshi Henshū Iinkai, ed. *Kakunodateshi: Kita-ke jidai 2* (The history of Kakunodate: The northern household period 2). Vol 4. Tokyo: Daiichi Hōki Shuppan Kabushikigaisha, 1970.

Kaufman, Les, and Kenneth Mallory, eds. *The Last Extinction.* Cambridge: MIT Press, 1986.

Kawada, Minoru. *The Origin of Ethnography in Japan: Yanagita Kunio and His Times.* Translated by Toshiko Kishida-Ellis. London: Kegan Paul International, 1993.

Kayano Shigeru. *Our Land Was a Forest: An Ainu Memoir.* Boulder: Westview Press, 1980.

Kellert, Stephen R. "Attitudes, Knowledge, and Behavior toward Wildlife among the Industrial Superpowers: United States, Japan, and Germany." *Journal of Social Issues* 49, no. 1 (1993): 53–69.

Kete, Kathleen. *The Beast in the Boudoir: Petkeeping in Nineteenth-Century Paris.* Berkeley and Los Angeles: University of California Press, 1994.

Kikuchi Isao. *Kinsei no kikin* (Early modern famines). Tokyo: Yoshikawa Kōbunkan, 1997.

Knight, John. "On the Extinction of the Japanese Wolf." *Asian Folklore Studies* 56 (1997): 130–59.

———. *Waiting for Wolves in Japan: An Anthropological Study of People-Wildlife Relations.* Oxford: Oxford University Press, 2003.

Kobayashi Takiji. *"The Factory Ship" and "The Absentee Landlord."* Translated by Frank Motofuji. Seattle: University of Washington Press, 1973.

Kōchi shinbun (Kōchi newspaper), September 19, 2002, evening edition. See www.kochinews.co.jp/0209/020919evening04.htm.

Koganezawa Masaaki. "Nikkō ni okeru shika no zōka to shinrin seitaikei e no eikyō, soshite ōkami dōnyū no hitsuyōsei" (The link between the rise in deer numbers and forest ecology in Nikkō and the necessity of wolf introduction). *Foresuto Kōru* (Forest call) 6 (May 1999): 4–5.

Kokushi daijiten (The comprehensive dictionary of Japanese history). Vols. 1–15. Edited by Kokushi Daijiten Henshū Iinkai. Tokyo: Yoshikawa Kōbunkan, 1979–98.

Koschmann, J. Victor, Ōiwa Keibō, and Yamashita Shinji, eds. *International Perspectives on Yanagita Kunio and Japanese Folklore Studies*. Ithaca: Cornell University East Asia Program, 1985.

Kouwenhoven, Arlette, and Matthi Forrer. *Siebold and Japan: His Life and Work*. Leiden: Hotei Publishing, 2000.

LaFleur, William R. "Saigyō and the Buddhist Value of Nature." In *Nature in Asian Traditions of Thought: Essays in Environmental Philosophy*, edited by J. Baird Callicott and Roger T. Ames. Albany: State University of New York Press, 1989.

Latour, Bruno, and Steve Woolgar. *Laboratory Life: The Construction of Scientific Facts*. Thousand Oaks: Sage Publications, 1979. Reprint, Princeton: Princeton University Press, 1986.

Law, Jane Marie. "Violence, Ritual Reenactment, and Ideology: The *Hōjō-e* (Rite for Release of Sentient Beings) of the Usa Hachiman Shrine in Japan." *History of Religions* 33, no. 4 (1994): 325–57.

Lawrence, Barbara, and William H. Bossert. "Multiple Character Analysis of *Canis lupus, latrans,* and *familiaris,* with a Discussion of the Relationships of *Canis niger.*" *American Zoologist* 7, no. 1 (February 1967): 223–32.

Lawrence, R. D. *In Praise of Wolves*. New York: Ballantine Books, 1986.

Lawton, John H., and Robert M. May, eds. *Extinction Rates*. Oxford: Oxford University Press, 1995.

Leakey, Richard, and Roger Lewin. *The Sixth Extinction: Patterns of Life and the Future of Humankind*. New York: Doubleday, 1995.

Lensen, George Alexander. *The Russian Push toward Japan: Russo-Japanese Relations, 1697–1875*. Princeton: Princeton University Press, 1959.

Lopez, Barry Holstun. *Of Wolves and Men*. New York: Simon and Schuster, 1978.

Machida Hiroshi. "Hikage no kazura no mabushii midori" (The radiant green of the shade vine). *Mitsumine-san* (Mount Mitsumine) 168 (April 1992): 6.

Maki, John M. *William Smith Clark: A Yankee in Hokkaido*. Sapporo: Hokkaido University Press, 1996.

Marceau, Lawrence E. *Takebe Ayatari: A Bunjin Bohemian in Early Modern Japan*. Ann Arbor: Center for Japanese Studies, University of Michigan, 2002.

Martin, Paul S., and Richard G. Klein, eds. *Quaternary Extinctions: A Prehistoric Revolution*. Tucson: University of Arizona Press, 1984.

Masson, Jeffrey Moussaieff. *Dogs Never Lie about Love: Reflections on the Emotional World of Dogs.* New York: Crown Publishers, 1997. Reprint, New York: Three Rivers Press, 1997.

Masson, Jeffrey Moussaieff, and Susan McCarthy. *When Elephants Weep: The Emotional Lives of Animals.* New York: Delacorte Press, 1995.

Mayer, Fanny Hagin, trans. and ed. *The Yanagita Kunio Guide to the Japanese Folk Tale.* Bloomington: Indiana University Press, 1948.

McClelland, Peter D. *Sowing Modernity: America's First Agricultural Revolution.* Ithaca: Cornell University Press, 1997.

McIntosh, Robert P. *The Background of Ecology: Concept and Theory.* Cambridge: Cambridge University Press, 1985.

McIntyre, Rick. *A Society of Wolves: National Parks and the Battle over the Wolf.* Stillwater, MN: Voyageur Press, 1993.

———, ed. *War against the Wolf: America's Campaign to Exterminate the Wolf.* Stillwater, MN: Voyageur Press, 1995.

McNamee, Thomas. *The Return of the Wolf to Yellowstone.* New York: Henry Holt and Co., 1997.

McNeill, J. R. *Something New under the Sun: An Environmental History of the Twentieth-Century World.* New York: W. W. Norton and Co., 2000.

Mech, L. David. *The Arctic Wolf: Ten Years with the Pack.* Stillwater, MN: Voyageur Press, 1988.

———. *The Way of the Wolf.* Stillwater, MN: Voyageur Press, 1991.

———. *The Wolf: The Ecology and Behavior of an Endangered Species.* Minneapolis: University of Minnesota Press, 1970.

———. *The Wolves of Isle Royale.* Washington, DC: U.S. Government Printing Office, 1966.

———. *Wolves of the High Arctic.* Stillwater, MN: Voyageur Press, 1992.

Mech, L. David, Layne G. Adams, Thomas J. Meier, John W. Burch, and Bruce W. Dale. *The Wolves of Denali.* Minneapolis: University of Minnesota Press, 1998.

Mech, L. David, and Luigi Boitani, eds. *Wolves: Behavior, Ecology, and Conservation.* Chicago: University of Chicago Press, 2003.

Mitman, Gregg. *The State of Nature: Ecology, Community, and American Social Thought, 1900–1950.* Chicago: University of Chicago Press, 1992.

"Miyagi-ken" (Miyagi Prefecture). In *Kadokawa Nihon chimei daijiten* (The Kadokawa dictionary of Japanese place-names), vol. 4, edited by Takeuchi Rizō. Tokyo: Kadokawa Shoten, 1979.

Miyake Hitoshi. "Mitsumine-san no shugen" (The mountain asceticism of Mount Mitsumine). *Mitsumine-san* (Mount Mitsumine) 144 (April 1994): 6–7.

———. *Shugendō: Essays on the Structure of Japanese Folk Religion.* Ann Arbor: Center for Japanese Studies, University of Michigan, 2001.

Miyoshi, Masao. *As We Saw Them: The First Japanese Embassy to the United States.* New York: Kōdansha International, 1979.

Morris, Ivan. *The World of the Shining Prince: Court Life in Ancient Japan.* New York: Kōdansha International, 1964.

Morris-Suzuki, Tessa. "Creating the Frontier: Border, Identity, and History in Japan's Far North." *East Asian History* 7 (June 1994): 18–23.

———. *A History of Japanese Economic Thought.* London: Routledge, 1989.

———. *Re-inventing Japan: Time, Space, Nation.* New York: M. E. Sharpe, 1998.

Morse, Ronald A. *Yanagita Kunio and the Folklore Movement: The Search for Japan's National Character and Distinctiveness.* New York: Garland Publishing, 1990.

Murie, Adolph. *The Wolves of Mount McKinley.* Seattle: University of Washington Press, 1985.

Murray, John A., ed. *Out among the Wolves: Contemporary Writings on the Wolf.* Vancouver: Whitecap Books, 1993.

Nagata Hōsei. *Hokkaidō Ezogo chimei kai* (Understanding Hokkaido's Ainu-language place-names). 2 vols. Sapporo: Hokkaidō Chō, 1891.

Nakamura Teiri. *Dōbutsutachi no reiryoku* (The spiritual powers of animals). Tokyo: Chikuma Shobō, 1989.

———. *Kitsune no Nihonshi* (The fox in Japanese history). Tokyo: Nihon Editā Sukūru Shuppanbu, 2001.

———. *Tanuki to sono sekai* (The raccoon-dog and its world). Tokyo: Asahi Shinbunsha, 1990.

Nara Kenshi Henshū Iinkai, ed. *Nara kenshi: Dōbutsu-shokubutsu* (The history of Nara Prefecture: Animals and plants). Vol. 2. Tokyo: Meicho Shuppan, 1990.

Nelson, John K. *A Year in the Life of a Shinto Shrine.* Seattle: University of Washington Press, 1996.

Nibbashi Kazuaki. "Nihon no dōbutsuen no rekishi" (The history of Japanese zoos). In *Dōbutsuen to iu media* (The media called zoos), ed. Watanabe Morio et al. Tokyo: Seikyūsha, 2000.

Nie, Martin A. *Beyond Wolves: The Politics of Wolf Recovery and Management.* Minneapolis: University of Minnesota Press, 2003.

"Niikappu bokujō no keikyō" (The situation at the Niikappu ranch). *Hokkai tsūshi* (Northern sea interpreter) 7 (July 1880): 4–9.

Nish, Ian, ed. *The Iwakura Mission in America and Europe: A New Assessment.* Richmond, Surrey, UK: Japan Library, 1998.

Nishimura Saburō. *Bunmei no naka no hakubutsugaku: Seiō to Nihon* (Natural history in civilizations, west and east). 2 vols. Tokyo: Kinokuniya Shoten, 2000.

Notehelfer, F. G. "Japan's First Pollution Incident." *Journal of Japanese Studies* 1, no. 2 (1975): 351–83.

Nowak, Ronald M. "Another Look at Wolf Taxonomy." In *Ecology and Conservation of Wolves in a Changing World,* edited by Ludwig N. Carbyn, Steven H. Fritts, and Dale R. Seip. Edmonton: Canadian Circumpolar Institute, University of Alberta, 1995.

Obeyesekere, Gananath. *Medusa's Hair: An Essay on Personal Symbols and Religious Experience*. Chicago: University of Chicago Press, 1984.

Ōga Tetsuo, ed. *Nihon daihyakka zensho* (The complete encyclopedia of Japan). Vol. 22. Tokyo: Shōgakukan, 1988.

Ohnuki-Tierney, Emiko. *The Monkey as Mirror: Symbolic Transformations in Japanese History and Ritual*. Princeton: Princeton University Press, 1987.

———. *Rice as Self: Japanese Identities through Time*. Princeton: Princeton University Press, 1993.

Okada Akio. *Dōbutsu: Nihonshi shōhyakka* (Animals: An encyclopedia of Japanese history). Tokyo: Kondō Shuppansha, 1979.

Ono, Sokyo. *Shinto: The Kami Way*. Rutland, VT: Charles E. Tuttle Co., 1962.

Ōta Shinya. *Karasu wa machi no ōsama da* (Crows are kings of the city streets). Fukuoka: Ashishobō Yūgengaisha, 1999.

Parker, Heidi G., Lisa V. Kim, Nathan B. Sutter, Scott Carlson, Travis D. Lorentzen, Tiffany B. Malek, Gary S. Johnson, Hawkins B. DeFrance, Elaine A. Ostrander, and Leonid Kruglyak. "Genetic Structure of the Purebred Domestic Dog." *Science* 304, no. 5674 (May 21, 2004): 1160–64.

Phillips, Michael K., and Douglas W. Smith. *The Wolves of Yellowstone*. Stillwater, MN: Voyageur Press, 1996.

Piggott, Joan R. *The Emergence of Japanese Kingship*. Stanford: Stanford University Press, 1997.

Pittau, Joseph, S.J. *Political Thought in Early Meiji Japan, 1868–1889*. Cambridge: Harvard University Press, 1967.

Pollan, Michael. *The Botany of Desire: A Plant's-Eye View of the World*. New York: Random House, 2002.

Ponting, Clive. *A Green History of the World: The Environment and the Collapse of Great Civilizations*. New York: Penguin Books, 1991.

Ritvo, Harriet. *The Animal Estate: The English and Other Creatures in the Victorian Age*. Cambridge: Harvard University Press, 1987.

———. *The Platypus and the Mermaid and Other Figments of the Classifying Imagination*. Cambridge: Harvard University Press, 1997.

"Saitama-ken" (Saitama Prefecture). In *Kadokawa Nihon chimei daijiten* (The Kadokawa dictionary of Japanese place-names), vol. 11, edited by Takeuchi Rizō. Tokyo: Kadokawa Shoten, 1980.

Sarashina Genzō. *Hoppō dōbutsuki* (Northern animal chronicle). Sapporo: Hokkaidō Raiburarii 2, 1977.

Sarashina Genzō and Sarashina Kō. *Kotan seibutsuki: Yajū-kaijū-gyozoku* (A biological chronicle of Ainu villages: Volume on animals, marine mammals, and fishes). 3 vols. Tokyo: Hōsei Daigaku Shuppankyoku, 1976.

Schach, Paul. "Russian Wolves in Folktales and Literature of the Plains: A Question of Origins." *Great Plains Quarterly* 3, no. 2 (Spring 1983): 67–78.

Schullery, Paul, ed. *The Yellowstone Wolf: A Guide and Sourcebook.* Worland, WY: High Plains Publishing Co., 1996.

Schwartz, Marion. *A History of Dogs in the Early Americas.* New Haven: Yale University Press, 1997.

Shibusawa Keizō, comp. *Japanese Life and Culture in the Meiji Era.* Translated by Charles S. Terry. Tokyo: Ōbunsha, 1958.

Shively, Donald H. "Tokugawa Tsunayoshi, the Genroku Shogun." In *Personality in Japanese History,* edited by Albert M. Craig and Donald H. Shively. Berkeley and Los Angeles: University of California Press, 1970. Reprint, Ann Arbor: Center for Japanese Studies, University of Michigan, 1995.

Sibley, William F. *The Shiga Hero.* Chicago: University of Chicago Press, 1979.

Siddle, Richard. *Race, Resistance and the Ainu of Japan.* London: Routledge, 1996.

Singer, Peter. *Animal Liberation.* New York: Avon Books, 1991.

Skabelund, Aaron. "Civilizing the Streets and Defending National Territory: The Creation of the 'Japanese' Dog, 1853–1941." Paper presented at the Japanimals Symposium, Columbia University, 2001.

———. "Loyalty and Civilization: A Canine History of Japan, 1850–2000." Ph.D. diss., Columbia University, 2004.

Smyers, Karen A. *The Fox and the Jewel: Shared and Private Meanings in Contemporary Japanese Inari Worship.* Honolulu: University of Hawai'i Press, 1999.

Sonoda Minoru. "Shinto and the Natural Environment." In *Shinto in History: Ways of the Kami,* edited by John Breen and Mark Teeuwen. Honolulu: University of Hawai'i Press, 2000.

Souyri, Pierre François. *The World Turned Upside Down: Medieval Japanese Society.* Translated by Käthe Roth. New York: Columbia University Press, 2001.

Stegner, Wallace. *Wolf Willow: A History, a Story, and a Memory of the Last Plains Frontier.* New York: Viking Press, 1962. Reprint, New York: Penguin Books, 1990.

Steinhart, Peter. *The Company of Wolves.* New York: Vintage Books, 1995.

Strong, Kenneth. *Ox against the Storm: A Biography of Tanaka Shozo—Japan's Conservationist Pioneer.* Richmond, Surrey, UK: Japan Library, 1977.

Sugimoto Isao. *Itō Keisuke* (Itō Keisuke). Tokyo: Yoshikawa Kōbunkan, 1960.

Sugimoto, Masayoshi, and David L. Swain. *Science and Culture in Traditional Japan, A.D. 600–1854.* Cambridge: MIT Press, 1978.

Sutō Isao. *Yama no hyōteki: Inoshishi to yamabito no seikatsushi* (Target in the mountains: A history of the daily lives of wild boars and mountain people). Tokyo: Miraisha, 1991.

Suzuki, Wajiro, Hiroshi Tanaka, and Tohru Nakashizuka. "Long Term Ecological Research at Senju-ga-hara Forest Reserve in Nikko, Central Japan." See http://ss.ffpri.affrc.go.jp/labs/femnet/nikko/nikko.htm.

Taniguchi Kengo. *Inu no Nihonshi: Ningen to tomo ni ayunda ichiman nen no monogatari* (The dog in Japanese history: A tale of walking as a friend of humans for ten thousand years). Tokyo: PHP Kenkyūjo, 2000.

Tawara Hiromi. *Hokkaidō no shizen hogo* (Protecting Hokkaido's nature). Sapporo: Hokkaidō Daigaku Tosho Kankōkai, 1979.

Thomas, Julia Adeney. "Naturalizing Nationhood: Ideology and Practice in Early Twentieth-Century Japan." In *Japan's Competing Modernities: Issues in Culture and Democracy, 1900–1930*. Edited by Sharon A. Minichiello. Honolulu: University of Hawai'i Press, 1998.

———. *Reconfiguring Modernity: Concepts of Nature in Japanese Political Ideology*. Berkeley and Los Angeles: University of California Press, 2001.

Tierkel, Ernest S. "Canine Rabies." In *The Natural History of Rabies*, vol. 2, edited by George M. Baer. New York: Academic Press, 1975.

Toby, Ronald P. *State and Diplomacy in Early Modern Japan: Asia in the Development of the Tokugawa Bakufu*. Princeton: Princeton University Press, 1984. Reprint, Stanford: Stanford University Press, 1991.

———. "Why Leave Nara? Kanmu and the Transfer of the Capital." *Monumenta Nipponica* 40, no. 3 (Autumn 1985): 331–47.

Tōkyō To, ed. *Ueno dōbutsuen hyakunenshi* (One hundred year history of the Ueno zoo). 2 vols. Tokyo: Daiichi Hōki Shuppan Kabushikigaisha, 1982.

Totman, Conrad. *The Green Archipelago: Forestry in Pre-industrial Japan*. Berkeley and Los Angeles: University of California Press, 1989. Reprint, Athens: Ohio University Press, 1998.

———. *A History of Japan*. Malden, MA: Blackwell Publishers, 2000.

Toyohara Shōji. "Unmemke ni okeru senkō ichi ni tsuite" (On the positioning of the puncture in the *unmemke*). *Kan Ohōtsuku* (Circle Okhotsk) 3 (1995): 63–69.

Tsukamoto Manabu. *Edo jidaijin to dōbutsu* (The people of the Edo period and animals). Tokyo: Nihon Editā Sukūru Shuppanbu, 1995.

———. *Shōrui o meguru seiji* (The politics of the clemency for living creatures). Tokyo: Heibonsha, 1993.

Tsurumi, E. Patricia. *Factory Girls: Women in the Thread Mills of Meiji Japan*. Princeton: Princeton University Press, 1990.

Tucker, Mary Evelyn. *Moral and Spiritual Cultivation in Japanese Neo-Confucianism: The Life and Thought of Kaibara Ekken (1630–1714)*. Albany: State University of New York Press, 1989.

Tyler, Royall. "Kōfuku-ji and Shugendō." *Japanese Journal of Religious Studies* 16, nos. 2–3 (June–September 1989): 143–80.

Ue Toshikatsu. *Yamabito no dōbutsushi: Kishū Hatenashi Sanmyaku Haruaki* (The animal journal of a mountain person: Spring and fall in the Kishū and the Hatenashi Mountain Range). Tokyo: Shinshuku Shobō, 1998.

Ueno Masuzō. *Nihon dōbutsugakushi* (The history of Japan's zoology). Tokyo: Yasaka Shobō, 1987.

Vaporis, Constantine. *Breaking Barriers: Travel and the State in Early Modern Japan*. Cambridge: Harvard University Press, 1994.

Varley, John D., Wayne G. Brewster, Sarah E. Broadbent, and Renee Evanoff, eds. *Wolves for Yellowstone? A Report to the United States Congress*. Vol. 4, *Research and Analysis*. Billings, MT: National Park Service, Yellowstone National Park, 1992.

von Den Driesch, Angela. *A Guide to the Measurement of Animal Bones from Archaeological Sites*. Cambridge: Peabody Museum of Archaeology and Ethnology, Harvard University, 1976.

Walker, Brett L. "Commercial Growth and Environmental Change in Early Modern Japan: Hachinohe's Wild Boar Famine of 1749." *Journal of Asian Studies* 60, no. 2 (May 2001): 329–51.

———. *The Conquest of Ainu Lands: Ecology and Culture in Japanese Expansion, 1590–1800*. Berkeley and Los Angeles: University of California Press, 2001.

———. "Foreign Affairs and Frontiers in Early Modern Japan: A Historiographical Essay of the Field." *Early Modern Japan: An Interdisciplinary Journal* 10, no. 2 (Fall 2002): 44–62.

———. "The History and Ecology of the Extinction of the Japanese Wolf." *Japan Foundation Newsletter* 29, no. 1 (October 2001): 10–13.

———. "Shinto." In *Encyclopedia of World Environmental History*, vol. 3., ed. Shepard Krech III, John R. McNeill, and Carolyn Merchant. New York: Routledge, 2004.

Walthall, Anne. *The Weak Body of a Useless Woman: Matsuo Taseko and the Meiji Restoration*. Chicago: University of Chicago Press, 1998.

Weiner, Jonathan. *The Beak of the Finch*. New York: Vintage Books, 1994.

White, Richard. "Animals and Enterprise." In *The Oxford History of the American West*, edited by Clyde A. Milner II, Carol A. O'Connor, and Martha A. Sandweiss. New York: Oxford University Press, 1994.

Williams, Duncan Ryūken. "Animal Liberation, Death, and the State: Rites to Release Animals in Medieval Japan." In *Buddhism and Ecology: The Interconnection of Dharma and Deeds*, edited by Mary Evelyn Tucker and Duncan Ryūken Williams. Cambridge: Harvard University Press, 1997.

Wise, Steven M. *Drawing the Line: Science and the Case for Animal Rights*. Cambridge, MA: Perseus Books, 2002.

———. *Rattling the Cage: Toward Legal Rights for Animals*. Cambridge, MA: Perseus Books, 2001.

Worster, Donald. *Nature's Economy: A History of Ecological Ideas*. Cambridge: Cambridge University Press, 1977.

Yanai Kenji. *Maboroshi no Nihon ōkami* (The elusive Japanese wolf). Urawashi: Sakitama Shuppankai, 1993.

Yokoyama Haruo. "Honzanha shugen Mitsumine-san no kōryū" (The rise of

Honzan school mountain asceticism at Mount Mitsumine). *Kokugakuin zasshi* (Journal of Kokugakuin University) 80, no. 10 (October 1979): 284–96.

Yonemoto, Marcia. *Mapping Early Modern Japan: Space, Place, and Culture in the Tokugawa Period (1603–1868)*. Berkeley and Los Angeles: University of California Press, 2003.

Yoshida Noriyuki. "Robo People: And Robo Cats and Dogs." *Look Japan* 47, no. 547 (October 2001): 6–11.

INDEX

Italicized page numbers refer to figures and tables.

LIBRARY OF CONGRESS
CATALOGING-IN-PUBLICATION DATA

Walker, Brett L., 1967–
The lost wolves of Japan / Brett L. Walker ;
foreword by William Cronon.—1st ed.
 p.   cm.—(Weyerhaeuser environmental books)
Includes bibliographical references and index.
ISBN 0-295-98492-9 (hardback : alk. paper)
ISBN 13: 978-0-295-98492-6

1. Wolves—Japan.   2. Extinct animals—Japan.
I. Title.   II. Weyerhaeuser environmental book.
QL737.C22W33   2005   599.773'0952—dc22   2004031040